E. J. LeCuyer

Introduction to College Mathematics with A Programming Language

Springer-Verlag

New York   Heidelberg   Berlin

Edward J. LeCuyer
Western New England College
Department of Mathematics
Springfield, MA 01119
USA

AMS Subject Classifications: 00-01, 00A05, 26-01, 26-04, 26A06, 26A09

Library of Congress Cataloging in Publication Data

LeCuyer, E   J
  Introduction to college mathematics with A Programming Language

  (Undergraduate texts in mathematics)
  Includes indexes.
  1. Mathematics—1961–   2. Mathematics—Data processing.
3. APL (computer program language)   1. Title.
QA39.2.L4   510   78-2054

ISBN 0-387-90280-5   Springer-Verlag New York
ISBN 3-540-90280-5   Springer-Verlag Berlin Heidelberg

Undergraduate Texts in Mathematics

*To my wife, Carol, and children, Karen,
Michael, and Todd, whose love, patience,
and encouragement made this work possible.
To Howard Peelle for all of his help.*

Preface

The topics covered in this text are those usually covered in a full year's course in finite mathematics or mathematics for liberal arts students. They correspond very closely to the topics I have taught at Western New England College to freshmen business and liberal arts students. They include set theory, logic, matrices and determinants, functions and graphing, basic differential and integral calculus, probability and statistics, and trigonometry. Because this is an introductory text, none of these topics is dealt with in great depth. The idea is to introduce the student to some of the basic concepts in mathematics along with some of their applications. I believe that this text is self-contained and can be used successfully by any college student who has completed at least two years of high school mathematics including one year of algebra. In addition, no previous knowledge of any programming language is necessary.

The distinguishing feature of this text is that the student is given the opportunity to learn the mathematical concepts via **A** **P**rogramming Language (APL). APL was developed by Kenneth E. Iverson while he was at Harvard University and was presented in a book by Dr. Iverson entitled *A Programming Language*[1] in 1962. He invented APL for educational purposes. That is, APL was designed to be a consistent, unambiguous, and powerful notation for communicating mathematical ideas. In 1966, APL became available on a time-sharing system at IBM. Today, APL is gaining wide acceptance in such fields as business, insurance, scientific research, and education. The reason for this is that APL is one of the most concise, versatile, and powerful computer programming languages yet developed. Programs requiring several steps in other computer languages become very

---

[1]*A Programming Language* by Kenneth E. Iverson, New York: John Wiley and Sons, (1962).

concise in APL, if a program is needed at all. This is both because many primitive functions are available directly on the APL keyboard and because such APL operations as + and × can be applied to arrays of any size (as well as to scalars). Yet, in spite of power and sophistication of APL, it is not a difficult language to learn. One can use APL to solve mathematical problems immediately after only a few minutes of instruction.

Conventional mathematical notation and APL notation are presented *in parallel* throughout the text. Thus, if one desires, it is possible to ignore the APL and still use this text as a standard survey-of-mathematics text. Alternatively, one may use the text in conjunction with an APL terminal. APL notation corresponds closely to standard mathematical notation, and many mathematical processes are executed very easily in APL. By using the computer, the student can save a great deal of time doing tedious calculations and can concentrate more on the principles and concepts of the mathematics. In addition, the APL programs tend to reinforce these principles and concepts. It is my experience that by using APL, the student may learn the mathematical concepts better while finding the learning of mathematics meaningful and enjoyable. As an important bonus, he will be learning a powerful programming language which he will then be able to use in many other courses as well as in the "real world."

The mathematical concepts and the APL notation are presented in parallel throughout the text because I believe that the APL can best be learned as needed in the development of the mathematics rather than as a separate topic. However, it might also be quite useful to have an APL reference for those who have not previously been exposed to the APL language. Therefore, I have included as an appendix an introduction to APL, including the writing and revising of APL programs. This appendix can be quickly perused at the start of the course and then referred to as needed throughout the course.

Finally, I would like to express my appreciation to Dr. Howard A. Peelle of the University of Massachusetts for his encouragement and his numerous valuable suggestions on ways to improve upon this text. Also, I would like to thank the many students at the University of Massachusetts and at Western New England College who used the preliminary versions of this test for their preserverance, encouragement, and suggestions.

July, 1977   Edward J. LeCuyer, Jr.

Contents

Contents

Contents

Set theory 1

When one thinks of the new math introduced in the public schools about 1960, the first mathematical notion that comes to mind is that of sets. Using the notion of set, elementary school teachers are supposed to be able to better explain the basic ideas of arithmetic. Thus, one could conclude that every adult should know some set theory in order to carry on an intelligent conversation with elementary school children (about mathematics). Sets do provide a good foundation for many topics in mathematics. Therefore, set theory is an ideal topic to begin a survey of mathematics.

1.1 Sets

A *set* is a collection of objects. The objects in the set are called *elements*.

Notation

Sets are designated by capital letters. In conventional mathematical notation, the elements of a set are enclosed in braces. For example,

$$A = \{1, 3, 5, 7\}$$

$$B = \{x \mid x \text{ is a student in this class}\}.$$

The second form of a set above is known as set builder notation. The symbol | is read as "such that."

In APL, the set A above is designated by

$$A \leftarrow 1\ 3\ 5\ 7.$$

To express the fact that an element x "belongs to" a set A, we write $x \in A$. The symbol $\in$ is read "belongs to." Thus, $3 \in A$ is true. (3 belongs to the set A.) However $4 \notin A$. (4 does not belong to A.) Notice that $\in$ is a primitive function on the APL keyboard. It yields a 1 (for true) or a 0 (for false). Thus, consider the following examples in APL:

```
A←1 3 5 7
```

```
3∈A
```
1   3 does belong to A.

```
4∈A
```
0   4 does not belong to A.

```
   1 2 3 4∈A
```
1 0 1 0   1 and 3 do belong to A, but 2 and 4 do not belong to A. Note that you get a set (of 1's and 0's) when you ask (set)∈(set)?

The empty set

It is possible for a set to have no elements. Such a set is called the *empty set* (or *null set*). In mathematics, the empty set is symbolized by ∅. For example, if

$$A = \{1,3,5,7\} \quad \text{and} \quad B = \{2,4,6,8\},$$

then the set of elements common to A and B is the empty set ∅.

In APL, one can express the empty set by $\iota 0$. The symbol ι (iota) is located above the I on the APL keyboard. If N is a nonnegative integer, then ιN yields the set of positive integers up to and including N. Thus, $\iota 0$ yields the set of positive integers up to and including 0. Since there are no such positive integers, $\iota 0$ is the empty set. If one enters $\iota 0$ on the terminal and then pushes the RETURN key, the computer prints nothing. In other words, it yields the empty set.

```
A←ι0
```
                               The name A is given to $\iota 0$.
```
A
```
                               The value of A is requested.
                               The computer responds with nothing.

Subset

Given two sets A and B, A is a *subset* of B if every element belonging to A also belongs to B. In mathematics, this is symbolized by $A \subset B$. For example, if

$$A = \{1,3,5,7\} \quad \text{and} \quad B = \{2,4,6,8\} \quad \text{and} \quad C = \{1,2,3,4,5,6,7,8\},$$

then $A \subset C$ and $B \subset C$.

Let us now consider an APL program for determining whether or not a set A is a subset of a set B. (For a general discussion of programs, refer to the appendix.)

Program 1.1   SUBSET

```
∇IS←A SUBSET B
```
                               The result of this program, *IS*, will be either 1 (yes) or 0 (no).
```
[1]  IS←(∧/A∈B)
     ∇
```

To understand this program, consider the following examples:

```
A←1 3 5 7
B←1 2 3 4
C←1 2 3 4 5 6 7 8
```

$A \in B$ 1 1 0 0	$A \in B$ yields a vector of 1's and 0's. It tests each element of A to see if it belongs to B. 1 and 3 do, but 5 and 7 don't. $\wedge /1\ 1\ 0\ 0$ yields a 0. $\wedge /$ is a logical operator. It will yield 1 only if all of the numbers following it are 1's.
$\wedge /A \in B$ 0	
$A\ SUBSET\ B$ 0	A is not a subset of B.
$A \in C$ 1 1 1 1	Every element belonging to A also belongs to C.
$\wedge /A \in C$ 1	Since $A \in C$ is a complete vector of 1's, then $\wedge /A \in C$ is 1.
$A\ SUBSET\ C$ 1	A is a subset of C.

Equal sets

Two sets A and B are said to be *equal* if both $A \subset B$ and $B \subset A$. In other words, A and B have exactly the same elements. In conventional mathematical notation, the symbol used to express the fact that a set A equals a set B is $A = B$. For example, if

$$A = \{1,3,5,7,9\} \quad \text{and} \quad B = \{5,7,3,1,9\},$$

then $A = B$, since $A \subset B$ and $B \subset A$.

An APL program for the equality of two sets which uses the above program *SUBSET* as a subprogram[1] follows:

Program 1.2 EQUAL

∇ *IS← A EQUAL B*

[1] *IS←(A SUBSET B)*$\wedge$*(B SUBSET A)* ∇	If (*A SUBSET B*) is 1 (yes), and also (*B SUBSET A*) is 1 (yes), then *IS* is 1.

---

[1]A subprogram is a program used within another program.

Otherwise *IS* is 0 (no).) ($1 \land 1$ yields 1, but $1 \land 0$, $0 \land 1$ and $0 \land 0$ all yield 0.

Examples

```
A←1 3 5 7 9
B←5 7 3 1 9
C←1 2 3 4 5
A EQUAL B
```
1   True.
```
A EQUAL C
```
0   False.
```
D←1 1 3 5 5 7 9 9
A EQUAL D
```
1

Since the sets A and D have the same elements, they are equal sets. We do not list an element more than once in a set, since by so doing, we do not create a new set.

EXERCISES

1. Let $A = \{1,3,5,7,9\}$, $B = \{2,4,6,8\}$, $C = \{1,2,3,4\}$, $D = \{6,2,8,4\}$, $E = \emptyset$, and $F = \{1,2,3,4,5,6,7,8,9\}$. Determine whether the following are true or false:
   (a) $5 \in A$
   (b) $5 \in B$
   (c) $A \subset F$
   (d) $C \subset A$
   (e) $E \subset B$
   (f) $B \subset D$
   (g) $B = D$
   (h) $B = C$

2. Repeat Exercise 1 on an APL terminal, using the programs *SUBSET* and *EQUAL* where appropriate.

3. List the elements of the following sets:
   (a) The subset of the set F in Exercise 1 consisting of elements divisible by 3.
   (b) The set of vowels in the word "mathematics."
   (c) The set of months in a year.
   (d) The set of colors in the rainbow.

4. Let
   $A \leftarrow$ 'APL'
   $B \leftarrow$ '$MATE$'
   $M \leftarrow$ '$MATHEMATICS$'
   $S \leftarrow$ '$SETTHEORY$'
   $P \leftarrow$ 'PAL'

4

Evaluate the following on an APL terminal:
(a) *'E' ∈ S*
(b) *'L' ∈ S*
                         Refer to the appendix for a discussion of
                         representing literals in APL.

(c) *B ∈ M*
(d) *∧/B ∈ M*
(e) *B SUBSET M*
(f) *A SUBSET B*
(g) *A ∈ B*
(h) *∧/A ∈ B*
(i) *A EQUAL B*
(j) *A EQUAL P*

5. List all subsets of the set $A = \{a,b,c,d\}$. How many subsets are there?

6. In general, if a set has n elements, how many subsets does it have?

7. If $\{x,x^2,y\} = \{1,2,4\}$, find x and y.

8. Let

    A←1 3 5 7

    B←1 2 3 4

    C←2 4 6 8

Evaluate the following on an APL terminal and see if you can figure out what they do:
(a) *A ∈ B*
                         ∨ is the logical function "or." 1∨0,
                         0∨1, 1∨1 all yield 1, while 0∨0 yields
                         0. Also, ∨/A yields 1 if at least one 1
                         appears in A, where A is a vector of all
                         0's and 1's. ~ is the logical operator
                         "complement." ~1 yields 0 and ~0
                         yields 1. Also, ~A changes the 1's in A to
                         0's and the 0's in A to 1's.

(b) *∨/A ∈ B*
(c) *A ∈ C*
(d) *∨/A ∈ C*
(e) *~A ∈ C*
(f) *∨/1 0 1 0*
(g) *∨/0 0 0 0*
(h) *∨/~A ∈ C*

1.2 Operations with sets

There are various operations that can be used to create new sets from old sets. The first of these is intersection.

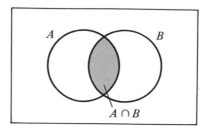

Figure 1.1   A Venn diagram.

Intersection

Let A and B be two sets. The *intersection* of A and B, symbolized by $A \cap B$, is the set of elements common to A and B. Using set builder notation,

$$A \cap B = \{ x \mid x \in A \text{ and } x \in B \}.$$

It will be helpful for us to visualize the sets formed by certain set operations. To do this, we shall use Venn diagrams.[2] Let the elements of A and B be schematically represented by the points inside the circles labeled A and B in Figure 1.1. Then, the intersection $A \cap B$ is represented by the points in the shaded region.

Examples

Let $A = \{1,3,5,7\}$, $B = \{1,2,3,4\}$, and $C = \{2,4,6,8\}$. Then, $A \cap B = \{1,3\}$ and $A \cap C = \varnothing$.

If A and C are two sets such that $A \cap C = \varnothing$, then A and C are said to be *disjoint* or *mutually exclusive*. If A and C are disjoint, then the Venn diagram would consist of two nonoverlapping circles.

We now have the following program for intersection. This program takes a set A on the left and a set B on the right and creates a new set, called *COMMON*, since the elements in the intersection are those common to both A and B.

Program 1.3 INTERSECT

∇ *COMMON* $\leftarrow A$ *INTERSECT B*

[1]   *COMMON* $\leftarrow (A \in B)/A$
$\quad \nabla$

To understand how this program works, consider the following examples:

$A \leftarrow 1$ 3 5 7
$B \leftarrow 1$ 2 3 4
$C \leftarrow 2$ 4 6 8

[2]Venn diagrams are named after their inventor, John Venn (1834–1923), and English logician.

A∈B 1 1 0 0	**A∈B** yields a vector of 0's and 1's. The 1's correspond to the elements in *A* which are also in *B*. The 0's correspond to the elements in *A* which are not in *B*.
1 1 0 0/A 1 3	**1 1 0 0/A** picks out the elements of *A* corresponding to the 1's in **1 1 0 0**.
A INTERSECT B 1 3	To execute the program, type **A IN-TERSECT B**. The result is the same as that of **(A∈B)/A**.
B∈A 1 0 1 0	Can you explain this result?
1 0 1 0/B 1 3	Can you explain this?
B INTERSECT A 1 3	
A∈C 0 0 0 0	
0 0 0 0/A	This result is the empty set. Why?
A INTERSECT C	The empty set. *A* and *C* are disjoint.

Set difference

The *difference* between two sets *A* and *B*, conventionally symbolized by $A - B$, is the set of elements of *A* which are not in *B*:

$$A - B = \{ x \mid x \in A \text{ but } x \notin B \}.$$

The Venn diagram for $A - B$ is shown in Figure 1.2

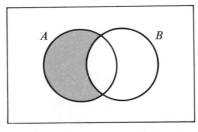

Figure 1.2   A Venn diagram. $A-B$ is the shaded region.

7

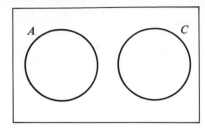

Figure 1.3   A Venn diagram of two disjoint sets.

Examples

Let $A = \{1,3,5,7\}$, $B = \{1,2,3,4\}$, and $C = \{2,4,6,8\}$. Then, $A - B = \{5,7\}$ and $B - C = \{1,3\}$ and $A - C = \{1,3,5,7\} = A$. Notice that if $A \cap C = \emptyset$, then $A - C = A$. To see why, study the Venn diagram of Figure 1.3. Also, for any set, $A - A = \emptyset$. Let us consider a program for set difference.

Program 1.4 DIFFERENCE

```
   ∇D←A DIFFERENCE B
[1]  D←(~A∈B)/A
   ∇
```

To understand how this program works, consider the examples:

```
   A←1 3 5 7
   B←1 2 3 4
```

```
   A∈B
1 1 0 0                          As before.
```

```
   ~A∈B                          ~(1 1 0 0) changes 0's to 1's and
0 0 1 1                          1's to 0's.
```

```
     0 0 1 1/A
5 7
```

```
     A DIFFERENCE B
5 7
```

The universal set

The *universal set*, U, in any discussion, consists of all of the elements under consideration in the discussion. When considering the complement of a set, it is important to know the universal set.

The complement of a set

The *complement* of a set A, symbolized by A', is the set of elements that are not in A but are still in the universal set U. Thus, $A' = U - A$.

In Venn diagrams, the universal set is all of the points in the rectangle.

8

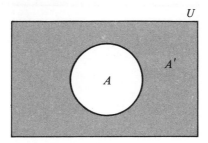

Figure 1.4   Venn diagram showing $A' = U - A$.

All other sets in the discussion are subsets of U. They are denoted by circles inside of U. The Venn diagram for A' is given in Figure 1.4 by the shaded region.

Example

Let $A = \{1, 3, 5, 7\}$. To get A', we need to know the universal set U. If $U = \{1, 2, 3, 4, 5, 6, 7, 8\}$, then $A' = \{2, 4, 6, 8\}$. But, if $U = \{1, 3, 5, 7, 9\}$, then $A' = \{9\}$. If $A = U$, then $A' = \varnothing$. If $A = \varnothing$, then $A' = U$.

The following program for complement is quite obvious.

Program 1.5 COMPLEMENT

```
      ∇C←COMPLEMENT A
[1]   C←U DIFFERENCE A ∇
```
The program *DIFFERENCE* is used as a subprogram.

Example

```
      U←1 2 3 4 5 6 7 8
      A←1 3 5 7

      U DIFFERENCE A
2 4 6 8

      COMPLEMENT A
2 4 6 8

      (U DIFFERENCE A) EQUAL (COMPLEMENT A)
1                                    True.
```

Union

The *union* of two sets A and B, symbolized by $A \cup B$, is the set of all elements appearing in A or in B or in both.

$$A \cup B = \{ x \mid x \in A \text{ or } x \in B \}.$$

(Here, "or" means one or the other or both.) The Venn diagram for $A \cup B$ is shown in Figure 1.5.

9

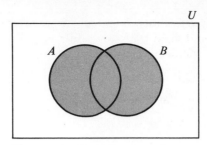

Figure 1.5   Venn diagram of $A \cup B$.

Example

If $A = \{1,3,5,7\}$ and $B = \{1,2,3,4\}$, then $A \cup B = \{1,2,3,4,5,7\}$. [*Note:* $A - B = \{5,7\}$ and $B \cup (A - B) = \{1,2,3,4,5,7\} = A \cup B$.]

In the Venn diagram in Figure 1.6., $A - B$ is shaded in using horizontal lines, while B is shaded in using vertical lines. Notice that $A \cup B = B \cup (A - B)$, and that $B \cap (A - B) = \varnothing$, or B and $A - B$ are disjoint.

Before considering a program for union, let us consider the use of the comma in APL. The operation of placing a comma between two sets is referred to as *catenation*.

Examples

```
        A←1 3 5 7
        B←1 2 3 4
        C←2 4 6 8

        A,B
1 3 5 7 1 2 3 4

        A,C
1 3 5 7 2 4 6 8
```

The comma between two sets (catenation) just chains the elements of the second set onto the end of the first set. If the sets are disjoint, as are A

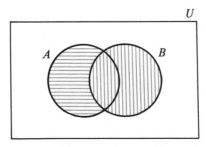

Figure 1.6   Venn diagram of $A \cup B$.

and C, the result is $A \cup C$. Since B and $A - B$ are always disjoint, then **B, A DIFFERENCE B** should yield the same result as $A \cup B$.

This suggests the following program for union:

Program 1.6 **UNION**

   ▽ *EITHER ← A UNION B*

[1]   *EITHER ← B, A DIFFERENCE B* ▽

Example

   A←1 3 5 7
   B←1 2 3 4

   B, A DIFFERENCE B
1 2 3 4 5 7

   A UNION B
1 2 3 4 5 7

Collectively exhaustive sets

Two sets are said to be *collectively exhaustive* if their union is the universal set.

Example

Let $A = \{1,3,5,7\}$ and $B = \{2,4,6,8\}$ and let the universal set be $U = \{1,2,3,4,5,6,7,8\}$. Then A and B are collectively exhaustive since $A \cup B = U$.

For any set A, A and its complement A' are always collectively exhaustive since $A \cup A' = U$. They are also mutually exclusive, since $A \cap A' = \emptyset$.

Symmetric difference of two sets

One other operation on two sets is the symmetric difference. The *symmetric difference* of a set A and a set B is the set of elements that are in A or in B but not in $A \cap B$. The standard mathematical symbol for this operation is $A \triangle B$. Notice that $A \triangle B = (A - B) \cup (B - A)$. A Venn diagram for this operation is shown in Figure 1.7.

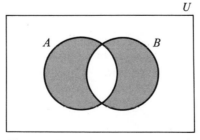

Figure 1.7   Venn diagram of $A \triangle B$.

11

Example

Let $A=\{1,3,5,7\}$ and $B=\{1,2,3,4\}$. Then, $A-B=\{5,7\}$ and $B=\{2,4\}$, so $(A-B)\cup(B-A)=\{2,4,5,7\}$, or $A\Delta B=\{2,4,5,7\}$.

It will be left as an exercise for the student to write a program for the symmetric difference of A and B.

EXERCISES

1. Consider the following universal set U and subsets A, B, C, and D.

    $U\leftarrow1\ 2\ 3\ 4\ 5\ 6\ 7\ 8\ 9\ 0$

    $A\leftarrow0\ 3\ 6\ 9$

    $B\leftarrow2\ 4\ 6\ 8$

    $C\leftarrow0\ 1\ 2\ 3\ 4$

    $D\leftarrow5\ 6\ 7\ 8\ 9$

Find the following using the definitions in this section and then check your answers using the APL terminal and the programs in this section. [*Note*: the exercises are first stated using conventional mathematical symbolism then using the APL program symbolism.]

. (a) A pair of disjoint sets.
  (b) A pair of collectively exhaustive sets.
  (c) $A\cap C$ *(A INTERSECT C)*
  (d) $D-A$ *(D DIFFERENCE A)*
  (e) $U-A$ *(U DIFFERENCE A)*
  (f) A' *(COMPLEMENT A)*
  (g) $A\cup D$ *(A UNION D)*
  (h) $(A\cup C)\cap D$ *(A UNION C) INTERSECT D*
  (i) $(A\cap D)\cup(C\cap D)$ *(A INTERSECT D) UNION (C INTERSECT D)*
  (j) $C\cap(D-A)$ *(C INTERSECT (D DIFFERENCE A))*
  (k) $(A\cap B)'$ *(COMPLEMENT (A INTERSECT B))*
  (l) $A'\cup B'$ *(COMPLEMENT A) UNION (COMPLEMENT B)*

2. Consider the following universal set U and subsets A, B, and C.

    $U\leftarrow$'ABCDEFGHIJKLMNOPQRSTUVWXYZ'

    $A\leftarrow$'AMPLE'

    $B\leftarrow$'METRIC'

    $C\leftarrow$'HELP'

Find the following using the definitions in this section. Then check your answers using the APL terminal and the programs in this section.

  (a) $A\cap B$ *(A INTERSECT B)*
  (b) $A\cup C$ *(A UNION C)*
  (c) $A-B$ *(A DIFFERENCE B)*
  (d) B' *(COMPLEMENT B)*
  (e) $C\cap(A-B)$ *(C INTERSECT A DIFFERENCE B)*

3. Consider the following universal set U and subsets A, S, and F:

$$U = \{\text{all cards in an ordinary deck of 52 playing cards}\}$$
$$A = \{\text{all of the aces}\}$$
$$S = \{\text{all of the spades}\}$$
$$F = \{\text{all of the face cards}\}$$

Describe the following sets in words:

(a) *A INTERSECT S*
(b) *A UNION F*
(c) *COMPLEMENT S*
(d) *A INTERSECT F*
(e) *S DIFFERENCE F*
(f) *S INTERSECT F*

4. Write a program for the symmetric difference of a set A and a set B.

1.3 A set theory drill and practice program (optional)

In this section, we present a program *SETTHEORY* which can be used by a student to practice the operations of intersection, union and difference. This is presented to illustrate the use of an APL program in drill and practice. It is an interactive program in which the student and the computer carry on a dialog. If it is saved in a workspace, then it can be used by a student to practice set theory. In any event, it might be worthwhile studying this program as a prototype of an APL drill and practice program.

```
        ∇ SETTHEORY
[1]     A←5?9
[2]     B←5?9
[3]     A;'INTERSECT';B;' =?'
[4]     GUESS:ANSWER←□
[5]     →(ANSWER EQUAL A INTERSECT B)/NEXT
[6]     'NO TRY AGAIN '
[7]     →GUESS
[8]     NEXT:A;' UNION';B;' =?'

[9]     TRY:RESPONSE←□
[10]    →(RESPONSE EQUAL A UNION B)/LAST
[11]    'SORRY TRY AGAIN'

[12]    →TRY

[13]    LAST:A;' DIFFERENCE';B;' =?'
[14]    SAY:REPLY←□
[15]    →(REPLY EQUAL A DIFFERENCE B)/END
```

[16]   *'WRONG TRY AGAIN'*
[17]   *→SAY*
[18]   *END:'WANT ANOTHER?'*
[19]   *'ENTER Y FOR YES, N FOR NO'*
[20] $→('Y' \in \Box)/1$
[21]   *'O.K. , GOODBYE'*
         ∇

In this program, two sets, A and B, consisting of 5 random digits from 1 to 9 are selected. The computer then prints out a request for the student to enter the intersection of these two sets. If the student correctly computes the intersection, then the computer requests the union of these two sets. If the student incorrectly answers the intersection question, then the computer prints *NO, TRY AGAIN* followed by $\Box$, and the student can try again. When he finally answers the intersection question correctly, he is given the union question. If he misses it, the computer prints *SORRY, TRY AGAIN*. When he answers the union question, he is asked for the difference of A and B. If he misses this, the computer prints *WRONG, TRY AGAIN*. When he gets the difference correct, the computer asks if he would like another problem. [*Note*: If *A INTERSECT B or A DIFFERENCE B* is empty, the student should enter $\iota0$.]

To run this program, type *SETTHEORY*. For example:

```
    SETTHEORY
5 9 2 7 8 INTERSECT 1 7 2 8 4 = ?
□ :
      2 7 8
5 9 2 7 8 UNION 1 7 2 8 4  = ?
□ :
      5 9 2 7 8 1 4
5 9 2 7 8 DIFFERENCE 1 7 2 8 4  = ?
□ :
      5 9
WANT ANOTHER?
ENTER Y FOR YES, N FOR NO
Y
1 8 3 7 4 INTERSECT 5 4 1 8 9  = ?
□ :
      1 8
NO TRY AGAIN
□ :
      1 8 4
1 8 3 7 4 UNION 5 4 1 8 9  = ?
□ :
      1 8 3 7 4 5 9
```

```
1 8 3 7 4 DIFFERENCE 5 4 1 8 9 =?
□ :
    3 7
WANT ANOTHER?
ENTER Y FOR YES, N FOR NO
N
O.K., GOODBYE
```

1.4 Boolean algebra

An *algebraic system* is a collection of objects, numbers, or sets together with one or more operations on these objects to create new objects in the collection, plus some laws concerning these operations. A *Boolean algebra*, named for George Boole, one of the originators of set theory, is any algebraic system similar to the system of subsets of a universal set U with operations of intersection, union, and complementation and the laws listed below. In this section, we shall consider the laws of Boolean algebra. These laws are listed both in conventional mathematical notation and in APL notation using our programs.

The laws of Boolean algebra

The idempotent laws

$A \cap A = A$ *(A INTERSECT A) EQUAL A*

$A \cup A = A$ *(A UNION A) EQUAL A*

The commutative laws

$A \cap B = B \cap A$ *(A INTERSECT B) EQUAL (B INTERSECT A)*

$A \cup B = B \cup A$ *(A UNION B) EQUAL (B UNION A)*

The associative laws

$A \cap (B \cap C) = (A \cap B) \cap C$
    (A INTERSECT (B INTERSECT C)) EQUAL ((A INTERSECT B) INTERSECT C)

$A \cup (B \cup C) = (A \cup B) \cup C$
    (A UNION (B UNION C)) EQUAL ((A UNION B) UNION C)

The distributive laws

$A \cap (B \cup C) = (A \cap B) \cup (A \cap C)$
    (A INTERSECT (B UNION C)) EQUAL ((A INTERSECT B) UNION (A INTERSECT C))

$A \cup (B \cap C) = (A \cup B) \cap (A \cup C)$
    (A UNION (B INTERSECT C)) EQUAL ((A UNION B) INTERSECT (A UNION C))

15

Operations with the universal set

$A \cap U = A$ (A INTERSECT U) EQUAL A
$A \cup U = U$ (A UNION U) EQUAL U

Operations with the empty set

$A \cap \varnothing = \varnothing$ (A INTERSECT ι0) EQUAL (ι0)
$A \cup \varnothing = A$ (A UNION ι0) EQUAL A

Laws of complements

$A'' = A$ (COMPLEMENT (COMPLEMENT A)) EQUAL A
$A \cup A' = U$ (A UNION COMPLEMENT A) EQUAL U
$A \cap A' = \varnothing$ (A INTERSECT COMPLEMENT A) EQUAL (ι0)
$U' = \varnothing$ (COMPLEMENT U) EQUAL (ι0)
$\varnothing' = U$ (COMPLEMENT ι0) EQUAL U

DeMorgan's laws

$(A \cup B)' = A' \cap B'$
(COMPLEMENT A UNION B) EQUAL ((COMPLEMENT A) INTERSECT (COMPLEMENT B))

$(A \cap B)' = A' \cup B'$
(COMPLEMENT A INTERSECT B) EQUAL ((COMPLEMENT A) UNION (COMPLEMENT B))

In mathematics, most laws are discovered by first considering particular examples. If a mathematician notices that a statement seems to be true for several particular examples, he then *conjectures* that perhaps that statement is always true. Then, he sets out to *prove* that the statement is always true. If he can do this, the conjecture becomes a theorem or a law.

The computer is very helpful in showing that statements are true or false with particular examples. In this respect, the computer is very valuable in doing mathematical research. Consider the following examples in which we test some of the laws of Boolean algebra on particular examples using APL.

Example 1

Let us test the distributive law.

```
A←1 3 5 7
B←3 4 5 6
C←1 2 3 4
B UNION C
1 2 3 4 5 6

A INTERSECT (B UNION C)
1 3 5

A INTERSECT B
3 5
```

A INTERSECT C

1 3

(A INTERSECT B) UNION (A INTERSECT C)

1 3 5

(A INTERSECT (B UNION C))
EQUAL (A INTERSECT B) UNION (A INTERSECT C)*

1

The 1 above stands for "true." Thus, this particular example seems to support the distributive law.

Example 2

Now, we'll test the law *(A UNION COMPLEMENT A) EQUAL U.*

A←1 3 5 7
U←1 2 3 4 5 6 7 8

COMPLEMENT A

2 4 6 8

A UNION COMPLEMENT A

1 2 3 4 5 6 7 8

(A UNION COMPLEMENT A) EQUAL U

1

Thus, we have verified this law of complements with this particular example.

Example 3

Finally, we shall test the DeMorgan law *(COMPLEMENT A INTERSECT B) EQUAL ((COMPLEMENT A) UNION (COMPLEMENT B)).*

U←1 2 3 4 5 6 7 8
A←1 3 5 7
B←1 2 3 4

A INTERSECT B

1 3

COMPLEMENT A INTERSECT B

2 4 5 6 7 8

COMPLEMENT A

2 4 6 8

COMPLEMENT B

5 6 7

*Due to space limitations this instruction has been printed on two lines. In reality it must be entered on one line.

(COMPLEMENT A) UNION (COMPLEMENT B)
2 4 5 6 7 8

(COMPLEMENT A INTERSECT B)
EQUAL ((COMPLEMENT A) UNION (COMPLEMENT B))*
1

Thus, this particular example helps us to believe DeMorgan's law.

Now that we have seen that these laws are valid for the particular examples above, we seek a method of proving that they are true in general. In order to do this, we shall use Venn diagrams. Two sets are considered to be equal if they have the same Venn diagrams. Let's draw Venn diagrams for the examples above.

Example 1

Since, as shown in Figure 1.8, the Venn diagrams are the same, then these sets are equal. Thus, $A \cap (B \cup C) = (A \cap B) \cup (A \cap C)$.

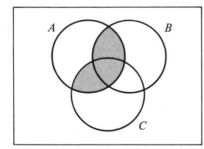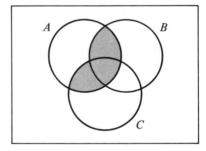

Figure 1.8 Left: Diagram of $A \cap (B \cup C)$. Right: Diagram of $(A \cap B) \cup (A \cap C)$.

Example 2

$A \cup A' = U$. In Figure 1.9, U is the whole rectangle. Obviously, the elements in A, represented by the horizontal lines, unioned with the elements in A', represented by the vertical lines, fill up the entire rectangle, U.

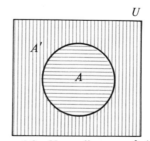

Figure 1.9 Venn diagram of $A \cup A'$.

*Due to space limitations this instruction has been printed on two lines. In reality it must be entered on one line.

18

Example 3

$(A \cap B)' = A' \cup B'$. Since, as shown in Figure 1.10, the Venn diagrams are the same, the sets are equal.

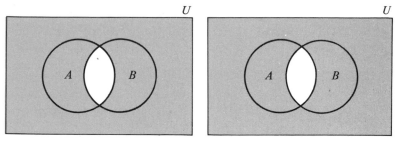

Figure 1.10 Left: Diagram of $A' \cup B'$. Right: Diagram of $(A \cap B)'$.

Exercises

1. Let

$U \leftarrow 0\ 1\ 2\ 3\ 4\ 5\ 6\ 7\ 8\ 9$

$A \leftarrow 0\ 3\ 6\ 9$

$B \leftarrow 0\ 1\ 2\ 3\ 4\ 5$

$C \leftarrow 4\ 5\ 6\ 7\ 8\ 9$

Test the validity of the following laws of Boolean algebra with the above sets on an APL terminal:

(a) $A \cap (B \cap C) = (A \cap B) \cap C$
   (A INTERSECT (B INTERSECT C)) EQUAL ((A INTERSECT B) INTERSECT C)

(b) $A \cup (B \cap C) = (A \cup B) \cap (A \cup C)$
   (A UNION B INTERSECT C) EQUAL (A UNION B) INTERSECT (A UNION C)

(c) $A \cap A' = \emptyset$
   (A INTERSECT COMPLEMENT A) EQUAL (ι0)

(d) $(A \cup B)' = A' \cap B'$
   (COMPLEMENT A UNION B) EQUAL (COMPLEMENT A) INTERSECT (COMPLEMENT B)

2. Verify the laws of Boolean algebra in Exercise 1 by drawing Venn diagrams.

3. There are many other properties of the operations of set theory not included in our list of laws of Boolean algebra. Test the validity of the following properties with the sets in Exercise 1 on an APL terminal. [Note: They may not all be true.]

(a) $A \cap B' = A - B$
   (A INTERSECT COMPLEMENT B) EQUAL (A DIFFERENCE B)

(b) $(A - B) \cup (A \cap C) = A - (B \cup C)$
   ((A DIFFERENCE B) UNION (A INTERSECT C)) EQUAL (A DIFFERENCE B UNION C)

19

(c) $A \cup B = (A - B) \cup (B - A) \cup (A \cap B)$
   (A UNION B) EQUAL ((A DIFFERENCE B) UNION (B DIFFERENCE A))
   UNION (A INTERSECT B)

4. Prove or disprove the properties in Exercise 3 by drawing Venn diagrams.

1.5 The number of elements in a set

Given a set A, the APL symbol for the number of elements in the set A is ρA. The letter ρ is located above the R on the keyboard.

Examples

   U←1 2 3 4 5 6 7 8 9
   A←1 3 5 7 9
   B←1 2 3 4 5
   C←'APL IS A PROGRAMMING LANGUAGE'

   ρA
5
   ρB
5
   ρA INTERSECT B
3
   ρA UNION B
7
   ρA DIFFERENCE B
2
   ρ COMPLEMENT A
4
   ρC
31   [*Note*: In literal data, ρ counts spaces too.]

In this text, we shall always denote the number of elements in a set A by ρA. There is no standard conventional symbol for the number of elements in a set.

Set theory can often be used to clarify otherwise complicated problems and to aid in solving them. One useful application of set theory and Venn diagrams is in counting the number of elements in the intersection, union, difference, and complement of various sets. The following examples are illustrations of this.

Example 1

Find the number of cards in an ordinary deck of 52 playing cards which are either face cards (jacks, queens, or kings) or spades. If we let F be the set of face cards and S be the set of spades, then we want the number of elements in the set $F \cup S$, or $\rho(F \cup S)$. We cannot just merely add ρF and ρS, because then we would be including the jack, queen, and king of spades twice in our sum, since they are in both F and S. In other words, they are in $F \cap S$. In order to make sure that we count each card exactly once, we could use the following formula:

$$\rho(F \cup S) = (\rho F) + (\rho S) - \rho(F \cap S).$$

Since ρF is 12, ρS is 13, and $\rho(F \cap S)$ is 3, then $\rho(F \cup S)$ is 22. The same result could have been arrived at by using the Venn diagram shown in Figure 1.11. In fact, the Venn diagram actually helps to clarify the situation.

The next example illustrates even better the use of a Venn diagram in finding the number of elements in a set.

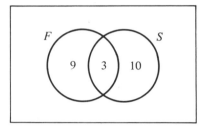

Figure 1.11   Venn diagram of the sets F and S.

Example 2

A survey was taken of 1000 citizens in a town to see how many read each of three magazines X, Y, and Z. It was found that 200 read X, 250 read Y, and 150 read Z. It was also found that 100 read both X and Y, 50 read both X and Z, and 50 read both Y and Z. In addition, 25 read all three magazines. The following questions were asked:

(a) How many of the citizens read at least one of the magazines?
(b) How many read none of them?
(c) How many read only X?

These questions would be quite difficult to answer without the aid of a Venn diagram. The Venn diagram in Figure 1.12 illustrates and illuminates the situation quite clearly however. The circles X, Y, and Z divide the universal set of 1000 citizens into 8 mutually exclusive, collectively exhaustive regions. In each region, we can list the exact number of people belonging exclusively to that region. Then, a little arithmetic will answer our questions. It is easiest to list the numbers of elements in each region if one starts with the intersection of the three sets X, Y, and Z and then works outward.

(a) The number who read at least one is $25 + 25 + 25 + 75 + 125 + 75 + 75 = 425$.

21

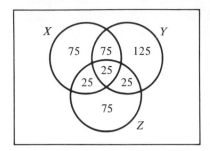

Figure 1.12   Venn diagram of sets X, Y, and Z.

(b) The number who read none is $1000 - 425 = 575$.
(c) The number who read only X is 75.

The following example illustrates the use of a table in finding the number of elements in a set.

Example 3

At a certain college, it is desired to learn the following information:

(a) How many students are either seniors or have grade-point averages above 3.00?
(b) How many students are either freshmen or have grade-point averages below 2.00?

The registrar furnishes us with the following information:

Grade-point averages

Academic year	Under 2.00	2.00–2.50	2.51–3.00	Over 3.00
Freshmen	75	170	130	25
Sophomores	60	120	100	20
Juniors	40	110	100	25
Seniors	25	100	70	30

Using this table, we can easily answer the above questions.

(a) Adding across the seniors row, we can see that there are 225 seniors. Adding down the over 3.00 column, there are 100 students over 3.00. From the intersection of the seniors row and the over 3.00 column, there are 30 people who are both seniors and are over 3.00. Thus, there are $225 + 100 - 30 = 295$ students who are either seniors or have grade-point averages over 3.00.

(b) Adding across the freshmen row, there are 400 freshmen. Adding down the 2.00 column, there are 200 students under 2.00. From the intersection of the freshmen row and the under 2.00 column, there are 75 people who are both freshmen and are under 2.00. Thus, there are

$400 + 200 - 75 = 525$ students who are either freshmen or have grade-point averages under 2.00.

Exercises

1. Let

    $U \leftarrow 0\ 1\ 2\ 3\ 4\ 5\ 6\ 7\ 8\ 9$
    $A \leftarrow 0\ 3\ 6\ 9$
    $B \leftarrow 0\ 1\ 2\ 3\ 4$
    $C \leftarrow 5\ 6\ 7\ 8\ 9$

   Find the following at an APL terminal:
   (a) ρA
   (b) ρB
   (c) $\rho(A\ INTERSECT\ B)$
   (d) $\rho(A\ UNION\ B)$
   (e) $\rho(COMPLEMENT\ C)$
   (f) $\rho(A\ DIFFERENCE\ B)$

2. Let

    $U \leftarrow {}'ABCDEFGHIJKLMNOPQRSTUVWXYZ'$
    $A \leftarrow {}'COMPUTER'$
    $B \leftarrow {}'TERMINAL'$

   Find the following at an APL terminal:
   (a) ρA
   (b) ρB
   (c) $\rho(A\ INTERSECT\ B)$
   (d) $\rho(A\ UNION\ B)$
   (e) $\rho(A\ DIFFERENCE\ B)$
   (f) $\rho(COMPLEMENT\ A)$

3. Find the number of cards in an ordinary deck of 52 playing cards which are
   (a) Either red or face cards.
   (b) Either kings or aces.
   (c) Neither diamonds nor aces.

4. In a certain class of 100 students, 15 got A in math, 10 made the Dean's list, and 5 got A in math and made the Dean's list. How many neither got A in math nor made the Dean's list?

5. A secretary phoned the 80 members of a club to call a meeting. The day of the meeting had to be either Wednesday or Friday. She found that 25 people were free on Wednesday only, 15 people were free on Friday only, and 20 people were free on both Wednesday and Friday. How many were free on neither day?

6. A college student is paid $1 for each person he interviews about his likes and dislikes for two types of deodorants, A and B. He finds that 30 like A, 25 like B, and 15 like both, and 10 like neither. How much should he be paid?

23

7. Five hundred women are interviewed about which sports they like. It is found that 185 like baseball, 135 like football, 110 like hockey, 50 like baseball and football, 45 like baseball and hockey, 35 like football and hockey, and 20 like all three sports.
   (a) How many like at least one of these sports?
   (b) How many like none of these sports?
   (c) How many like only baseball?
   (d) How many like football and hockey but not baseball?
   (e) How many like baseball or hockey but not football?

8. In trying to decide on the main course for a dinner, a chef finds that of 25 people who will be at the dinner, 14 like steak, 12 like lobster, and 11 like chicken. Also, 5 like steak and lobster, 5 like steak and chicken, 4 like lobster and chicken, and 2 like all three.
   (a) How many like steak only?
   (b) How many like lobster only?
   (c) How many like chicken only?
   (d) If they couldn't get lobster, how many people would be disappointed?

9. A poll is taken to see whether or not some people believe that a college education is necessary for a youth today. Their responses are tabulated below:

Sex	Yes	No	Not sure
Men	300	240	60
Women	200	160	40

   (a) How many men said no?
   (b) How many were either men or said no?
   (c) How many were either women or said yes?
   (d) How many people were included in the poll?

10. A poll is taken to relate a person's political preference to his income bracket with the following results:

Income bracket	Democrat	Republican	Independent
High	70	90	40
Middle	180	140	80
Low	50	70	80

   (a) How many are Democrats?
   (b) How many of the high income people are Republicans?
   (c) How many are either middle income or Independents?
   (d) How many are not Independents?

Logic 2

Logic is another application of Boolean algebra. The operations and laws used in logic are analogous to those of set theory in many respects. Knowledge of logic is often quite helpful in the deductive thinking process used in making decisions.

2.1 Statements and logical operations

In logic, a *statement* or a *proposition* is an assertion that can be either true or false but not both. This doesn't mean that everyone must have the same opinion of the truth value of the statement. Two people might disagree as to the truth value of the statement. However, for any given person at a given time, the statement is either true or false but not both. The following are examples of statements:

1. Learning mathematics using APL is fun.
2. Learning mathematics using APL is fun and easy.
3. Learning mathematics using APL is fun or hard.
4. Learning mathematics using APL is fun if it is easy.

The first example above is an example of a simple statement. A *simple statement* is a statement that makes just one assertion. The other statements above are examples of compound statements. A *compound statement* is a statement that makes more than one assertion. Compound statements are made up of two or more simple statements joined by "connectives" such as *and, or, if...then*. We shall consider methods for determining whether compound statements are true or false. This will depend on whether the simple statements making up the compound statements are true or false, as well as on the rules governing the particular connective or connectives being used in the statement. We now consider the rules governing certain logical connectives.

Conjunction

Let A and B be two statements. The *conjunction* of A and B, denoted by $A \wedge B$, and read as A "and" B, is defined by the following "truth table":

A	B	$A \wedge B$
T	T	T
T	F	F
F	T	F
F	F	F

A *truth table* is a table that defines the truth values of a compound statement based upon the truth values of the simple statements comprising it. $\wedge$ is defined so that $A \wedge B$ is true only if A and B are both true.

Logical conjunction, $\wedge$, is a keyboard operation in APL. Recall that in APL 1 can be interpreted to mean "true" and 0 "false." Consider the following uses of $\wedge$ on the APL terminal:

Examples

$1 \wedge 1$	1 "and" 1 is true.
1	
$1 \wedge 0$	1 "and" 0 is false.
0	
$0 \wedge 1$	0 "and" 1 is false.
0	
$0 \wedge 0$	0 "and" 0 is false.
0	
$A \leftarrow 1\ 1\ 0\ 0$	These correspond to the truth values
$B \leftarrow 1\ 0\ 1\ 0$	of A and B in the above truth table.
$A \wedge B$	The corresponding elements of A
1 0 0 0	and B are compared using the operation $\wedge$.

The following examples illustrate the use of the operation $\wedge$:

Examples

1. "In 1974, Johnny Bench batted over .300 *and* hit more than 30 home runs." Is this compound statement true or false? Let A be the statement that Bench batted over .300. Let B be the statement that he hit more than 30 home runs. The above statement can be symbolically represented as $A \wedge B$. Since Bench actually batted .280 with 33 home runs, A is false and B is true. Therefore, $A \wedge B$ is false.

2. "APL is a powerful language, *but* it is easy to learn." The word "but" here has the same meaning as the word "and." Let A be the statement that APL is a powerful language. Let B be the statement that APL is easy to learn. The above statement is symbolically represented as $A \wedge B$. Of course, we shall take both of the above statements A and B as true. Therefore, $A \wedge B$ is true. If the reader believes either A or B to be false, then for him, $A \wedge B$ is false. (Such a person is a member of the minority, we sincerely hope.)

Disjunction

Let A and B be two statements. The *disjunction* of A and B, denoted by $A \vee B$ and read as A "or" B, is defined by the following truth table:

A	B	$A \vee B$
T	T	T
T	F	T
F	T	T
F	F	F

The only time $A \vee B$ is false is if both A and B are false. The "or" here is the inclusive "or." In other words, it means one or the other or both. Logical disjunction is a keyboard function in APL.

Examples

$1 \vee 1$
1   1 "or" 1 is true.

$1 \vee 0$
1   1 "or" 0 is true.

$0 \vee 1$
1   0 "or" 1 is true.

$0 \vee 0$
0   0 "or" 0 is false.

$A \leftarrow 1\ 1\ 0\ 0$
$B \leftarrow 1\ 0\ 1\ 0$

$A \vee B$ The corresponding elements of A
1 1 1 0   and B are compared using the operation $\vee$.

The following examples illustrate the use of the operation $\vee$:

Examples

1. "In 1974, Johnny Bench batted over .300 *or* hit more than 30 home runs." As before, let A be the statement that Bench batted over .300. Let B be the statement that he hit more than 30 home runs. The above

27

statement can be symbolically represented by $A \lor B$. Since A is false, but B is true, then $A \lor B$ is true.

2. "The President is a good speaker *or* he is a dictator." Let A be the statement that the President is a good speaker. Let B be the statement that he is a dictator. Then, the above statement can be represented as $A \lor B$. Whether or not the above statement is true would depend on the person's assessment of the truth values of A and B. The only people for which this statement would be false are those who believe that the President is not a good speaker and that he is not a dictator.

3. "The candidate will either win the election *or* he will lose it." Let A be the statement that the candidate will win the election. Let B be the statement that he will lose the election. The above statement $A \lor B$ is always true, since at least one of the statements A or B must be true.

Negation

Let A be a statement. The logical negation of A, denoted by $\sim A$ and read as "not" A, is defined by the following truth table:

A	$\sim A$
T	F
F	T

Thus, the truth value of $\sim A$ is just the opposite of that of A. Logical negation is also a keyboard operation in APL.

Examples

~ 1 The opposite of true is false.

0

~ 0 The opposite of false is true.

1

$C \leftarrow 1\ 0$

$\sim C$ $\sim$ negates each element of C.

0 1

If A is a statement, then $\sim A$ is used to reflect the statement with meaning opposite to that of A. Consider the following examples:

Examples

1. "The Edsel was not a successful automobile." Let A be the statement that the Edsel was a successful automobile. The above statement can be symbolized by $\sim A$. Since A is false, then $\sim A$ is true.

2. "Tomorrow it will neither rain nor be colder." Let R denote the statement that tomorrow it will rain. Let C denote the statement that tomorrow it will be colder. The above statement can be symbolized by

$\sim(R\vee C)$. Let us investigate the truth table for this statement.

$R\leftarrow1\ 1\ 0\ 0$
$C\leftarrow1\ 0\ 1\ 0$

$R\vee C$
1 1 1 0

$\sim(R\vee C)$
0 0 0 1

Thus, this statement will be true only if R and C are both false. Another operation in APL that accomplishes the same objective as $\sim(R\vee C)$ is the "nor" operation $\veebar$ obtained by overstriking the $\vee$ and the $\sim$. Thus,

$R\veebar C$
0 0 0 1

3. "Tomorrow it will not rain and be colder." This can be symbolized by $\sim(R\wedge C)$. Let us investigate the truth table for this example.

$R\wedge C$
1 0 0 0

$\sim(R\wedge C)$
0 1 1 1

Thus, this statement will be true unless R and C are both true. Another operation in APL that accomplishes the same objective as $\sim(R\wedge C)$ is the "nand" operation $\barwedge$ obtained by overstriking the $\wedge$ and the $\sim$. Thus,

$R\barwedge C$
0 1 1 1

4. "Tomorrow it will not rain and it will not be colder." The logical symbolism for this statement is $(\sim R)\wedge(\sim C)$. The truth table is as follows:

$(\sim R)\wedge(\sim C)$

0 0 0 1

Notice that this statement has the same truth table as that of Example 2. Actually, the statements in Examples 2 and 4 have exactly the same meaning. They are logically equivalent statements. This topic of logically equivalent statements will be discussed in detail in Section 2.3.

5. "The Princess is neither a beauty nor a charmer, but she is loaded with money." Symbolically, this would be represented as $(\sim(A\vee B))\wedge C$, where A is that she has beauty, B is that she is a charmer, and C is that she is loaded with money. In order to construct a truth table for a

statement containing three simple statements, it is necessary to include all possible combinations of truth values for the three statements. This is illustrated below:

$A \leftarrow 1\ 1\ 1\ 1\ 0\ 0\ 0\ 0$
$B \leftarrow 1\ 1\ 0\ 0\ 1\ 1\ 0\ 0$
$C \leftarrow 1\ 0\ 1\ 0\ 1\ 0\ 1\ 0$

$A \lor B$
1 1 1 1 1 1 0 0
$\sim (A \lor B)$
0 0 0 0 0 0 1 1
$(\sim (A \lor B)) \land C$
0 0 0 0 0 0 1 0

Thus, this statement is true only if A and B are false and C is true.

Exclusive disjunction

Let A and B be two statements. The *exclusive disjunction* of A and B, usually denoted by $A \underline{\lor} B$, is defined by the following truth table:

A	B	$A \underline{\lor} B$
T	T	F
T	F	T
F	T	T
F	F	F

The symbol $\underline{\lor}$ is also read as "or." However, it is the exclusive "or." It is used when the meaning is A or B, but not both. In other words, when it is not possible for both A and B to be true at the same time. In APL, the symbol $\neq$ is used for the exclusive disjunction.

Example

"Sam will either wear his blue suit or his brown suit." Let A denote the statement that Sam will wear his blue suit. Let B denote the statement that Sam will wear his brown suit. Since he can't wear both suits, this statement is denoted as $A \underline{\lor} B$.

$A \leftarrow 1\ 1\ 0\ 0$
$B \leftarrow 1\ 0\ 1\ 0$

$A \neq B$
0 1 1 0

Since $1 \neq 1$ is false, $1 \neq 0$ is true, $0 \neq 1$ is true, and $0 \neq 0$ is false.

EXERCISES

1. Construct truth tables for the following compound statements:
    (a) $(\sim(A \vee B)) \wedge B$
    (b) $(A \,\underline{\vee}\, B) \wedge B$
    (c) $B \wedge \sim(A \vee B)$
    (d) $\sim B \vee A$
    (e) $\sim A \wedge B$
    (f) $(\sim A) \vee (\sim B)$
    (g) $\sim(A \wedge B)$
    (h) $\sim A \neq B$

2. Check your answers to Exercise 1 at an APL terminal.

3. Construct truth table for the following compound statements:
    (a) $(A \wedge B) \vee C$
    (b) $(A \vee C) \wedge (B \vee C)$
    (c) $\sim((\sim A) \wedge (\sim B)) \wedge C$
    (d) $(A \vee B) \wedge C$
    (e) $((\sim A) \,\underline{\wedge}\, (\sim B)) \wedge C$
    (f) $(A \vee B \vee C) \wedge \sim(A \vee B \vee C)$
    (g) $(A \wedge B \wedge C) \vee \sim(A \wedge B \wedge C)$
    (h) $A \neq (B \neq C)$

4. Check your answers to Exercise 3 at an APL terminal.

5. Let A be the statement, "We are in a period of inflation." Let B be, "The standard of living is rising." Let C be, "The economy is sound."
   Express each of the following compound statements in symbolic form, and find their truth tables:
    (a) We are in a period of inflation, and the economy is not sound.
    (b) We are neither in a period of inflation nor is the economy sound.
    (c) We are in a period of inflation or the standard of living is rising.
    (d) The economy is sound and the standard of living is rising, and we are not in a period of inflation.
    (e) The economy is sound, but we are in a period of inflation or the standard of living is rising.
    (f) We are not in a period of inflation and the standard of living is not rising, but the economy is sound.

6. Suppose that for Mrs. L, statement A is true, B is true, and C is false. Then, what truth values should Mrs. L assign to each of the statements in Exercise 5?

7. Let A be the statement, "Prices are rising." Let B be, "There is a great deal of unemployment." Let C be, "People are discouraged."
   Express each of the following compound statements in symbolic form, and find their truth tables:
    (a) Prices are rising and people are discouraged.
    (b) Prices are not rising or there is not a great deal of unemployment.
    (c) Prices are rising or there is a great deal of unemployment, but people are not discouraged.

(d) People are discouraged, but prices are not rising and there is not a great deal of unemployment.
(e) There is a great deal of unemployment. However, prices are neither rising nor are people discouraged.
(f) Prices are rising, but there is not a great deal of unemployment and people are not discouraged.

8. Suppose that for Mr. E, A is true, B is true, and C is false. Determine the truth values that Mr. E should assign to each of the statements in Exercise 7.

2.2 Conditional statements

A very important logical connective in mathematics and in logic is the conditional.

Conditional

Let A and B be two statements. The *conditional* statement "if A then B," denoted by $A \Rightarrow B$, is defined by the following truth table:

A	B	$A \Rightarrow B$
T	T	T
T	F	F
F	T	T
F	F	T

In the statement $A \Rightarrow B$, A is called the *antecedent* and B is called the *consequent*. Notice that $A \Rightarrow B$ is true in all cases except for the one in which the antecedent is true and the consequent false.

Ordinarily, one doesn't try to justify definitions in mathematics. However, perhaps this truth table doesn't appear very obvious to the reader. Therefore, we shall attempt to justify it with the following example:

Example

Suppose someone said, "If the Yankees play on television, then I will watch the game." Let A be the statement: "The Yankees play on television," and let B be the statement: "I will watch the game."

Possibility 1: A true and B true

In this case, the statement is true, since the person did what he said he would do. He was telling the truth.

Possibility 2: A true and B false.

In this case, we would all agree that the person lied. The Yankees were on television, but he didn't watch the game. Thus, the statement $A \Rightarrow B$ is false in this case.

Possibilities 3 and 4: A false and B true or false.

We really couldn't call this person a liar in either of these cases, since he didn't say whether or not he would watch the game if the Yankees were not on television. Thus, he was telling the truth in these cases, as far as we know.

The conditional symbol, $\Rightarrow$, does not appear on the APL keyboard. (A similar symbol $\rightarrow$ does appear. However, this symbol is used for branching in programs and is not used for the conditional.) In APL, the symbol $\leqslant$ can be used to compare the truth values of two statements A and B. The resulting truth values are exactly the same as those in the truth table for the conditional.

Example

$$A \leftarrow 1 \ 1 \ 0 \ 0$$
$$B \leftarrow 1 \ 0 \ 1 \ 0$$

$\qquad A \leqslant B$ Since $1 \leqslant 1$ is true, $1 \leqslant 0$ is false, $0 \leqslant 1$
$1 \ 0 \ 1 \ 1$ is true, and $0 \leqslant 0$ is true.

Implications

A statement A is said to *imply* a statement B if B must be true whenever A is true. If A implies B, then the second row of the truth table for $A \Rightarrow B$ is not possible, since we can not have A true and B false. Thus, if A implies B, the conditional $A \Rightarrow B$ is always true. Such a statement which is always true is called a *logically true statement* or a *tautology*.

In mathematics, implications are very important, since all theorems and definitions are implications. If A implies B, then the conditional $A \Rightarrow B$ can be read in one of the following ways:

If A then B.
A implies B.
B if A.
B whenever A.
A is a sufficient condition for B.

Consider the following examples:

1. "If ψ is a mathematical symbol, then ψ appears on the APL keyboard." This is *not* an implication, since it is possible for the consequent to be false while the antecedent is true. For example, ψ is a mathematical symbol which does not appear on the APL keyboard.
2. "If N is an integer, then N is a rational number." This is an implication, because whenever the antecedent is true, the consequent must be true, since the set of integers is a subset of the set of rational numbers.
3. "If r is a root of a polynomial, then $x - r$ can be factored out of the polynomial." This is not only an implication but a theorem in algebra.

4. "If $x - r$ can be factored out of a polynomial, then r is a root of the polynomial." This is also a theorem in algebra. It is the converse of the previous theorem. Therefore, it is also an implication.

Biconditional

Let A and B be two statements. The biconditional statement "A if and only if B," symbolized by $A \Leftrightarrow B$, is defined by the following truth table:

A	B	$A \Leftrightarrow B$
T	T	T
T	F	F
F	T	F
F	F	T

Thus, $A \Leftrightarrow B$ is true whenever A and B have the same truth values. In fact, we could use a biconditional statement to describe the biconditional truth table: "A is true if and only if B is true; and A is false if and only if B is false." In APL, the biconditional can be conveyed by the $=$ symbol between the two statements.

Example

```
A←1 1 0 0
B←1 0 1 0

A=B
1 0 0 1
```
Since $1=1$ is true, $1=0$ is false, $0=1$ is false, and $0=0$ is true.

```
(A⩽B)∧(B⩽A)
1 0 0 1
```
These illustrate that the biconditional is really the conjunction of two conditionals.

$$(A=B)=(A⩽B)\wedge(B⩽A)$$

Double implications

If A implies B, and also B implies A, then we have a *double implication*. In mathematics, definitions are double implications, since if A is defined by B, then A is just a shorter way to say B, and A is true if and only if B is true. Many theorems are also double implications. A theorem is a double implication if its converse is also a theorem. In the previous examples of implications, Example 3 is a double implication. In fact, Examples 3 and 4 can be combined into one theorem: r is a root of a polynomial if and only if $x - r$ can be factored out of the polynomial. This can also be stated as: A necessary and sufficient condition that r be a root of a polynomial is that $x - r$ can be factored out of the polynomial. Not every theorem is a double implication, however. Example 2 is not a double implication. There are an infinite number of rational numbers which are not integers.

Before leaving this section, let us consider some examples of the truth tables of some more complex statements involving the conditional and the biconditional.

Examples

1. "If the weather is pleasant tomorrow, then Mr. E will play golf or go fishing." Let A be the statement that the weather is pleasant tomorrow, B be the statement that Mr. E will play golf, and C be the statement that he will go fishing. The above compound statement can be symbolized by $A \Rightarrow (B \lor C)$.

$A \leftarrow 1\ 1\ 1\ 1\ 0\ 0\ 0\ 0$
$B \leftarrow 1\ 1\ 0\ 0\ 1\ 1\ 0\ 0$
$C \leftarrow 1\ 0\ 1\ 0\ 1\ 0\ 1\ 0$

$B \lor C$
1 1 1 0 1 1 1 0

$A \leqslant (B \lor C)$ Recall that in APL, $\Rightarrow$ is represented
1 1 1 1 0 1 1 1 1   by $\leqslant$.

Thus, the only way for this statement to be false is for A to be true and B and C to both be false. In other words, it would have to be pleasant and Mr. E would have to not play golf and not fish.

2. "Mr. E will play golf or go fishing if and only if the weather is pleasant." This statement can be symbolized by $(B \lor C) \Leftrightarrow A$.

$(B \lor C) = A$ Recall that in APL, $\Leftrightarrow$ is represented
1 1 1 0 0 0 0 1   by $=$.

This statement can be false in many ways. It will be false if the weather is pleasant and Mr. E fails to play golf or fish. It will also be false if he plays golf or fishes and the weather is not pleasant.

3. "If the weather is pleasant, then Mr. E will play golf. But, the weather is pleasant. Therefore, Mr. E will play golf." This can be symbolized as $((A \Rightarrow B) \land A) \Rightarrow B$.

$A \leftarrow 1\ 1\ 0\ 0$
$B \leftarrow 1\ 0\ 1\ 0$

$A \leqslant B$
1 0 1 1

$(A \leqslant B) \land A$
1 0 0 0

$((A \leqslant B) \land A) \leqslant B$
1 1 1 1

This statement is always true. It is a logically true statement. In fact, it is a sneak preview of a "valid" argument. This will be considered in detail in Section 2.4.

EXERCISES

1. Construct truth tables for the following compound statements:
   (a) $(A \vee (\sim A)) \Rightarrow B$
   (b) $A \Rightarrow (B \Rightarrow A)$
   (c) $A \Leftrightarrow \sim B$
   (d) $((\sim B) \Rightarrow (\sim A)) \Rightarrow (A \Rightarrow B)$
   (e) $(A \vee B) \Leftrightarrow C$
   (f) $((A \Rightarrow B) \wedge \sim B) \Rightarrow \sim A$

2. Check your answers to Exercise 1 at an APL terminal using $\leqslant$ and $=$ for $\Rightarrow$ and $\Leftrightarrow$.

3. Let A be the statement, "We are in a period of inflation." Let B be the statement, "The standard of living is rising." Let C be the statement, "The economy is sound."
   Express each of the following statements in symbolic form, and find their truth tables.
   (a) The economy is sound if and only if the standard of living is rising.
   (b) If the economy is sound, then the standard of living is rising and we are not in a period of inflation.
   (c) If we are in a period of inflation, then the standard of living is not rising and the economy is not sound.
   (d) The economy is sound if and only if the standard of living is rising and we are not in a period of inflation.

4. Suppose that for Mrs. L, statement A is true, B is true, and C is false. Then, what truth values should Mrs. L assign to each of the compound statements in Exercise 3?

5. Let A be the statement, "Prices are rising." Let B be the statement, "There is a great deal of unemployment." Let C be the statement, "People are discouraged."
   Express each of the following statements in symbolic form and find their truth tables.
   (a) If prices are rising and there is a great deal of unemployment, then people are discouraged.
   (b) Prices are rising and there is a great deal of unemployment if and only if people are discouraged.
   (c) If prices are rising or there is a great deal of unemployment, then people are discouraged.
   (d) If there is not a great deal of unemployment and prices are not rising, then people are not discouraged.

6. Suppose that for Mr. E, A is true, B is true, and C is false. Determine the truth values that Mr. E should assign to each of the compound statements in Exercise 5.

7. Which of the following are implications? Double implications?
   (a) If $x = 2$, then $x^2 = 4$.
   (b) If people are discouraged, then prices are rising.
   (c) If the standard of living is rising, then the economy is sound.

(d) If one is from Boston, then he is a Red Sox fan.
(e) If x is an element of both A and B, then $x \in A \cap B$.
(f) If the sides of a quadrilateral are all equal, then it is a square.
(g) If the pairs of opposite sides of a quadrilateral are parallel, then it is a parallelogram.
(h) If the Yankees win the pennant, they will play in the World Series.

2.3 Logical equivalence

Two statements A and B are said to be *logically equivalent*, symbolized by $A \equiv B$, if and only if they have the same truth tables. If A and B are logically equivalent, then if the truth tables for A and B are compared using the relation $=$, the result will be all 1's (trues).

Example

$((\sim A) \vee B))$ is logically equivalent to $(A \Rightarrow B)$, since

$$A \leftarrow 1 \ 1 \ 0 \ 0$$
$$B \leftarrow 1 \ 0 \ 1 \ 0$$

$$\sim A$$
$$0 \ 0 \ 1 \ 1$$

$$(\sim A) \vee B$$
$$1 \ 0 \ 1 \ 1$$

$A \leqslant B$	Recall that $\leqslant$ is used in place of $\Rightarrow$
$1\ 0\ 1\ 1$	in APL.
$((\sim A) \vee B) = (A \leqslant B)$	
$1\ 1\ 1\ 1$	Since $1 = 1$, $0 = 0$, $1 = 1$, and $1 = 1$ are all true.

In the beginning of this chapter, we stated that, like set theory, symbolic logic is a Boolean algebra. If this is so, then the logical operations should satisfy the laws of Boolean algebra. To illustrate this, we shall replace the set theory operation intersection with the logical operation conjunction, the set theory operation union with the logical operation disjunction, and the set theory operation complement with the logical operation negation. Also, the universal set U will be replaced by a logically true statement U consisting entirely of 1's, and the empty set $\varnothing$ will be replaced by a logically false statement $\varnothing$ consisting entirely of 0's. Then, the laws of Boolean algebra in logic are as follows.

The laws of Boolean algebra

The idempotent laws

$$(A \wedge A) \equiv A \quad \text{and} \quad (A \vee A) \equiv A.$$

37

The commutative laws

$$(A \wedge B) \equiv (B \wedge A) \quad \text{and} \quad (A \vee B) \equiv (B \vee A).$$

The associative laws

$$(A \wedge (B \wedge C)) \equiv ((A \wedge B) \wedge C) \quad \text{and} \quad (A \vee (B \vee C)) \equiv ((A \vee B) \vee C).$$

The distributive laws

$$(A \wedge (B \vee C)) \equiv ((A \wedge B) \vee (A \wedge C)).$$
$$(A \vee (B \wedge C)) \equiv ((A \vee B) \wedge (A \vee C)).$$

Operations with the logically true statement U

$$(A \wedge U) \equiv A \quad \text{and} \quad (A \vee U) \equiv U.$$

Operations with the logically false statement ∅

$$(A \wedge \varnothing) \equiv \varnothing \quad \text{and} \quad (A \vee \varnothing) \equiv A.$$

Laws of negation

$$(\sim(\sim A)) \equiv A.$$
$$(A \vee \sim A) \equiv U.$$
$$(A \wedge \sim A) \equiv \varnothing.$$
$$(\sim U) \equiv \varnothing.$$
$$(\sim \varnothing) \equiv U.$$

DeMorgan's laws

$$(\sim(A \vee B)) \equiv ((\sim A) \wedge (\sim B))$$
$$(\sim(A \wedge B)) \equiv ((\sim A) \vee (\sim B))$$

Let us verify some of these laws of Boolean algebra in APL.

Example 1

Let us verify the law $(A \wedge U) \equiv A$.

```
A←1 0
U←1 1
```
The logically true statement U (all 1's).

```
   A∧U
1 0
```
Since $A \wedge U$ has the same truth table as A, then $(A \wedge U) \equiv A$.

Example 2

Let us now verify that $(A \wedge \sim A) \equiv \emptyset$.

A←1 0
~A
0 1

A∧~A
0 0

Since $A \wedge \sim A$ has the same truth table as the logically false statement (all 0's), then $(A \wedge \sim A) \equiv \emptyset$.

Example 3

We now verify the associative law $(A \wedge (B \wedge C)) \equiv ((A \wedge B) \wedge C)$.

A←1 1 1 1 0 0 0 0
B←1 1 0 0 1 1 0 0
C←1 0 1 0 1 0 1 0

B∧C
1 0 0 0 1 0 0 0

A∧(B∧C)
1 0 0 0 0 0 0 0

A∧B
1 1 0 0 0 0 0 0

(A∧B)∧C
1 0 0 0 0 0 0 0

Since $(A \wedge (B \wedge C))$ and $((A \wedge B) \wedge C)$ have the same truth tables, they are logically equivalent.

Example 4

We now verify the distributive law $(A \wedge (B \vee C)) \equiv ((A \wedge B) \vee (A \wedge C))$.

B∨C
1 1 1 0 1 1 1 0

A∧(B∨C)
1 1 1 0 0 0 0 0

A∧B
1 1 0 0 0 0 0 0

A∧C
1 0 1 0 0 0 0 0

(A∧B)∨(A∧C)
1 1 1 0 0 0 0 0

Note that the second and fifth sets of 1's and 0's are the same.

39

Example 5

Finally, we verify the DeMorgan law $(\sim(A\vee B))\equiv((\sim A)\wedge(\sim B))$.

      A←1 1 0 0
      B←1 0 1 0

      $A\vee B$
1 1 1 0

      $\sim(A\vee B)$ Note the second and fifth lines of 1's
0 0 0 1   and 0's are the same.

      $\sim A$

0 0 1 1

      $\sim B$
0 1 0 1

      $(\sim A)\wedge(\sim B)$
0 0 0 1

Of course, there are many other logically equivalent statements not included in our list of laws of Boolean algebra.

Example 6

$(A\Leftrightarrow B)\equiv((A\Rightarrow B)\wedge(B\Rightarrow A))$.

      $A=B$
1 0 0 1

      $A\leqslant B$
1 0 1 1   Note that the first and last lines of
                                1's and 0's are the same.
      $B\leqslant A$
1 1 0 1

      $(A\leqslant B)\wedge(B\leqslant A)$
1 0 0 1

This last example shows that the biconditional is logically equivalent to the conjunction of the two conditionals.

EXERCISES

1. Check the validity of the following laws of Boolean algebra at an APL terminal:
    (a) $(A\wedge B)\equiv(B\wedge A)$
    (b) $(A\vee(B\vee C))\equiv((A\vee B)\vee C)$
    (c) $(A\vee(B\wedge C))\equiv((A\vee B)\wedge(A\vee C))$
    (d) $(A\vee U)\equiv U$
    (e) $(A\vee\sim A)\equiv U$
    (f) $(\sim(A\wedge B))\equiv((\sim A)\vee(\sim B))$

2. Determine whether or not the following pairs of statements are logically equivalent at an APL terminal:
(a) $((\sim A) \Rightarrow (\sim B))$ and $(B \Rightarrow A)$.
(b) $(A \wedge (A \Rightarrow B))$ and $(A \Rightarrow B)$
(c) $((A \Rightarrow B) \wedge (B \Rightarrow C))$ and $(A \Rightarrow C)$
(d) $(((A \wedge \sim B) \vee (A \wedge C)) \Rightarrow (B \wedge C))$ and $(A \Rightarrow B)$

2.4   Arguments

An *argument* is an assertion that from a set of one or more statements, called the *premises* or the *hypotheses*, one can deduce another statement, called the *conclusion*.

If the statements comprising the hypothesis are denoted by $A_1, A_2, \ldots, A_n$, and the conclusion is denoted by C, then the argument can be expressed as a conditional: $(A_1 \wedge A_2 \wedge \ldots \wedge A_n) \Rightarrow C$.

One of the major applications of symbolic logic is in determining whether arguments are *valid* or *invalid*. If an argument is invalid, it is often called a *fallacy*. An argument is valid if it is a logically true statement or a tautology. In other words, it is valid if its truth table consists entirely of 1's (trues). Thus, it is the symbolic form of the argument, rather than the particular facts making up the statements in the hypothesis and conclusion, that determines whether it is valid or a fallacy. We will consider several examples now.

Example 1

Consider the following argument: "If you brush with toothpaste X, you will have fewer cavities. But, you do not brush with toothpaste X. Therefore, you will not have fewer cavities." Let A be the statement that you brush your teeth with toothpaste X, and let B be the statement that you will have fewer cavities. The above argument can then be symbolically represented as

$$((A \Rightarrow B) \wedge \sim A) \Rightarrow (\sim B).$$

Let us use APL to test this argument for validity.

```
    A←1 1 0 0
    B←1 0 1 0

    A≤B
1 0 1 1

    ~A
0 0 1 1

    (A≤B)∧~A
0 0 1 1

    ~B
0 1 0 1
```

$$((A \leqslant B) \wedge \sim A) \Rightarrow (\sim B)$$
1 1 0 1

Since we don't get all 1's in the truth table, this argument is not valid. It is a fallacy. In fact, fallacies of this form are quite common.

Example 2

"If you brush with toothpaste X, you will have fewer cavities. You do brush with toothpaste X. Thus, you will have fewer cavities." Letting A and B be as in Example 1, this argument can be symbolically represented as $((A \Rightarrow B) \wedge A) \Rightarrow B$. Let's test it for validity.

$A \leftarrow$ 1 1 0 0
$B \leftarrow$ 1 0 0 0

$A \leqslant B$
1 0 1 1

$(A \leqslant B) \wedge A$
1 0 0 0

$((A \leqslant B) \wedge A) \leqslant B$
1 1 1 1

Since this is a logically true statement (all 1's in the truth table), then the argument is valid. [*Note*: This does not mean that toothpaste X *caused* you to have fewer cavities, nor does it mean that you actually have fewer cavities. It only means that the above argument was formed in such a way as to make it valid. Any good advertizing firm would be very careful to present valid arguments for advertizing its products.]

Example 3

"If $\sqrt{2}$ is rational, then it can be expressed in the form a/b, where a and b are integers and $b \neq 0$. However, $\sqrt{2}$ cannot be expressed in this form. Therefore, $\sqrt{2}$ not rational." Let A be the statement that $\sqrt{2}$ is rational. Let B be the statement that $\sqrt{2}$ can be expressed in the form a/b, where a and b are integers and $b \neq 0$. The above argument has the form $((A \Rightarrow B) \wedge \sim B) \Rightarrow \sim A$. Testing it for validity:

$A \leftarrow$ 1 1 0 0
$B \leftarrow$ 1 0 1 0

$A \leqslant B$
1 0 1 1

$\sim B$
0 1 0 1

$(A \leqslant B) \wedge \sim B$
0 0 0 1

$$\sim A$$
0 0 1 1

$$((A \leqslant B) \wedge \sim B) \leqslant \sim A$$
1 1 1 1

Thus, the argument is valid.

Example 4

"If you do not do the work, you will not pass this course. Thus, if you pass this course, you did the work." Let A be the statement that you do the work. Let B be the statement that you will pass this course. The symbolic representation of this argument is $((\sim A) \Rightarrow (\sim B)) \Rightarrow (B \Rightarrow A)$.

$$A \leftarrow 1\ 1\ 0\ 0$$
$$B \leftarrow 1\ 0\ 1\ 0$$

$$(\sim A) \leqslant (\sim B)$$
1 1 0 1

$$B \leqslant A$$
1 1 0 1

$$((\sim A) \leqslant (\sim B)) \leqslant (B \leqslant A)$$
1 1 1 1

The argument is valid.

Example 5

"If you do the work, you will pass the course. If you pass the course, you will be very happy. Therefore, if you do the work, you will be very happy." Let A and B be as in Example 4, and let C be the statement that you will be very happy. We now check this argument for validity. It can be symbolized by $((A \Rightarrow B) \wedge (B \Rightarrow C)) \Rightarrow (A \Rightarrow C)$.

$$A \leftarrow 1\ 1\ 1\ 1\ 0\ 0\ 0\ 0$$
$$B \leftarrow 1\ 1\ 0\ 0\ 1\ 1\ 0\ 0$$
$$C \leftarrow 1\ 0\ 1\ 0\ 1\ 0\ 1\ 0$$

$$A \leqslant B$$
1 1 0 0 1 1 1 1

$$B \leqslant C$$
1 0 1 1 1 0 1 1

$$(A \leqslant B) \wedge (B \leqslant C)$$
1 0 0 0 1 0 1 1

$$A \leqslant C$$
1 0 1 0 1 1 1 1

$$((A \leqslant B) \wedge (B \leqslant C)) \leqslant (A \leqslant C)$$
1 1 1 1 1 1 1 1

The argument is valid. Happiness is doing your work...especially at an APL terminal.

1. Test the following symbolic representations of arguments for validity at an APL terminal:
   (a) $(A \Rightarrow \sim B) \Rightarrow (B \Rightarrow \sim A)$
   (b) $((A \Rightarrow B) \wedge B) \Rightarrow A$
   (c) $((A \vee B) \wedge \sim A) \Rightarrow B$
   (d) $((A \vee B) \wedge B) \Rightarrow \sim A$
   (e) $((\sim(A \wedge B)) \wedge A) \Rightarrow \sim B$
   (f) $((A \Leftrightarrow B) \wedge (B \vee C)) \Rightarrow (A \vee C)$

In each of the following exercises, form the symbolic representation of the argument and determine whether it is valid or a fallacy:

2. If you are mathematically inclined, then you will have no difficulty learning APL. Therefore, if you have difficulty learning APL, then you are not mathematically inclined.

3. The President is a good speaker or he is a diplomat. He is not a good speaker. Thus, he is a diplomat.

4. If it rains tomorrow, I won't play golf. If I don't play golf, I will be angry. So, if it rains tomorrow, I will be angry.

5. The standard of living is rising or we are in a period of inflation. The standard of living is rising. Therefore, we are not in a period of inflation.

6. The standard of living is rising if and only if the economy is sound. The standard of living is rising. So, we can conclude that the economy is sound.

7. If prices are rising, then people are discouraged. But, prices are not rising. Thus, people are not discouraged.

8. If prices are falling, then there is a great deal of unemployment. If there is a great deal of unemployment, then people are discouraged. Therefore, if prices are falling, people are discouraged.

9. If you are not careful, then you will be hurt. You didn't get hurt. So, you must have been careful.

10. If we don't help country X, they will lose their independence. Thus, if we do help country X, they will not lose their independence.

11. If you like Merle Haggard, then you like country music. But, you don't like Merle Haggard. So, you don't like country music.

12. If you don't believe in yourself, then you won't be successful. Therefore, if you are successful, you must believe in yourself.

Vectors and matrices 3

A matrix is a rectangular array of numbers arranged in rows and columns. As we shall see, a matrix is a convenient device for organizing data that would otherwise require several pages. In addition, many mathematical problems can be expressed much more concisely and solved much more easily using matrix notation and matrix operations. For example, a system of 10 linear equations with 10 unknowns can be expressed as a simple matrix equation $A \cdot X = B$, and solved by a matrix equation $X = A^{-1} \cdot B$. For reasons such as these, matrix algebra has become a required topic for students of business administration and many branches of science and engineering. In this chapter, we will concern ourselves with some of the basic theory of matrix algebra and explore the use of APL in working with vectors and matrices. In the next two chapters, we will consider some applications of matrices. As we shall see, APL is very well suited for handling arrays such as vectors and matrices.

3.1 Vectors

Definition of a vector

A vector with n components is an ordered array of n real numbers.

Examples

1. An example of a 4-component row vector is

$$R = (1, 2, 3, 4).$$

2. An example of a 4-component column vector is

$$C = \begin{bmatrix} 1 \\ 2 \\ 3 \\ 4 \end{bmatrix}.$$

Unless otherwise stated, we shall always use row vectors. As with sets, a vector is denoted as follows in APL:

> $R \leftarrow 1\ 2\ 3\ 4$ The name of the vector is R. R has 4 real components: 1, 2, 3, and 4.

> R The command to print out R.
> 1 2 3 4

Unlike sets, however, an element in a vector may be repeated. Also, unlike sets, the order of the components in a vector is significant. A vector is an *ordered* array of numbers.

Example

For contrast, consider the following sets:

> $A \leftarrow 1\ 2\ 3\ 4$
> $B \leftarrow 2\ 1\ 3\ 4$

> A EQUAL B
>
> 1   As sets, A does equal B, since they have the same elements.

> $A = B$
> 0 0 1 1 $1 = 2$, $2 = 1$ are false, but $3 = 3$, $4 = 4$ are true.

Equal vectors

Two vectors, A and B, are equal, denoted by $A = B$, if and only if they have the same number of components and all corresponding components are equal.

Examples

> $A \leftarrow 1\ 2\ 3\ 4$
> $B \leftarrow 1\ 2\ 3\ 4\ 5$
> $C \leftarrow 2\ 1\ 3\ 4$
> $D \leftarrow \iota\ 4$ Recall that $\iota 4$ yields the positive integers from 1 to 4.

> $A = B$
> LENGTH ERROR A and B do not have the same number of components.

$A = C$
0 0 1 1

$\wedge/A = C$
0

A does not equal C, since all *corresponding* components are not equal.

$A = D$
1 1 1 1

$\wedge/A = D$
1

A does equal D, since they have the same number of components and corresponding components are equal.

Thus, the APL operation to determine whether or not two vectors are equal is the "and" reduction, denoted as $\wedge/V = W$, where V and W are the vectors being compared.

An application of vectors

Vectors are often used as a convenient way of representing data. For example, suppose that a company manufactures a product called a "gadget." Suppose that a gadget is made up of parts categorized by the parts numbers 051, 035, 068, and 047, in the following quantities: 3 of part 051, 5 of part 035, 2 of part 068, and 1 of part 047. This information can be conveniently conveyed using the following parts vector G (for "gadgets"):

$$G = (\overset{051}{3} , \overset{035}{5} , \overset{068}{2} , \overset{047}{1})$$

The position of the component tells the employee the part to which it corresponds. This company has the convention that the first component is the number of 051's, the second component is the number of 035's, the third component is the number of 068's, and the fourth component is the number of 047's. If everyone in the company knows of this convention, then the parts vector for a gadget can be denoted in APL as simply

$G \leftarrow 3\ 5\ 2\ 1$.

If this company also produces "widgets," which use the same parts, and if the parts vector for a widget is

$W \leftarrow 4\ 6\ 1\ 2$,

then the number of 051's in a widget is 4, the number of 035's is 6, the number of 068's is 1, and the number of 047's is 2.

47

Indexing with vectors

In APL, the *i*th component of a vector *G* is denoted by *G*[I].

Examples

G←3 5 2 1

G [1]   The first component of *G*.
3
G [3]   The third component of *G*.
2
G [5]
INDEX ERROR   *G* has no 5th component.

G [2.5]
INDEX ERROR   The index must be a positive integer.

G [1 3]   The 1st and 3rd components.
3 2
G [4 3 1 2]   Rearranging the components in order
1 2 3 5   of magnitude.

Altering a vector

To replace a component of a vector with a new number, do as follows:

G [2]←4   Replace the second component of *G*
                               by 4.
G
3 4 2 1

G [4]←3   Replace the fourth component of *G*
                               by 3.
G
3 4 2 3

Catenation

The operation of catenation with vectors is the same as with sets. To chain a component or a vector onto the end of a vector in APL, just use the comma.

Examples

G←3 5 2 1
W←4 6 1 2

G,4   Chaining 4 onto the end of *G*.
3 5 2 1 4

G, W   Chaining *W* onto the end of *G*.
3 5 2 1 4 6 1 2

The size of a vector

As with sets, the ρ ("rho") applied to a vector computes the size of the vector. In other words, it computes the number of components in the vector.

$R \leftarrow 10\ 3\ 15\ 7\ 986$

ρR The number of components in R.
5

$\rho\rho R$ $\rho\rho R$ gives the number of dimensions
1   of R.

Since a row is 1 dimensional, then the number of dimensions of a vector is 1. $\rho\rho R$ is often referred to as the *rank* of R.

3.2 Operations with vectors

We now consider some APL operations with vectors. These operations provide very good illustrations of the power of APL as applied to arrays. First we consider ways of operating on a vector with a single number.

Examples

$R \leftarrow 2\ 1\ 4\ 3\ 5$

$R+1$ Addition.
3 2 5 4 6

$R-1$ Subtraction.
1 0 3 2 4

$R\times 3$ Multiplication.
6 3 12 9 15

$R\div 2$
1 0.5 2 1.5 2.5   Division.

$R*2$ Exponentiation.
4 1 16 9 25

Notice that in each example above, the operation is applied to the number together with each component of R.

Parallel processing

The following examples illustrate operations with two vectors. They use the notion of "parallel processing" in which the operation is applied to all of the corresponding components of the two vectors. In order to perform these operations, of course, the vectors must have the same number of components.

49

Examples

```
R←2 1 4 3 5
V←3 ¯1 2 1 0
W←2 1 3
```

```
R+V
5 0 6 4 5
```
The corresponding components are added.

```
R−V
¯1 2 2 2 5
```
The corresponding components are subtracted.

```
R×V
6 ¯1 8 3 0
```
The corresponding components are multiplied.

```
R+W
LENGTH ERROR
```
R and *W* cannot be added, since they do not have the same number of components.

Reduction

It is often useful to add up or multiply the components of a vector. This can be done with the use of the reduction symbol /. The general form is operation/vector. It reduces the vector to a single number by applying the operation to the successive components of the vector from right to left.

Examples

```
V←1 2 3 4
```

```
+/V
10
```
This is sum reduction. The components of *V* are added.

```
×/V
24
```
This is times reduction. The components of *V* are multiplied.

Inner products

A very useful and important operation with vectors is that of inner product. If $V=(v_1, v_2, \ldots, v_n)$ and $W=(w_1, w_2, \ldots, w_n)$ are two n component vectors, then the *inner product* of V and W is the number $v_1 \cdot w_1 + v_2 \cdot w_2 + \ldots + v_n \cdot w_n$. Let us look at a program for computing inner products. Notice that the header of this program is designed so as to create a dyadic function called *INNER*. It combines the two vectors V and W, and creates the explicit value *PRODUCT*.

Program 3.1 INNER

   ∇ *PRODUCT← V INNER W*

[1]   *PRODUCT← + / V × W*   The corresponding components of *V*
       ∇   and *W* are multiplied and the results
                                           added up yielding the result *PROD-
                                           UCT*.

Example

   V ←1 2 3 4
   W←2 ̄3 4 1

   V INNER W
12

This operation of inner product can also be accomplished directly on the APL keyboard by using *V+.×W*. In general, if α and ω are operations in APL, then the expression of the form *Vα.ωW* is called an *inner product*. The result is that ω is applied to the corresponding components of *V* and *W*, followed by α reduction applied to the result. Let us do the above example using this notation:

   V+.×W
12

This operation of inner product will be used later in multiplying matrices.

An application of vectors

Recall the company in the previous section which produces two products, gadgets and widgets. These products consist of parts called 051, 035, 068, and 047. The number of each of these parts in each product is given by the vectors:

   G←3 5 2 1
   W←4 3 1 2

1. In order to produce a "gidget," one merely fastens a gadget to a widget. Find a vector for the number of each part needed to produce one gidget.

   G+ W   The sum of the vectors *G* and *W*
7 8 3 3   accomplishes this.

2. If one received an order for 5 gadgets and 10 widgets, find a vector for the number of each part needed to fill the order.

   (5 × G)+(10 × W)
55 55 20 25   Need 55 051's, 55 035's, 20 068's, 25
                                        047's.

51

3. If 051's cost $.50 each, 035's cost $1.00 each, 068's cost $0.75 each, and 047's cost $1.50 each, use vectors to find the cost of producing a gadget.

    C←.50 1.00 .75 1.50   This is the unit cost vector.

    C+.×*G*   The inner product of the unit cost
9.50   vector with the parts vector for a
                                    gadget yields the total cost, $9.50, of
                                    a gadget.

4. If gadgets sell for $12.00 each, find the profit per gadget.

    12.00−9.50   Profit is revenue minus cost.
2.50

5. An employee has discovered that a better widget can be produced if one uses 2 068's instead of 1, and 3 051's instead of 4. Write an APL expression to make these changes in *W*.

    W[1 3]←3 2   The first component of *W* is replaced
                              by 3 and the third component by 2.

    W
3 3 2 2   The new value of *W*.

6. Another employee has discovered that both products will be better if one part 072 is included. Write APL expressions to chain this new part onto the end of the parts vectors for gadgets and widgets.

    G←*G*,1   The use of catenation.

    W←*W*,1

    G   The new *G*.
3 5 2 1 1
    W   The new *W*.
4 3 1 2 1

EXERCISES

1. Consider the vectors

    S←2 3 ‾1 5

    T←4 0 1 ‾2

    U←‾2 6 7

    V←8 0 1

Evaluate the following (Do them by hand first; then, check your answers at an APL terminal):

(a) 3×*S*+*T*	(e) *S*,*U*	(i) +/*S*×*T*
(b) (3×*S*)+*T*	(f) *S*+3	(j) (+/*U*∗2)∗.5
(c) *S*×*T*	(g) *S*+*U*	(k) *S*[3 2]
(d) *U*∗2	(h) *T*×‾1	(l) *T*[3]

2. Compute the inner products of S and T and of U and V.

3. A candy vendor sells five brands of candy called brand A, B, C, D, and E respectively. He records his daily sales of each brand in a vector called S. On a certain day, S is as follows:

    $S \leftarrow 92\ 81\ 35\ 49\ 57$

(a) Write an APL expression for the number of different brands.
(b) Write an APL expression for the number of packages of brand D in S.
(c) Write an APL expression for the subvector V of S whose brands are vowels.
(d) Write an APL expression for the total number of packages sold on this day.
(e) Write an APL expression for changing the number of brand C from 35 to 38.

4. The candy vendor in Exercise 3 has been selling candy for three days and has the following three sales vectors for each of these days:

    $S1 \leftarrow 92\ 81\ 38\ 49\ 57$

    $S2 \leftarrow 120\ 68\ 19\ 25\ 75$

    $S3 \leftarrow 67\ 50\ 37\ 29\ 63$

(a) Find the total sales vector for the three days.
(b) Find the total number of packages of candy sold in the three days.
(c) Suppose the prices of the candies are 0.10 for brand A, 0.15 for B, 0.20 for C, 0.10 for D, and 0.15 for E. Express these prices as a price vector.
(d) Write an APL expression for computing the revenue for each of the three days, and find these revenues.

3.3 Matrices

In Section 3.1., it was noted that when the symbol ρ is used as a monadic operator, as in ρR, then ρR computes the "size" of R. In other words, it computes the number of components in R. If ρ is used as a dyadic operator, then ρ arranges the elements on the right according to the structure on the left. Consider the following examples:

    $5\ \rho 3$
$3\ 3\ 3\ 3\ 3$ Five 3's.

    $5\ \rho'*'$
$*\ \ *\ \ *\ \ *\ \ *$ Five $*$'s.

    $6\ \rho 89$
$89\ 89\ 89\ 89\ 89\ 89$ Six 89's.

    $6\ \rho 8\ 9\ 10$
$8\ 9\ 10\ 8\ 9\ 10$ Six numbers from 8 9 10 respectively.

```
      3 3ρ5
5 5 5                                  Three rows, 3 columns of 5's.
5 5 5
5 5 5

      3 2ρ1 2 3 4 5 6
1 2
3 4                                    Three rows, 2 columns.
5 6

      3 3ρι9
1 2 3                                  Three rows, 3 columns.
4 5 6
7 8 9
```

Matrix of order $m \times n$

A matrix of order $m \times n$ (m by n) is a rectangular array of numbers arranged in m rows and n columns.

The following is a 2×3 matrix A:

$$A = \begin{bmatrix} 4 & 2 & 0 \\ 3 & 1 & 5 \end{bmatrix}$$

In APL, this matrix would be represented as follows:

```
    A←2 3ρ4 2 0 3 1 5
    A
4 2 0
3 1 5
```

Note the following monadic uses of ρ with matrices:

```
    ρA
2 3                                    The order of A.

    (ρA)[1]
2                                      The number of rows in A.

    (ρA)[2]
3                                      The number of columns in A.

    ρρA
2                                      The rank of a matrix is 2, since a
                                       matrix has two dimensions, rows and
                                       columns.
```

Equal matrices

Two matrices A and B are equal, usually denoted by $A = B$, if and only if they have the same orders and their corresponding elements are equal.

Examples

```
    A←3 3ρ1 2 3 4 5 6 7 8 9
    A
1 2 3
4 5 6
7 8 9
```

```
    B←3 3ρι 9

    B
1 2 3
4 5 6
7 8 9
```

```
    A=B
1 1 1
1 1 1
1 1 1
```
The relation = compares the corresponding elements of *A* and *B*. If the matrices are equal, the result will be all 1's (trues).

```
    C←2 3ρ1 2 3 4 5 6
    C
1 2 3
4 5 6
```

```
    D←3 2ρ1 2 3 4 5 6
    D
1 2
3 4
5 6
```

```
    C=D
LENGTH ERROR
```
C and *D* do not have the same orders.

Indexing with matrices

We now illustrate the use of indices with matrices:

Examples

```
    A←2 3ρ4 2 0 3 1 5
    A
4 2 0
3 1 5
```

```
    A[2;1]
3
```
The element in the second row, first column.

```
    A[1;2]
2
```
The element in the first row, second column.

55

A[2;]	The second row.
3 1 5	

A[;2]	The second column. Note that it is
2 1	expressed as a row vector.

Altering a matrix

The following examples illustrate the ways in which to make changes in a matrix:

A←3 3ρ ι9	Creating a matrix.
A	
1 2 3	
4 5 6	
7 8 9	

A[2;3]←0	Change the element in the second
A	row, third column to 0.
1 2 3	
4 5 0	
7 8 9	

A[2;]←3 1 5	Change the second row of *A*.
A	
1 2 3	
3 1 5	
7 8 9	

A[;3]←2 3 0	Change the third column of *A*.
A	
1 2 2	
3 1 3	
7 8 0	

Catenation with matrices

It might also be useful to be able to chain one or more rows onto a matrix. Let us illustrate this also:

Examples

A←2 3ρ1 2 3 4 5 6	
A	
1 2 3	
4 5 6	

B←7 8 9	A new row to be attached to *A*.

```
      A←A,[1]B                    Augment A by this new row, B.
      A                          B is chained onto the first dimension
1 2 3                            of A, its rows.
4 5 6
7 8 9
```

```
      C←1 0 2                    A new column to be attached to A.

      A←A, C                     Augment A by this new column, C.
      A                                A,C is an abbreviated notation
1 2 3 1                          for A,[2]C.
4 5 6 0
7 8 9 2
```

Transposing a matrix

The *transpose* of an $m \times n$ matrix A is a new matrix of order $n \times m$ obtained by making the rows of A into columns (in the same order). Of course, this will also make the columns of A into rows. The transpose of A is accomplished easily in APL by entering ⍉A. (The symbol ⍉ is made by typing $\circ$, located above the letter O on the keyboard, then backspacing and overstriking the \ .) We shall use the symbol ⍉A to denote the transpose of A throughout this text. Other texts commonly use the symbol A^t or $Tr(A)$ for the transpose of A. To illustrate this operation, we shall transpose the matrix A above:

```
      ⍉A
1 4 7
2 5 8
3 6 9
1 0 2
```

The last change that we wish to consider making in a matrix at this time involves the ability to specify certain submatrices of a matrix.

The "take" and "drop" functions

The "take" function ↑ is located above the letter Y on the keyboard. The "drop" function ↓ is located above the U.

```
      B←4 4⍴⍳ 16
      B
 1  2  3  4
 5  6  7  8
 9 10 11 12
13 14 15 16
```

```
      2 2↑B                      Take the first two rows and two
1 2                              columns of B.
5 6
```

2 ⁻2↑B 3 4 7 8	Take the first two rows and last two columns of B.
⁻2 ⁻2↑B 11 12 15 16	Take the last two rows and columns of B.
2 2↓B 11 12 15 16	Drop the first two rows and columns of B.
⁻2 ⁻2↓B 1 2 5 6	Drop the last two rows and columns of B.

EXERCISES

1. Let

$$M = \begin{bmatrix} 2 & 1 & 4 & 3 \\ 5 & 0 & 2 & 1 \\ 3 & ^-1 & 7 & 0 \end{bmatrix}$$

Do the following exercises at an APL terminal:
(a) Enter M on the APL terminal.
(b) Find the order of M.
(c) Find the number of columns of M.
(d) Change the element in the second row third column to 8.
(e) Change the third column to 3 ⁻1 5.
(f) Change the second row to 1 0 0 0.
(g) Transpose M.
(h) Augment M by a new row 0 0 0 1.
(i) Drop the last row and first column of M.
(j) Take the first two rows and last two columns of M.

2. Let

$$C = \begin{bmatrix} 1 & 3 & 2 & 0 \\ 5 & ^-1 & 0 & 4 \\ 0 & 5 & ^-2 & 3 \end{bmatrix}$$

Compute the following:
(a) ρC (f) $(\rho C)[1]$
(b) $\rho(\rho C)$ (g) $(\rho C)[2]$
(c) $C[2;]$ (h) $2 \times C[3;]$
(d) $C[;2]$ (i) $C[2;] + (^-5) \times C[1;]$
(e) $C[2;3]$ (j) $C[;3\ 1]$

3. A company produces four products called A, B, C, and D. Each product is made up of five parts called a, b, c, d, and e. The number of each part needed

for each product is contained in the following matrix P:

$$P = \begin{array}{c c} & \begin{array}{c c c c c} a & b & c & d & e \end{array} \\ \begin{array}{c} A \\ B \\ C \\ D \end{array} & \left[\begin{array}{c c c c c} 2 & 1 & 3 & 4 & 2 \\ 3 & 1 & 5 & 2 & 3 \\ 2 & 1 & 4 & 3 & 2 \\ 3 & 1 & 2 & 5 & 3 \end{array} \right] \end{array}$$

Express each of the following in APL notation and evaluate:
(a) The row of parts required for a product B.
(b) The column consisting of the numbers of part d in each product.
(c) The number of part c in a product B.
(d) It was decided that 3 d's are needed in a B. Make this change in B.
(e) It was decided to increase the number of e's in each product by 1. Make the appropriate change in the matrix P.
(f) A new product E has been added to production, consisting of 4 a's, 2 b's, 0 c's, 5 d's, and 2 e's. Include this new product E in an augmented matrix P.
(g) An order is received for 5 product C's. Write an APL expression to extract from P the number of each part needed to fill the order.

3.4 Operations with matrices

We now consider some APL operations with matrices. These operations are very similar to the operations with vectors in Section 3.2. in that the operations apply to each element of the array, and they further illustrate the power of APL as applied to arrays. First we consider operating on a matrix with a number.

Examples

```
      A←2 3ρ4 2 0 3 1 5
      A
4 2 0
3 1 5
```

```
      A+1
5 3 1                          Add 1 to each element.
4 2 6
```

```
      A−1
3 1 ¯1                         Subtract 1 from each element.
2 0 4
```

```
      A×2
8 4 0                          Multiply each element by 2.
6 2 10
```

```
      A÷2
  2   1   0                    Divide each element by 2.
1.5 0.5 2.5
```

```
     A*2
16 4  0
 9 1 25
```
Raise each element to the power 2.

Parallel processing

The following examples illustrate the addition and subtraction of two matrices. In order to perform these operations, the matrices must have the same orders.

Examples

```
    A←3 3ρι 9
    A
1 2 3
4 5 6
7 8 9
    B←3 3ρ2 3 0 ¯1 2 5 0 4 ¯2
    B
 2 3  0
¯1 2  5
 0 4 ¯2
```

```
    A+B
3  5 3
3  7 11
7 12 7
```
Corresponding elements are added.

```
    A-B
¯1 ¯1  3
 5  3  1
 7  4 11
```
Corresponding elements are subtracted.

```
    A×B
 2  6   0
¯4 10  30
 0 32 ¯18
```
Corresponding elements are multiplied.

Reduction with matrices

By using the idea of reduction, one can perform an operation down the columns or across the rows of a matrix. The following examples illustrate how this is done with addition and multiplication.

Examples

```
    A←3 3ρι 9
    A
1 2 3
4 5 6
7 8 9
```

$+/[1]A$ 12 15 18	Adds the rows of A—vertically (down the columns).
$+/[2]A$ 6 15 24	Adds the columns of A—horizontally (across the rows).
$+/(+/[1]A)$ 45	The sum of 12, 15, and 18. The sum of all elements in A.
$+/(+/[2]A)$ 45	The sum of 6, 15, and 24. Also, the sum of all elements in A.
$\times/[1]A$ 28 80 162	Multiplies the rows of A.
$\times/[2]A$ 6 120 504	Multiplies the columns of A.

Matrix multiplication

Definition

The product of a matrix A of order $m \times k$ and a matrix B of order $k \times n$ is the matrix P of order $m \times n$ such that $P[I;J]$ is the inner product of the Ith row of A and the Jth column of B. In conventional mathematics, P is denoted by $A \cdot B$.

Example

Let

$$A = \begin{bmatrix} 1 & 2 & 3 \\ 4 & 5 & 6 \end{bmatrix} \quad \text{and} \quad B = \begin{bmatrix} 1 & 2 \\ 0 & 3 \\ 2 & 0 \end{bmatrix}.$$

Since A is a 2×3 matrix and B is a 3×2 matrix, then the product P is a 2×2 matrix. The elements of P are computed as follows:

$$P[1;1] = A[1;] + . \times B[;1] = (1\ 2\ 3) + . \times (1\ 0\ 2) = 7$$
$$P[1;2] = A[1;] + . \times B[;2] = (1\ 2\ 3) + . \times (2\ 3\ 0) = 8$$
$$P[2;1] = A[2;] + . \times B[;1] = (4\ 5\ 6) + . \times (1\ 0\ 2) = 16$$
$$P[2;2] = A[2;] + . \times B[;2] = (4\ 5\ 6) + . \times (2\ 3\ 0) = 23$$

Therefore,

$$P = A \cdot B = \begin{bmatrix} 7 & 8 \\ 16 & 23 \end{bmatrix}.$$

In order for this matrix multiplication to be possible, it is necessary that the number of columns of the left matrix, A, be equal to the number of rows of the right matrix, B. Otherwise, we will not have corresponding elements in the Ith row of A and the Jth column of B necessary to compute the vector inner product of this row and column.

Let us now consider an APL program which explicitly performs the multiplications of the two matrices A and B, step by step.

Program 3.2 MULTIPLY (Optional)

    ∇ *PRODUCT← A MULTIPLY B; I; J; ELEMENTS*

> *MULTIPLY* is a dyadic function which assigns to A and B the value *PRODUCT. I, J,* and *ELEMENTS* are local variables.

[1]   →((ρA)[2]=(ρB)[1])/OK

> Line 1 checks to see if multiplication is possible. If it is, the program branches to the line labeled OK. Otherwise, it prints *IMPOSSIBLE*, and branches to 0; thus ending the program.

[2]   *PRODUCT←'IMPOSSIBLE '*

[3]   →0

[4]   *OK: ELEMENTS←ι0*

[5]   *I←0*

> Lines 4, 5, and 7 initialize the values of I, J, and *ELEMENTS*. Lines 6 and 8 increment the values of I and J.

[6]   *NEXTROW: I←I+1*

[7]   *J←0*

[8]   *NEXTCOL: J←J+1*

> Line 9 forms the row of *ELEMENTS* of *PRODUCT*. It computes the inner product of the *I*th row of A and *J*th column of B, and chains it onto the previous *ELEMENTS*.

[9]   *ELEMENTS← ELEMENTS, A[I;]+.× B[; J]*

[10] →(J<(ρB)[2])/NEXTCOLUMN

[11] →(I<(ρA)[1])/NEXTROW

> Lines 10 and 11 make sure that all of the inner products are computed.

[12] *PRODUCT←((ρA)[1], (ρB)[2])ρ ELEMENTS*
      ∇

> Line 12 puts *ELEMENTS* into the appropriate size matrix called *PRODUCT*.

Let us consider some examples using this program *MULTIPLY:*

        A←2 3ρι 6
        A
1 2 3
4 5 6

        B←2 3ρ1 2 0 3 2 0
        B
1 2 0
3 2 0

        A MULTIPLY B
IMPOSSIBLE   Note that the number of columns of
                                A does not equal the number of rows
                                of *B*.

        B←⍉ *B*   Let *B* be the transpose of the previ-
        B   ous *B*.
1 3
2 2
0 0

        A MULTIPLY B
  5 7
14 22

It seems only fair to point out that this matrix multiplication can be
accomplished directly on the APL keyboard using the operation of inner
product by merely typing *A+.× B.*

Examples

        A+.× B   The example above.
  5 7
14 22

        B+.× A
13 17 21   Note that $(A+.\times B)\neq(B+.\times A)$ or
10 14 18 $A\cdot B\neq B\cdot A$. In fact, they are not even
  0 0 0   the same size.

        A←3 3ρι 9
        A
1 2 3
4 5 6
7 8 9
        B←3 3ρ1 3 5 2 6 7 1 0 1
        B
1 3 5
2 6 7
1 0 1

3 Vectors and matrices

```
    A+.×B
  8 15   22
 20 42   61
 32 69  100
    B+.×A
 48 57   66
 75 90  105
  8 10   12
```

Note, again, that $A \cdot B \neq B \cdot A$. Thus, matrix multiplication is *not* commutative.

```
    A←2 3ρι 6
    A
 1 2 3
 4 5 6

    B←2 3ρ5 6 4 3 2 1
    B
 5 6 4
 3 2 1

    A+.×B
 LENGTH ERROR
```

In this example, matrix multiplication is not possible, due to the orders of A and B.

An application of matrix multiplication

Suppose that a company produces three products called A, B, and C. Each product is composed of two subassemblies called I and II. The numbers of each subassembly in each product is given by the following matrix S:

$$S = \begin{matrix} & \text{I} & \text{II} \\ A & \begin{bmatrix} 1 & 1 \\ B & 2 & 1 \\ C & 1 & 2 \end{bmatrix} \end{matrix}$$

Now, each subassembly consists of three parts called a, b, and c. The numbers of each part in each subassembly is given by the parts matrix P:

$$P = \begin{matrix} & a & b & c \\ \text{I} & \begin{bmatrix} 3 & 2 & 1 \\ \text{II} & 2 & 3 & 2 \end{bmatrix} \end{matrix}$$

Find a matrix which gives the number of each part a, b, and c in each product A, B, and C. The solution is given by the matrix multiplication:

```
    S+.×P
 5 5 3
 8 7 4
 7 8 5
```

There are 5 a's in an A, 5 b's in an A, 3 c's in an A, 8 a's in a B, 7 b's in a B, 4 c's in a B, 7 a's in a C, 8 b's in a C, and 5 c's in a C.

EXERCISES

1. Consider the matrices

$$A = \begin{bmatrix} 2 & 1 & 3 \\ {}^-1 & 4 & 0 \end{bmatrix} \quad \text{and} \quad B = \begin{bmatrix} 5 & 0 & 1 \\ 0 & {}^-2 & 3 \end{bmatrix}.$$

Evaluate the following:
(a) $(2 \times A) + B$ (f) B, A
(b) $B - (3 \times A)$ (g) $A, [1]B$
(c) $B * 2$ (h) $+/[1]A$
(d) $A + 5$ (i) $\times/[2]A$
(e) $B - 3$ (j) $+/(+/[2]A)$

2. Check your answers to Exercise 1 at an APL terminal.

3. With pencil and paper, trace through the program *MULTIPLY* with the matrices

$$A = \begin{bmatrix} 1 & 2 & 3 \\ 4 & 5 & 6 \end{bmatrix} \quad \text{and} \quad B = \begin{bmatrix} 1 & 2 \\ 0 & 3 \\ 2 & 0 \end{bmatrix}.$$

4. Repeat Exercise 3 at an APL terminal using the *TRACE* command (see Appendix A.6).

5. Multiply the following matrices using pencil and paper:

(a) $A = \begin{bmatrix} 2 & 3 \\ 4 & 0 \end{bmatrix}$ and $B = \begin{bmatrix} 5 & {}^-1 \\ 2 & 1 \end{bmatrix}.$

(b) $A = \begin{bmatrix} 3 & 2 & 4 \\ 1 & 0 & 5 \end{bmatrix}$ and $B = \begin{bmatrix} 6 & 2 \\ 0 & 4 \\ 1 & 3 \end{bmatrix}.$

(c) B times A in Part (b).

(d) $A = \begin{bmatrix} 1 & 2 & 3 & 4 \\ 5 & 6 & 7 & 8 \end{bmatrix}$ and $B = \begin{bmatrix} 2 & 1 \\ 0 & 3 \\ {}^-1 & 4 \\ 5 & 0 \end{bmatrix}.$

6. Check your answers to Exercise 5 at an APL terminal.

7. Let

$$A = \begin{bmatrix} 1 & 2 \\ 3 & 4 \end{bmatrix} \quad \text{and} \quad B = \begin{bmatrix} 5 & 6 \\ 7 & 8 \end{bmatrix}.$$

Use the APL terminal for Parts (a)–(d).
(a) Compute $(A+B) + . \times (A-B)$ (i.e., $(A+B) \cdot (A-B)$ in conventional notation).
(b) Compute $(A + . \times A) - (B + . \times B)$ (i.e., $A^2 - B^2$ in conventional notation).
(c) Are the answers to Parts (a) and (b) the same?
(d) Do you have an explanation for this?

8. Let

$$A = \begin{bmatrix} 2 & 1 & 3 \\ 5 & 0 & 2 \\ 7 & 1 & 3 \end{bmatrix} \quad B = \begin{bmatrix} 3 & {}^-2 & 6 \\ {}^-1 & 7 & 0 \\ 5 & 4 & 9 \end{bmatrix} \quad C = \begin{bmatrix} 4 & 2 & 8 \\ 1 & 6 & 0 \\ 0 & 4 & 2 \end{bmatrix}.$$

3 Vectors and matrices

At an APL terminal, verify the fact that $A+.\times(B+C)$ equals $(A+.\times B)+(A+.\times C)$. This is called the distributive property. (In conventional notation, it would be written as $A\cdot(B+C)=A\cdot B+A\cdot C$.)

9. The following matrix consists of the numbers of subassemblies I, II, and III needed in the production of products A, B, and C.

$$S=\begin{bmatrix} 2 & 1 & 3 \\ 3 & 1 & 2 \\ 1 & 2 & 3 \end{bmatrix}$$

where row 1 is for A, row 2 for B, and row 3 for C. Column 1 is for I, column 2 for II, and column 3 for III. Each subassembly I, II, and III consists of parts a, b, and c as given in the following matrix:

$$P=\begin{bmatrix} 5 & 4 & 3 \\ 6 & 3 & 2 \\ 6 & 2 & 1 \end{bmatrix}$$

where row 1 is for I, row 2 for II, and row 3 for III. Column 1 is for a, column 2 for b, and column 3 for c. Find a matrix that will give the number of each part a, b, and c in each product A, B, and C.

10. A company produces widgets and gadgets. To produce each widget and gadget requires time on machines X, Y, and Z as given in the following matrix T:

$$T=\begin{matrix} \text{widgets} \\ \text{gadgets} \end{matrix} \begin{matrix} X & Y & Z \\ \begin{bmatrix} 1 & 2 & 1 \\ .5 & 3 & 1 \end{bmatrix} \end{matrix}.$$

These times are measured in hours. In other words, it takes 1 hour on machine X to produce a widget, 2 hours on machine Y to produce a widget, etc.

Set up an APL expression for the following problems and evaluate:
(a) An order is received for 10 widgets and 15 gadgets. Find a vector for the amount of time needed on each machine to fill the order.
(b) If the cost per hour on machine X is $5.00, on machine Y is $4.00, and on machine Z is $6.00, find the total cost of machine time in filling the order.

3.5 Properties of matrices

Now that we know how to add and multiply matrices, we will consider some properties of matrices using these operations. In this section, we will restrict our attention to square matrices of the same order. (Therefore, all of the operations will be defined.) The properties we will consider are analogous to the properties of the real number system. Rather than present formal proofs of these properties, we will illustrate them with the following matrices:

```
    A←3 3ρ1 ¯2 3 ¯4 5 ¯6 7 ¯8 9
    A
 1 ¯2  3
¯4  5 ¯6
 7 ¯8  9
```

66

```
      B←3 3ρ3 2 1 1 2 3 2 1 3
      B
3 2 1
1 2 3
2 1 3
      C←3 3ρ2 1 0 ¯1 4 3 5 ¯2 1
      C
 2  1 0
¯1  4 3
 5 ¯2 1
```

The reader will also be asked to test out these properties with some other particular examples, either by hand, or even better, at an APL terminal. The properties are listed in conventional notation.

1. Addition is commutative.
$$(A+B)=(B+A).$$

Example

```
      A+B
 4  0  4
¯3  7 ¯3
 9 ¯7 12
      B+A
 4  0  4
¯3  7 ¯3
 9  7 12
```

2. Addition is associative.
$$(A+(B+C))=((A+B)+C).$$

Example

```
      A+(B+C)
 6  1  4
¯4 11  0
14 ¯9 13
      (A+B)+C
 6  1  4
¯4 11  0
14 ¯9 13
```

3. There is an additive identity matrix Z.

   $(A + Z) = A$, where Z is a matrix completely filled with 0's.

   Example

```
      Z←3 3ρ0
      Z
0 0 0
0 0 0
0 0 0

      A+Z
 1 ¯2  3
¯4  5 ¯6
 7 ¯8  9

      A
 1 ¯2  3
¯4  5 ¯6
 7 ¯8  9
```

4. For any matrix A, there is an additive inverse $-A$.

   $$(A + -A) = Z.$$

   Example

```
      -A
¯1  2 ¯3
 4 ¯5  6
¯7  8 ¯9

      A+ -A
0 0 0
0 0 0
0 0 0
```

5. In general, multiplication is *not* commutative.

   $$(A \cdot B) \neq (B \cdot A), \quad \text{in general.}$$

   Example

```
      A+.×B
  7  1  4
¯19 ¯4 ¯7
 31  7 10

      B+.×A
 2  ¯4  6
14 ¯16 18
19 ¯23 27
```

6. Multiplication is associative.

$$(A \cdot (B \cdot C)) = ((A \cdot B) \cdot C).$$

Example

```
    A+.×(B+.×C)
 33   3    7
-69  -21  -19
105   39   31
```

```
    (A+.×B)+.×C
 33   3    7
-69  -21  -19
105   39   31
```

7. Multiplication is distributive over addition.

$$(A \cdot (B + C)) = (A \cdot B + A \cdot C)$$
$$((B + C) \cdot A) = (B \cdot A + C \cdot A).$$

Example

```
    A+.×(B+C)
 26  -12   1
-62   24   2
 98  -36  -5
```

```
    (A+.×B)+(A+.×C)
 26  -12   1
-62   24   2
 98  -36  -5
```

```
    (B+C)+.×A
  0  -3   6
 18 -18  18
 39 -51  63
```

```
    (B+.×A)+(C+.×A)
  0  -3   6
 18 -18  18
 39 -51  63
```

8. There is a multiplicative identity matrix *I*.

$$(A \cdot I) = (I \cdot A) = A, \quad \text{where } I \text{ is a matrix with 1's}$$
down the main diagonal and 0's elsewhere.

Example

```
I←3 3ρ1 0 0 0 1 0 0 0 1
I
1 0 0
0 1 0
0 0 1

    A+.×I
 1 ⁻2  3
⁻4  5 ⁻6
 7 ⁻8  9

     I+.×A
 1 ⁻2  3
⁻4  5 ⁻6
 7 ⁻8  9

    A
 1 ⁻2  3
⁻4  5 ⁻6
 7 ⁻8  9
```

9. The question of multiplicative inverses will be taken up in the next chapter.

EXERCISE

Let

$$A = \begin{bmatrix} 3 & 1 & 2 \\ 7 & -1 & 0 \\ 2 & 0 & 5 \end{bmatrix} \quad B = \begin{bmatrix} 0 & 2 & 3 \\ -1 & 6 & 2 \\ 5 & 0 & 4 \end{bmatrix} \quad C = \begin{bmatrix} 1 & 2 & 3 \\ 4 & 5 & 6 \\ 7 & 8 & 9 \end{bmatrix}.$$

Verify Properties 1–8 above for these matrices at an APL terminal using the proper APL notation.

Systems of linear equations

4

In this chapter, we shall discuss the use of matrices for representing and solving systems of linear equations. As we shall see, APL makes the job of solving certain systems of linear equations almost trivial.

4.1 Linear equations

A linear equation with n unknowns is an equation of the form

$$a_1 \cdot x_1 + a_2 \cdot x_2 + a_3 \cdot x_3 + \ldots + a_n \cdot x_n = B,$$

where $x_1, x_2, x_3, \ldots, x_n$ are the unknowns, and $a_1, a_2, a_3, \ldots, a_n$, and B are constants. a_i is called the *coefficient* of x_i for $i = 1, 2, \ldots, n$.

Solution of a linear equation

A *solution* to a linear equation $a_1 \cdot x_1 + a_2 \cdot x_2 + \ldots + a_n \cdot x_n = B$ is a vector $X = (x_1, x_2, \ldots, x_n)$ of values of the unknowns for which the equation is a true statement.

Examples (conventional notation)

1. $3x + 4y - 2z = 12$ (The unknowns are x, y, and z.) The vector $(2, 0, {}^-3)$ is a solution, since if x is replaced by 2, y by 0, and z by ${}^-3$, the result will be 12, so that the equation is a true statement for this vector. The vectors $(0, 3, 0)$ and $(3, 1, .5)$ are also solutions to this linear equation.
2. $3x + 2y = 6$ (The unknowns are x and y.) $(2, 0)$ and $(1, 1.5)$ are two solutions to this equation.
3. $2x_1 - x_2 + 3x_3 - 4x_4 = 8$ (The unknowns are x_1, x_2, x_3, x_4.) $(1, {}^-2, 0, {}^-1)$ and $(2, 2, 2, 0)$ are solutions to this equation.

Notice that for a linear equation, there can be more than one solution. In fact, there are an infinite number of solutions if $n > 1$.

An APL expression for a linear equation

If the coefficients $(a_1, a_2, \ldots, a_n)$ are expressed as a vector A, the unknowns $(x_1, x_2, \ldots, x_n)$ as a vector X, then the linear equation

$$a_1 \cdot x_1 + a_2 \cdot x_2 + \ldots + a_n \cdot x_n = B$$

can be expressed as the inner product

$(A + . \times X) = B.$

The following examples illustrate this point:

Examples

1. $3x + 4y - 2z = 12.$

   $A \leftarrow 3\ 4\ ^-2$
   $B \leftarrow 12$

   Let us use the APL notation to verify that $(3, 1, 0.5)$ is a solution to this equation.

   $X \leftarrow 3\ 1\ .5$

   $A + . \times X$

12

   $(A + . \times X) = B$

1   True. X is a solution.

2. $3x + 2y = 6.$
   Let us use APL to verify that $(1, 1.5)$ is a solution.

   $A \leftarrow 3\ 2$
   $B \leftarrow 6$
   $X \leftarrow 1\ 1.5$

   $(A + . \times X) = B$

1   True. X is a solution.

3. $2x_1 - x_2 + 3x_3 - 4x_4 = 8.$
   Let us show that $(2, 1, 2, 0)$ is *not* a solution to this equation, using APL.

   $A \leftarrow 2\ ^-1\ 3\ ^-4$
   $B \leftarrow 8$
   $X \leftarrow 2\ 1^-2\ 0$

   $A + . \times X$

9

   $(A + . \times X) = B$

0   False. X is not a solution.

Exercises

1. Use APL to determine whether or not the indicated vectors are solutions to the
   given linear equations.
   (a) $2x-3y=6$, $X=(1.5,\,^-1)$
   (b) $4x+y=7$, $X=(1,3)$
   (c) $3x+2y-z=8$, $X=(1,2,\,^-1)$.
   (d) $x_1-2x_2+3x_3-x_4=10$, $X=(2,1,3,1)$

2. Find two solutions to each of the following linear equations, and use APL to
   check your solutions.
   (a) $5x-3y=6$
   (b) $2x-6y+8z=24$
   (c) $3x_1-2x_2+5x_3-x_4=10$

3. A linear equation with one unknown is an equation of the form $A\cdot X=B$. (In
   APL notation, it is $(A\times X)=B$, where X is the unknown and A and B are
   constants with $A\neq0$.) Write an APL program to solve such an equation.

4. Solve the following equations:
   (a) $3\cdot X=7$ (b) $7\cdot X=5$ (c) $^-2\cdot Y=9$ (d) $4\cdot Z=\,^-3$

4.2 Two-by-two systems of linear equations

A *two-by-two system of linear equations* consists of two linear equations
with two unknowns, x_1 and x_2, as follows:

$$a_{11}\cdot x_1+a_{12}\cdot x_2=b_1$$
$$a_{21}\cdot x_1+a_{22}\cdot x_2=b_2.$$

[*Note*: The first subscript of a_{ij} denotes the equation to which it belongs,
while the second subscript denotes the variable to which it belongs.] A
solution to such a system is a vector $X=(x_1,x_2)$ for which *both* equations
are true statements.

Example

Consider the system

$$3x+2y=6$$
$$x-2y=4.$$

The vector (2.50, 0.75) is a solution, since both equations are true for this
vector. That is, if x is replaced by 2.50 and y by $^-0.75$ in both equations,
the resulting statements are true. The vector (2,0) is not a solution, since
only the first equation is true for this vector.

73

Matrix representation for a system of linear equations

A two-by-two system of linear equations can be represented as a matrix equation as follows: $A \cdot X = B$ (in conventional notation) or $(A + . \times X) = B$ (in APL notation), where

$$A = \begin{bmatrix} a_{11} & a_{12} \\ a_{21} & a_{22} \end{bmatrix}$$ The matrix of coefficients.

$X = (x_1, x_2)$ The vector of unknowns.

$B = (b_1, b_2)$ The vector of constants on the right.

Example

Let us use APL to verify that the vector $(2.50, {}^-0.75)$ is a solution to the system

$$3x + 2y = 6$$
$$x - 2y = 4.$$

```
      A←2 2ρ3 2 1 ¯2
      A
3  2
1 ¯2
      B←6 4
      X←2.50 ¯0.75

      A+.×X
6 4

      (A+.×X)=B
1 1                                  6=6 is true and 4=4 is true.

      ∧/(A+.×X)=B
1                                    Since both are true.
```

We now consider the question of how does one arrive at the solution to a two-by-two system of linear equations? We shall use a method known as *Gaussian elimination.*

Solving a two-by-two system by the method of Gaussian elimination

We shall make use of the following three operations on the equations in the system. These operations are permissible because they do not alter the solutions of the system. In fact, they may help us in finding the solutions.

Operation 1.

A multiple of one equation may be added to or subtracted from the other equation.

Operation 2.

An equation may be multiplied by a nonzero constant.

Operation 3.

Two equations may be interchanged.

In the method of elimination, we use these operations to eliminate a variable arriving at a single equation in a single unknown. This equation can then be solved for the other unknown. The following examples illustrate this method:

Examples

1.

$$3x + 2y = 6$$
$$x - 2y = 4.$$

Interchanging these two equations, we get the equivalent (two equations are "equivalent" if they have the same solutions) system

$$x - 2y = 4$$
$$3x + 2y = 6.$$

Adding the first equation to the second equation yields the equivalent system

$$x - 2y = 4$$
$$4x + 0y = 10.$$

So, since $4x = 10$, then $x = 2.50$. Replacing x by 2.50 in the first equation yields $2.50 - 2y = 4$, so that $y = {}^-0.75$ after a little algebra.

2.

$$x - 3y = 2$$
$$3x + 2y = {}^-5.$$

Multiplying the first equation by 3 yields the equivalent system

$$3x - 9y = 6$$
$$3x + 2y = {}^-5.$$

Adding to the second equation the multiple $^-1$ times the first equation yields

$$3x - 9y = 6$$
$$0x + 11y = {}^-11.$$

From the resulting second equation, we get $11y = {}^-11$ or $y = {}^-1$. Replacing y by $^-1$ in the first equation, we get $3x + 9 = 6$, or $3x = {}^-3$, or $x = {}^-1$.

So, the solution to the system is the vector ($^-$1,$^-$1). Check:

```
A←2 2ρ1 ¯3 3 2
B←2 ¯5
X←¯1 ¯1
A+.×X
```
2 ¯5   So, it checks.

3.
$$x+3y=4$$
$$2x+6y=6.$$

Adding the multiple $^-$2 times the first equation to the second equation one

$$x+3y= \ 4$$
$$0x+0y=^-2.$$

The second equation is now the ridiculous statement that $0=^-2$. Thus, the system has *no* solution. A system with no solutions is called an *inconsistent system*.

4.
$$x+3y=4$$
$$2x+6y=8.$$

Adding the multiple $^-$2 times the first equation to the second equation yields

$$x+3y=4$$
$$0x+0y=0.$$

The resulting second equation $0=0$ is certainly true. However, it isn't very useful in solving the system. The significance of this result is that it tells us that the system has an infinite number of solutions. Any solution to the first equation is automatically a solution to the second equation. A system with an infinite number of solutions is called a *redundant system*.

The above examples point out that a two-by-two system of equations may have one solution, no solutions, or an infinite number of solutions. The method of Gaussian elimination helps us to decide in which case the example lies. In the event that we are in the first or third case, it also helps us to find the solution or solutions.

Some applications of two-by-two systems of linear equations

The following examples illustrate the use of two-by-two systems of linear equations to solve everyday problems:

Example 1

A company produces two products: widgets and gadgets. To produce each widget requires 5 minutes on machine I and 3 minutes on machine II. To produce each gadget requires 4 minutes on each machine. Find the number of each product that can be produced in a day if machine I is operated for 6 hours and machine II for 5 hours.

Let x represent the number of widgets that can be produced in a day and y represent the number of gadgets that can be produced in a day.

For machine I, the total number of minutes spent on widgets is $5x$ and a total number of minutes spent on gadgets is $4y$. Since machine I is in operation for 6 hours, or 360 minutes, then the equation for time on machine I is

$$5x + 4y = 360.$$

For machine II, the total number of minutes spent on widgets is $3x$ and the total number of minutes spent on gadgets is $4y$. Since machine II is in operation for 5 hours, or 300 minutes, then the equation for time on machine II is

$$3x + 4y = 300.$$

Adding to the second equation, the multiple $^-1$ times the first equation yields the new equation $2x = 60$. Thus, $x = 30$. Therefore, from the first equation, $5 \cdot 30 + 4y = 360$, or $4y = 210$, or $y = 52.5$. Thus, in a day, the company can produce 30 widgets and 52.5 gadgets.

Example 2

A man deposits a total of $1000 in two banks, called bank A and bank B. The interest rate in bank A is 4 percent per year, and in bank B is 5 percent per year. His income from these deposits for the year was $42. How much did he deposit in each bank?

Let x represent the amount deposited in bank A and y represent the amount deposited in bank B. Then, the interest he received from the two banks is given by

$$0.04x + 0.05y = 42.$$

The total amount deposited is given by

$$x + y = 1000.$$

Adding to equation one the multiple $^-0.04$ times equation two yields

$$0.10y = 2.$$

Thus, $y = \$200$. Since the total amount deposited is $1000, then $x = \$800$. So, he deposited $800 in bank A and $200 in bank B.

EXERCISES

1. Express the following systems of linear equations as matrix equations $(A + . \times X) = B$, and check to see if the specified vectors, X, are solutions to the systems (do them at an APL terminal):

(a) $x + 2y = 5$
   $3x + 4y = 6$ $X = (^-4, 4.5)$
(b) $6x - y = 5$
   $4x + 2y = 6$ $X = (1, 1)$
(c) $4x + 3y = 2$
   $8x + 6y = 3$ $X = (^-0.25, 1)$
(d) $3x + 4y = 2$
   $6x + 8y = 4$ $X = (2, ^-1)$

77

2. Use the method of Gaussian elimination to solve the following systems of linear equations (if they have solutions):
   (a) $5x - y = 4$
       $3x + 2y = 5$
   (b) $2x + 4y = 3$
       $x + 3y = 2$
   (c) $x + 3y = 4$
       $2x + 6y = 8$
   (d) $2x + y = 2$
       $6x + 3y = 4$
   (e) $3x + 2y = 5$
       $4x + 3y = 2$

3. Set up systems of equations for and solve the following problems:
   (a) A carpenter builds bookcases and tables. Each bookcase requires 12 square feet of lumber and takes 2 hours to build. Each table requires 16 square feet of lumber and takes 1.5 hours to build. How many bookcases and tables can he build if he has 100 square feet of lumber and 12 hours?
   (b) A person wants to buy nuts and bolts. Each nut costs $.04 and each bolt costs $.06. He has $1.90. He needs 5 less than twice as many nuts as bolts. How many nuts and bolts should he buy?
   (c) A new diet restricts a person to 1300 calories a day, and 100 grams of protein per day. The dieter is allowed to only eat foods A and B on this diet. Each ounce of food A contains 100 calories and 8 grams of protein. Each ounce of food B contains 80 calories and 6 grams of protein. How many ounces of each food should this dieter eat to meet the exact amount of calories and protein in his diet?

4.3 Elementary row operations

In solving the two-by-two systems of linear equations in the previous section, we used three operations on the equations, which suggest the following operations on the rows of a matrix. These are called elementary row operations. They will be used in the next section to solve larger systems of linear equations. In future sections, we shall also use these operations to invert matrices and compute determinants.

Elementary row operations

1. A row can be multiplied by a nonzero real number.
2. A constant multiple of any row can be added to any other row.
3. Any pair of rows can be interchanged.

These operations are easily performed at an APL terminal. Recall that in APL, the Ith row of a matrix is denoted by $M[I;]$.

To multiply the Ith row of a matrix M by a nonzero constant C, simply replace $M[I;]$ by $C \times M[I;]$. This is done as follows:

$$M[I;] \leftarrow C \times M[I;].$$

Example

        M←3 3ρι9
        M
1 2 3
4 5 6
7 8 9
        M[2;]←3×M[2;]   Multiply row 2 by 3.
        M
 1   2   3   Note that you must request *M* to be
12 15 18   printed to see the change.
 7   8   9

To add to row *I* a multiple *C* of row *J*, replace *M[I;]* by *M[I;]* + *C*× *M[J;]*. This is done as follows: *M[I;]*←*M[I;]* + *C*× *M[J;]*.

Example

        M[2;]← M[2;]+(⁻12)× M[1;]   Add to row 2 the multiple ⁻12 times
        M   row 1.
1 2   3
0 ⁻9 ⁻18   Note that this change is made to the
7 8   9   latest version of *M*.

To interchange row *I* and row *J*, it is necessary to replace row *I* by row *J* and row *J* by row *I*. This is done as follows:

        M[I J;]←M[J I;].

Example

        M[1 3;]←M[3 1;]   Interchange rows 1 and 3.
        M
7 8   9
0 ⁻9 ⁻18
1 2   3

For each of these elementary row operations, there is a corresponding column operation. However, we shall have no need for these column operations in this text.

79

4.4 Larger systems of linear equations

We will now consider using elementary row operations to solve a system of m linear equations with n unknowns. Consider the system

$$a_{11} \cdot x_1 + a_{12} \cdot x_2 + \ldots + a_{1n} \cdot x_n = b_1$$
$$a_{21} \cdot x_1 + a_{22} \cdot x_2 + \ldots + a_{2n} \cdot x_n = b_2$$
$$\vdots \qquad \vdots \qquad\qquad \vdots \quad \vdots$$
$$a_{m1} \cdot x_1 + a_{m2} \cdot x_2 + \ldots + a_{mn} \cdot x_n = b_m.$$

This system can be represented as a matrix equation $A \cdot X = B$ (or $(A + . \times X)$ $= B$ in APL) where

$$A = \begin{bmatrix} a_{11} & a_{12} & \cdots & a_{1n} \\ a_{21} & a_{22} & \cdots & a_{2n} \\ \vdots & \vdots & & \vdots \\ a_{m1} & a_{m2} & \cdots & a_{mn} \end{bmatrix} \qquad X = \begin{bmatrix} x_1 \\ x_2 \\ \vdots \\ x_n \end{bmatrix} \qquad B = \begin{bmatrix} b_1 \\ b_2 \\ \vdots \\ b_m \end{bmatrix}.$$

To verify this, one only needs to do the indicated matrix multiplication. In fact, such a system can be expressed even more succinctly as a single matrix

$$C = A, B = \begin{bmatrix} a_{11} & a_{12} & \cdots & a_{1n} & b_1 \\ a_{21} & a_{22} & \cdots & a_{2n} & b_2 \\ \vdots & \vdots & & \vdots & \vdots \\ a_{m1} & a_{m2} & \cdots & a_{mn} & b_m \end{bmatrix} \qquad (C \leftarrow A, B \text{ in APL}).$$

One of the basic techniques for solving such a system is the method of Gaussian elimination which we know from solving the two-by-two systems. By this method, one attempts to reduce the system to one of the form

$$x_1 + c_{12} \cdot x_2 + c_{13} \cdot x_3 + \ldots + c_{1n} \cdot x_n = d_1$$
$$x_2 + c_{23} \cdot x_3 + \ldots + c_{2n} \cdot x_n = d_2$$
$$x_3 + \ldots + c_{3n} \cdot x_n = d_3$$
$$\ddots$$
$$x_n = d_n$$
$$0 = 0$$
$$\vdots$$
$$0 = 0.$$

Then, by solving the equations successively (from the bottom to the top), one can determine the values of the unknowns.

To reduce such a system of linear equations to this form (without altering the solutions), requires the use of a sequence of operations on the equations. The permissible operations are

1. Multiply an equation by a nonzero real number.
2. Add to an equation a constant multiple of another equation.
3. Interchange two equations.

These operations can also be performed as the analogous elementary row operations on the rows of the augmented matrix A, B. The following examples illustrate the technique:

Example 1

$$x+2y+3z=1$$ Conventional notation for a 3-by-3 linear system.
$$x+3y+5z=2$$
$$2x+5y+9z=3$$

$A\leftarrow3\ 3\rho1\ 2\ 3\ 1\ 3\ 5\ 2\ 5\ 9$ The matrix of coefficients.
$B\leftarrow1\ 2\ 3$ The vector of constants.

$C\leftarrow A,B$ The augmented matrix.
C
```
1 2 3 1
1 3 5 2
2 5 9 3
```

$C[2;]\leftarrow C[2;]+(^-1)\times C[1;]$ Row 2 is replaced by row 2 plus the
C multiple $^-1$ of row 1.
```
1 2 3 1
0 1 2 1
2 5 9 3
```

$C[3;]\leftarrow C[3;]+(^-2)\times C[1;]$ Row 3 is replaced by row 3 plus the
C multiple $^-2$ of row 1.
```
1 2 3 1
0 1 2 1
0 1 3 1
```

$C[3;]\leftarrow C[3;]+(^-1)\times C[2;]$ Row 3 is replaced by row 3 plus the
C multiple $^-1$ of row 2.
```
1 2 3 1
0 1 2 1
0 0 1 0
```

Thus, from row 3, $z=0$. From row 2, $y+2z=1$, so that $y=1$. From row 1, $x+2y+3z=1$. Substituting $z=0$ and $y=1$, we get $x=^-1$.

81

Check

    A+.×⁻1 1 0
    1 2 3   It checks. $A \cdot X = B$.

Example 2

$$2x + y - z = 4$$
$$x - 2y + z = 1$$
$$3x - y - 2z = 3$$

    A←3 3ρ2 1 ⁻1 1 ⁻2 1 3 ⁻1 ⁻2
    B←4 1 3
    C←A,B
    C
    2 1 ⁻1 4
    1 ⁻2 1 1
    3 ⁻1 ⁻2 3

    C[1 2;]←C[2 1;]   Interchange rows 1 and 2.
    C[2;]←C[2;]+(⁻2)×C[1;]   Replace row 2 by row 2 plus the
    C[3;]←C[3;]+(⁻3)×C[1;]   multiple ⁻2 of row 1. Replace row 3
    C   by row 3 plus the multiple ⁻3 of row
    1 ⁻2 1 1   1.
    0 5 ⁻3 2
    0 ⁻5 ⁻5 0

    C[2 3;]←C[3 2;]   Interchange rows 2 and 3.
    C[2;]←(÷5)×C[2;]   Multiply the new row 2 by 1/5.
    C
    1 ⁻2 1 1
    0 1 ⁻1 0
    0 5 ⁻3 2

    C[1;]←C[1;]+2×C[2;]   Replace row 1 by row 1 plus twice
                            row 2.
    C[3;]←C[3;]+(⁻5)×C[2;]   Replace row 3 by row 3 plus the
                            multiple ⁻5 of row 2.
    C
    1 0 ⁻1 1
    0 1 ⁻1 0
    0 0 2 2

    C[3;]←(÷2)×C[3;]   Multiply row 3 by 1/2.
    C
    1 0 ⁻1 1
    0 1 ⁻1 0
    0 0 1 1

$C[1;] \leftarrow C[1;] + C[3;]$ Replace row 1 by the sum of rows 1
$C[2;] \leftarrow C[2;] + C[3;]$ and 3. Replace row 2 by the sum of
C rows 2 and 3.

```
1 0 0 2
0 1 0 1
0 0 1 1
```

Thus, from row 1, $x = 2$; from row 2, $y = 1$; from row 3, $z = 1$.

$A + . \times 2\ 1\ 1$
```
4 1 3
```
It checks.

Example 3

$$x + y + z + w = 1$$
$$-x + z + 2w = 1$$
$$3x + 2y - w = 1$$
$$x + y + 2z + 2w = 1$$

$A \leftarrow 4\ 4\rho 1\ 1\ 1\ 1\ ^-1\ 0\ 1\ 2\ 3\ 2\ 0\ ^-1\ 1\ 1\ 2\ 2$
$B \leftarrow 1\ 1\ 1\ 1$
$C \leftarrow A,B$
C
```
 1 1 1  1 1
-1 0 1  2 1
 3 2 0 -1 1
 1 1 2  2 1
```

$C[2;] \leftarrow C[2;] + C[1;]$
$C[3;] \leftarrow C[3;] + (^-3) \times C[1;]$
$C[4;] \leftarrow C[4;] + (^-1) \times C[1;]$
C
```
1  1  1  1  1
0  1  2  3  2
0 -1 -3 -4 -2
0  0  1  1  0
```

$C[3;] \leftarrow C[3;] + C[2;]$
C
```
1 1  1  1 1
0 1  2  3 2
0 0 -1 -1 0
0 0  1  1 0
```

83

$$C[4;] \leftarrow C[4;] + C[3;]$$
$$C[3;] \leftarrow (^-1) \times C[3;]$$
$$C$$

```
1 1 1 1 1
0 1 2 3 2
0 0 1 1 0
0 0 0 0 0
```

Thus, row 4 yields $0 = 0$, indicating that we have a redundant system. From row 3, $z + w = 0$, or $z = -w$. From row 2, $y + 2z + 3w = 2$. Or, replacing z by $-w$, $y - 2w + 3w = 2$; or, $y + w = 2$; so, $y = 2 - w$. Finally, from row 1, $x + y + z + w = 1$. Replacing y by $2 - w$ and z by $-w$, we get $x + 2 - w - w + w = 1$; or, $x = w - 1$. This system has an infinite number of solutions, since w can be any real number. The general solution looks like $(w - 1, 2 - w, -w, w)$.

Example 4

$$x + 2y + 3z = 2$$
$$4x + 5z + 6z = 3$$
$$7x + 8y + 9z = 5$$

$$A \leftarrow 3 \; 3\rho\iota \; 9$$
$$B \leftarrow 2 \; 3 \; 5$$
$$C \leftarrow A, B$$
$$C$$

```
1 2 3 2
4 5 6 3
7 8 9 5
```

$$C[2;] \leftarrow C[2;] + (^-4) \times C[1;]$$
$$C[3;] \leftarrow C[3;] + (^-7) \times C[1;]$$
$$C$$

```
1  2   3   2
0 ¯3  ¯6  ¯5
0 ¯6 ¯12  ¯9
```

$$C[2;] \leftarrow (^-1) \times C[2;]$$
$$C$$

```
1  2   3   2
0  3   6   5
0 ¯6 ¯12  ¯9
```

$$C[3;] \leftarrow C[3;] + 2 \times C[2;]$$
$$C$$

```
1 2 3 2
0 3 6 5
0 0 0 1
```

Row 3 says that 0=1. This is ridiculous. Thus, this system has no solutions. It is an inconsistent system.

A system of linear equations has either one solution, no solutions, or an infinite number of solutions.

EXERCISES

1. Let

$$A = \begin{bmatrix} 3 & 2 & 5 \\ 1 & {}^-4 & 6 \\ 7 & 0 & 9 \end{bmatrix}.$$

Perform the following elementary row operations on A at an APL terminal.
(a) Interchange rows 1 and 2.
(b) Add to row 2 the multiple $^-3$ of row 1.
(c) Add to row 3 the multiple $^-7$ of row 1.
(d) Multiply row 2 by $1 \div 14$.

2. Using elementary row operations at an APL terminal, reduce the matrix

$$M = \begin{bmatrix} 1 & 2 & 3 \\ 4 & 5 & 6 \\ 7 & 8 & 9 \end{bmatrix} \quad \text{to} \quad \begin{bmatrix} 1 & 0 & {}^-1 \\ 0 & 1 & 2 \\ 0 & 0 & 0 \end{bmatrix}.$$

3. Using elementary row operations at an APL terminal, reduce the matrix

$$M = \begin{bmatrix} 2 & 1 & 3 & 4 \\ 3 & 1 & 5 & 2 \\ 2 & 1 & 4 & 3 \\ 3 & 1 & 2 & 5 \end{bmatrix} \quad \text{to} \quad \begin{bmatrix} 1 & 0 & 0 & 0 \\ 0 & 1 & 0 & 7 \\ 0 & 0 & 1 & {}^-1 \\ 0 & 0 & 0 & 0 \end{bmatrix}.$$

4. Use elementary row operations at an APL terminal to solve the following systems of linear equations:

(a) $2x+3y=6$
$5x-y=4$
(b) $x+2y+3z=3$
$4x+5y+6z=4$
$7x+8y+9z=5$
(c) $3x+y+2z=4$
$2x+3y+z=3$
$2x+y+z=2$

(d) $x+2y+3z+5w=5$
$2x+3y+5z+9w=4$
$3x+4y+7z+w=0$
$7x+6y+5z+4w=3$
(e) $x+2y+3z+4w=2$
$5x+6y+7z+8w=5$
$9x+10y+11z+12w=3$
$13x+14y+15z+16w=0$

4.5 Row reduced form

Using elementary row operations, there is a final form in which a matrix may be altered, called the *row reduced form*. The reduction of a matrix to this form has many applications in matrix algebra.

85

Row reduced form

A matrix M is in row reduced form if

1. The first K rows are nonzero vectors (vectors not containing all zeros), and the remaining rows are zero vectors.
2. The first nonzero entry in each nonzero row is a 1, and it occurs in a column to the right of the leading 1 in each preceding row
3. The first nonzero entry in each nonzero row is the only nonzero entry in its column.

The following are examples of matrices in row reduced form:

Examples

1.

$$\begin{matrix} 1 & 0 & 0 & 0 \\ 0 & 1 & 0 & 0 \\ 0 & 0 & 1 & 0 \\ 0 & 0 & 0 & 1 \end{matrix}$$

This is the 4-by-4 "identity" matrix. In reducing a matrix to row reduced form, one tries to make it as close to an identity matrix as possible.

2.

$$\begin{matrix} 1 & 0 & {}^-1 & 2 \\ 0 & 1 & 2 & 3 \\ 0 & 0 & 0 & 0 \\ 0 & 0 & 0 & 0 \end{matrix}$$

3.

$$\begin{matrix} 1 & 0 & 0 & 0 & 2 \\ 0 & 1 & 0 & 0 & 1 \\ 0 & 0 & 0 & 1 & 3 \\ 0 & 0 & 0 & 0 & 0 \\ 0 & 0 & 0 & 0 & 0 \end{matrix}$$

In Examples 2 and 3 above, it is not possible to make the matrices into identity matrices by further applications of elementary row operations. They are as close to the corresponding identities as possible.

Let us consider some examples of the method of reducing matrices to row reduced form.

Example 1

```
    M←3 3ρ4 3 0 1 3 2 ̄2 0 1
    M
4 3 0
1 3 2
̄2 0 1
```

$M[1\ 2;]\leftarrow M[2\ 1;]$ Interchange rows 1 and 2.
M
```
 1  3  2
 4  3  0
⁻2  0  1
```

$M[2;]\leftarrow M[2;]+(^-4)\times M[1;]$ Replace row 2 by row 2 plus ⁻4 times row 1.

$M[3;]\leftarrow M[3;]+2\times M[1;]$ Replace row 3 by row 3 plus 2 times row 1.

M
```
1   3   2
0  ⁻9  ⁻8
0   6   5
```

$M[2;]\leftarrow(1\div 9)\times M[2;]$ Multiply row 2 by ⁻1/9.
M
```
1 3 2
0 1 .8888888888
0 6 5
```

$M[1;]\leftarrow M[1;]+(^-3)\times M[2;]$ Replace row 1 by row 1 plus ⁻3 times row 2.

$M[3;]\leftarrow M[3;]+(^-6)\times M[2;]$ Replace row 3 by row 3 plus ⁻6 times row 2.

M
```
1  0  ⁻.6666666666
0  1   .8888888888
0  0  ⁻.3333333333
```

$M[3;]\leftarrow(^-3)\times M[3;]$ Multiply row 3 by ⁻3.
M
```
1  0  ⁻.6666666666
0  1   .8888888888
0  0  1
```

$M[1;]\leftarrow M[1;]+.6666666666\times M[3;]$

Making the elements in the third column above the 1 into 0's.

$M[2;]\leftarrow M[2;]+(^-.8888888888)\times M[3;]$
M
```
1  0  0
0  1  0
0  0  1
```

The final form of M is the 3×3 identity matrix.

Example 2

$$M \leftarrow 4 \ 4\rho\iota \ 16$$
$$M$$

```
 1  2  3  4
 5  6  7  8
 9 10 11 12
13 14 15 16
```

$M[2;] \leftarrow M[2;] + (^-5) \times M[1;]$ Making the elements in the first col-
$M[3;] \leftarrow M[3;] + (^-9) \times M[1;]$ umn below the 1 into 0's.
$M[4;] \leftarrow M[4;] + (^-13) \times M[1;]$
M

```
1   2    3    4
0  ⁻4   ⁻8  ⁻12
0  ⁻8  ⁻16  ⁻24
0 ⁻12  ⁻24  ⁻36
```

$M[2;] \leftarrow (1 \div ^-4) \times M[2;]$ Multiply row 2 by $^-1/4$ to create a 1
M in the second column.

```
1   2    3    4
0   1    2    3
0  ⁻8  ⁻16  ⁻24
0 ⁻12  ⁻24  ⁻36
```

$M[1;] \leftarrow M[1;] + (^-2) \times M[2;]$ Using the 1 in the second row and
$M[3;] \leftarrow M[3;] + 8 \times M[2;]$ column to make the rest of the sec-
$M[4;] \leftarrow M[4;] + 12 \times M[2;]$ ond column into 0's.
M

```
1 0 ⁻1 ⁻2
0 1  2  3
0 0  0  0
0 0  0  0
```

This is as close as we can make it to the identity matrix. This matrix cannot be made any simpler by using elementary row operations. This process of reducing a matrix to row reduced form is quite tedious. Therefore, we now consider a program which uses the elementary row operations to reduce a matrix to row reduced form.

Program 4.1 ROWFORM

```
    ∇R←ROWFORM M;K;L;H
[1]  K←0
[2]  RAISE:K←K+1
[3]  →K>(ρM)[2])/ANSWER
[4]  L←K
[5]  CHECK:→(M[K;K]≠0)/BEGIN
[6]  L←L+1
[7]  →(L>(ρM)[1])/RAISE
[8]  M[K,L;]←M[L,K;]
[9]  →CHECK
[10] BEGIN:H←0
[11] M[K;]←(1÷M[K;K])×M[K;]
[12] INCREASE: H←H+1
[13] →(H≠K)/NEXT
[14] →((K=(ρM)[1])∧(H=(ρM)[1]))/ANSWER
[15] H←H+1
[16] NEXT: M[H;]←M[H;]−M[H;K]×M[K;]
[17] →(H<(ρM)[1])/INCREASE
[18] →(K<(ρM)[1])/RAISE
[19] ANSWER: R←M
    ∇
```

Essentially, this program works as follows: Proceeding to line 5, if $M[1;1]$ is 0, row 1 of M is interchanged with row 2 (line 8). Then, branching back to *CHECK* (line 5), if $M[1;1]$ is still 0, rows 1 and 3 are interchanged. This continues until row 1 has been interchanged with all rows of M, $(L>(\rho M)[1])$, which means that the entire first column is all 0's, in which case the program branches back to *RAISE* (line 2), where K is increased to 2 and an analogous process is used on the second column. However, if we do get an $M[1;1]\neq0$, the program branches to *BEGIN* where $M[1;1]$ is made into a 1 by multiplying row 1 by $1÷M[1;1]$ (line 11). Then, all other elements in column 1 are made into 0's (line 16). In line 18, the program branches back to *RAISE* where K becomes 2.

Now, the same process is repeated in column 2. That is, by interchanging row 2 with those below it, the program attempts to find an $M[2;2]\neq0$. Then $M[2;2]$ is converted to 1 by multiplying row 2 by $1÷M[2;2]$. All other elements in column 2 are next converted to 0's. Then, we are branched back to *RAISE*, where K is made 3 and the same process is repeated with column 3. In other words, $M[3;3]$ is converted to 1, if possible, and all other elements in column 3 are converted to 0's. Then, to column 4, etc., until $K=(\rho M)[1]$, when we have all columns in the proper form. Then, this last form of M is printed out as our answer R in line 19.

This program does exactly as we would do if we were to convert M to row reduced form ourselves (by hand). Only, using the computer, it accomplishes this process much faster.

Let us consider some examples using this program *ROWFORM*.

Example 1

```
     M←4 4ρι 16
     M
 1   2   3   4
 5   6   7   8
 9  10  11  12
13  14  15  16
```

```
     ROWFORM  M
```
To run the program, type *ROW-FORM M.*

```
1 0  ̄1  ̄2
0 1   2   3
0 0   0   0
0 0   0   0
```

Example 2

```
     M←3 3ρ1 2 3 2 1 3 3 2 1
     M
1 2 3
2 1 3
3 2 1
```

```
     ROWFORM M
1 0 0
0 1 0
0 0 1
```
This is the best possible final form of a matrix, the identity.

Example 3

```
     M←3 4ρ1 2  ̄1 2 2 5  ̄2 3 1 2 1 2
     M
1 2  ̄1 2
2 5  ̄2 3
1 2   1 2
```

```
     ROWFORM M
1 0 0   4
0 1 0  ̄1
0 0 1   0
```

90

Example 4

```
M←3 3ρ1 2 3 2 4 8 1 2 5
M
```
1 2 3
2 4 8
1 2 5

```
ROW FORM M
```
1 2 0
0 0 1
0 0 0

EXERCISES

1. Use elementary row operations at an APL terminal to reduce the following matrices to row reduced form:

(a) $A = \begin{bmatrix} 1 & 2 & 3 \\ 1 & 3 & 5 \\ 2 & 5 & 9 \end{bmatrix}$

(b) $B = \begin{bmatrix} 1 & 1 & 1 \\ 2 & 3 & 4 \\ 1 & 3 & 5 \end{bmatrix}$

(c) $C = \begin{bmatrix} 1 & 2 & ^-1 & 4 \\ 2 & 4 & 3 & 5 \\ ^-1 & ^-2 & 6 & 0 \\ 5 & 0 & ^-3 & ^-1 \end{bmatrix}$

(d) $D = \begin{bmatrix} 0 & 1 & 2 & 3 \\ 1 & 1 & 1 & 5 \\ 2 & 1 & 0 & 7 \\ 1 & 0 & 1 & 2 \end{bmatrix}$

(e) $E = \begin{bmatrix} 1 & 2 & ^-1 & 4 \\ 2 & 4 & 3 & 5 \\ ^-1 & ^-2 & 6 & 0 \end{bmatrix}$

2. Use the program *ROWFORM* to check your answers to Exercise 1.

3. With a pencil and paper, trace the program *ROWFORM* with the matrix *D* above.

4. Using the *TRACE* command, trace the program *ROWFORM* with the matrix *D* above.

4.6 The inverse of a matrix

If a system of linear equations has the same number of equations as unknowns, and if the matrix of coefficients has as its row reduced form an identity matrix, then there is a very easy method for solving the system using the inverse of the matrix of coefficients. In this section, we shall consider the concept of the inverse of a matrix. In the next section, we shall apply the idea of matrix inversion to solving a linear system.

Recall, from Section 3.5., that if we are working with square *n* by *n* matrices, then there is a *multiplicative identity matrix I*. This matrix has the

property that if A is any n by n matrix, then
$$A \cdot I = I \cdot A = A.$$
I is the n by n matrix with 1's down the main diagonal and 0's elsewhere. In APL, the n by n identity matrix can be created as follows:

$I \leftarrow (\iota N) \circ . = (\iota N).$

This makes use of the concept of outer product. 1 is compared to each element of ιN using the logical operator, $=$, yielding the vector 1 0 0 $\cdots$ 0 as the first row. Then, 2 is compared to each element of ιN using $=$, yielding that vector 0 0 1 $\cdots$ 0 as the second row. Then, 3 is compared to each element of ιN using $=$, yielding the vector 0 0 1 $\cdots$ 0 as the third row, and so on through the nth row.

Example

$I \leftarrow (\iota 3) \circ . = (\iota 3)$
I
1 0 0
0 1 0
0 0 1

The diagram below might be helpful in understanding the outer product

$(\iota 3) \circ . = (\iota 3)$:

=	1	2	3
1	1	0	0
2	0	1	0
3	0	0	1

From now on, we shall refer to I as simply the *identity* matrix of appropriate size.

The inverse of a matrix

The *inverse* of an $n \times n$ matrix A, if it has an inverse, is the $n \times n$ matrix (usually denoted by A^{-1}) such that
$$(A \cdot A^{-1}) = (A^{-1} \cdot A) = I.$$

Examples

1. If
$$A = \begin{bmatrix} 2 & 3 \\ 4 & 7 \end{bmatrix},$$
then
$$A^{-1} = \begin{bmatrix} 3.5 & {}^-1.5 \\ {}^-2 & 1 \end{bmatrix},$$

since

$$(A \cdot A^{-1}) = (A^{-1} \cdot A) = \begin{bmatrix} 1 & 0 \\ 0 & 1 \end{bmatrix} = I.$$

2. If

$$A = \begin{bmatrix} 1 & 2 & 3 \\ 1 & 3 & 5 \\ 2 & 5 & 9 \end{bmatrix},$$

then

$$A^{-1} = \begin{bmatrix} 2 & -3 & 1 \\ 1 & 3 & -2 \\ -1 & -1 & 1 \end{bmatrix},$$

since

$$(A \cdot A^{-1}) = (A^{-1} \cdot A) = \begin{bmatrix} 1 & 0 & 0 \\ 0 & 1 & 0 \\ 0 & 0 & 1 \end{bmatrix} = I.$$

3. If

$$A = \begin{bmatrix} 1 & 2 \\ 3 & 6 \end{bmatrix},$$

then A has no inverse. In order to see this, suppose that

$$A^{-1} = \begin{bmatrix} x & y \\ z & w \end{bmatrix}.$$

Then,

$$(A \cdot A^{-1}) = \begin{bmatrix} x+2z & y+2w \\ 3x+6z & 3y+6w \end{bmatrix} = \begin{bmatrix} 1 & 0 \\ 0 & 1 \end{bmatrix}.$$

Therefore, $x+2z=1$ and $3x+6z=0$. However, this is impossible. So, A has no inverse.

Matrices that do not have inverses are said to be *singular*. Those with inverses are therefore *nonsingular*. Obviously, any matrix which is not square is singular, because if the matrix is not square, then $A \cdot A^{-1}$ and $A^{-1} \cdot A$ would not be the same size, let alone equal.

We now have three questions:

1. How does one decide whether a matrix is singular or nonsingular?
2. If the matrix is nonsingular, how does one compute its inverse?
3. Of what use is the inverse of a matrix?

In order to attempt to answer these questions, consider the following matrix equation:

$$(A \cdot X) = (I \cdot B) \qquad ((A + . \times X) = (I + . \times B) \text{ in APL notation}).$$

Using conventional notation, and multiplying both sides of this equation by A^{-1}, assuming A is nonsingular, one obtains

$$(A^{-1} \cdot (A \cdot X)) = (A^{-1} \cdot (I \cdot B)),$$

or using the associative property,

$$((A^{-1} \cdot A) \cdot X) = ((A^{-1} \cdot I) \cdot B), \text{ or } (I \cdot X) = X = (A^{-1} \cdot B).$$

Thus, if A^{-1} exists, it can be used to solve matrix equations like the one above. Consider the system of equations:

$$a_{11} \cdot x_1 + a_{12} \cdot x_2 + \ldots + a_{1n} \cdot x_n = 1 \cdot b_1 + 0 \cdot b_2 + \ldots + 0 \cdot b_n = b_1$$

$$a_{21} \cdot x_1 + a_{22} \cdot x_2 + \ldots + a_{2n} \cdot x_n = 0 \cdot b_1 + 1 \cdot b_2 + \ldots + 0 \cdot b_n = b_2$$

$$a_{n1} \cdot x_1 + a_{n2} \cdot x_2 + \ldots + a_{nn} \cdot x_n = 0 \cdot x_1 + 0 \cdot x_2 + \ldots + 1 \cdot x_n = b_n.$$

As a matrix equation, this would be $A \cdot X = I \cdot B$, where

$$A = \begin{bmatrix} a_{11} & a_{12} & \cdots & a_{1n} \\ a_{21} & a_{22} & \cdots & a_{2n} \\ \vdots & \vdots & & \vdots \\ a_{n1} & a_{n2} & \cdots & a_{nn} \end{bmatrix} \quad X = \begin{bmatrix} x_1 \\ x_2 \\ \vdots \\ x_n \end{bmatrix}$$

$$B = \begin{bmatrix} b_1 \\ b_2 \\ \vdots \\ b_n \end{bmatrix} \quad I = \begin{bmatrix} 1 & 0 & \cdots & 0 \\ 0 & 1 & \cdots & 0 \\ \vdots & \vdots & & \vdots \\ 0 & 0 & \cdots & 1 \end{bmatrix}.$$

If this system has a unique solution and has been solved by the method of elimination, then we would end up with a system like the following:

$$1 \cdot x_1 + 0 \cdot x_2 + \ldots + 0 \cdot x_n = x_1 = c_{11} \cdot b_1 + c_{12} \cdot b_2 + \ldots + c_{1n} \cdot b_n$$

$$0 \cdot x_1 + 1 \cdot x_2 + \ldots + 0 \cdot x_n = x_2 = c_{21} \cdot b_1 + c_{22} \cdot b_2 + \ldots + c_{2n} \cdot b_n$$

$$0 \cdot x_1 + 0 \cdot x_2 + \ldots + 1 \cdot x_n = x_n = c_{n1} \cdot b_1 + c_{n2} \cdot b_2 + \ldots + c_{nn} \cdot b_n.$$

Or, as a matrix equation, this would be

$$(I \cdot X) = X = (C \cdot B),$$

where $C = A^{-1}$, since $X = (A^{-1} \cdot B)$. Considering these two systems of linear equations, this process seems analogous to using elementary row operations to reduce the augmented matrix, A, I to the form I, A^{-1}.

An algorithm for matrix inversion

The above discussion yields the following algorithm for inverting a matrix:

1. Form the augmented matrix A, I.
2. Use elementary row operations on the rows of this augmented matrix to reduce it to the form I, A^{-1}.
3. Drop the matrix I from the resulting matrix.

[*Note*: This process also provides us with a test for the nonsingularity of a matrix. If the row reduced form of A is I, then A is nonsingular. Otherwise, A is singular.]

Let us apply the above algorithm to finding the inverses of the matrices in the previous examples:

Example 1

```
A←2 2ρ2 3 4 7
A
```
2 3
4 7
```
ROWFORM A
```
1 0   Since the row reduced form of A is
0 1   the identity, I, then A has an inverse. It is nonsingular.

$I←(\iota 2)\circ.=(\iota 2)$ The 2×2 identity matrix.

$M←A,I$ Step 1: forming the augmented matrix A, I.

```
ROWFORM M
```
1 0 3.5 ⁻1.5   Step 2: reducing the augmented
0 1 ⁻2   1   matrix to the form I, A^{-1}. Let the program ROWFORM do the work.

$INVERSE←2 \ ^{-}2\uparrow M$ Step 3: taking the inverse from the resulting matrix. We do this by
```
INVERSE
```
taking the first two rows and last two
3.5 ⁻1.5   columns of M.
⁻2   1

Thus,

$$A^{-1}=\begin{bmatrix} 3.5 & ^{-}1.5 \\ ^{-}2 & 1 \end{bmatrix}.$$

Let us check this using APL.

```
A+.×INVERSE
```
1 0
0 1   It checks.

Example 2

```
A←3 3ρ1 2 3 1 3 5 2 5 9
A
```
1 2 3
1 3 5
2 5 9

```
ROWFORM A
```
1 0 0   Since **ROWFORM A** is *I*, *A* is nonsin-
0 1 0   gular.
0 0 1

```
I←(ι3)∘.=(ι3)
```
               The 3×3 identity matrix.

```
M←A, I
M
```
               The augmented matrix
1 2 3 1 0 0
1 3 5 0 1 0
2 5 9 0 0 1

```
ROWFORM M
```
               Reducing *M* to row reduced form.
1 0 0   2 ‾3   1
0 1 0   1   3 ‾2
0 0 1 ‾1 ‾1   1

```
INVERSE←3 ‾3↑M
INVERSE
```
               Dropping *I*.

  2 ‾3   1   The inverse of *A*.
  1   3 ‾2
‾1 ‾1   1

```
A+.×INVERSE
```
1 0 0   Checking our result using the defini-
0 1 0   tion of inverse.
0 0 1   It checks.

Example 3

```
A←2 2ρ1 2 3 6
A
```
1 2
3 6

```
ROWFORM A
```
               Since **ROWFORM A** is not *I*, *A* is
1 2   singular. It has no inverse.
0 0

We now consider a simple APL program for finding the inverse of a
matrix. However, before using this program to find A^{-1}, first find

ROWFORM A using Program 4.1. If *ROWFORM A* is not the identity matrix, *I*, then *A* has no inverse. If *ROWFORM A* is the identity matrix, then *A* has an inverse which can be found using the following program.

Program 4.2 INVERSE

$\nabla INV \leftarrow INVERSE\ A; N; I;$

[1] $N \leftarrow (\rho A)[1]$

[2] $I \leftarrow (\iota N) \circ . = (\iota N)$

[3] $INV \leftarrow (N, -N) \uparrow ROWFORM\ (A, I)$
      ∇

If line 3 of this program is read from right to left, one can easily see that it is merely performing the 3 steps in the above algorithm.

Example 1

        $A \leftarrow 2\ 2\rho 2\ 3\ 4\ 7$
        A
    2 3
    4 7
        ROWFORM A
    1 0   Thus, *A* has an inverse.
    0 1
        INVERSE A
    3.5 ⁻1.5   Using the program *INVERSE* to com-
     ⁻2   1   pute A^{-1}.

Example 2

        $A \leftarrow 3\ 3\rho 1\ 2\ 3\ 1\ 3\ 5\ 2\ 5\ 9$
        A
    1 2 3
    1 3 5
    2 5 9
        ROWFORM A
    1 0 0   Thus, *A* has an inverse.
    0 1 0
    0 0 1

        INVERSE A
     2 ⁻3 1
     1 3 ⁻2
    ⁻1 ⁻1 1

97

EXERCISES

1. Use elementary row operations to find the inverses of the following matrices (if
they have inverses):

(a) $A = \begin{bmatrix} 3 & 4 \\ 1 & 2 \end{bmatrix}$

(b) $B = \begin{bmatrix} 2 & 1 & 3 \\ 3 & 1 & 2 \\ 1 & 2 & 3 \end{bmatrix}$

(c) $C = \begin{bmatrix} 1 & 2 & 3 \\ 4 & 5 & 6 \\ 7 & 8 & 9 \end{bmatrix}$

(d) $D = \begin{bmatrix} ^-1 & 4 & 2 \\ 7 & 2 & 0 \\ 0 & 3 & 5 \end{bmatrix}$

(e) $E = \begin{bmatrix} 2 & 1 & 3 & 4 \\ 3 & 1 & 5 & 2 \\ 2 & 1 & 4 & 3 \\ 3 & 1 & 2 & 5 \end{bmatrix}$

2. Use the program *INVERSE* to check your answers to Exercise 1.

3. Trace the program *INVERSE* using the *TRACE* command and the matrix B.

4. Prove that the inverse of a matrix is unique. [*Hint*: Suppose that there are two
inverses of A, called X and Y, and show that $X = Y$.]

5. Write an APL program for computing the $N \times N$ identity matrix I for any given
N.

4.7 Inverses in APL

It seems only fair to point out that if a matrix A is nonsingular, then its
inverse can be found directly on the APL system by simply typing $\boxdiv A$.
$\boxdiv$ is often called the "domino function." It is obtained by typing the
quad, $\square$, backspacing, and typing the $\div$ inside of the $\square$. If A is
singular, then, on most APL systems, an error message will result when
$\boxdiv A$ is entered.

Example 1

```
    A←2 2ρ 2 3 4 7
    A
2 3
4 7
    ⊞ A
3.5 ‾1.5
‾2   1
```

Example 2

```
    A←2 2ρ1 2 3 6
    A
1 2
3 6
    ⊞ A
DOMAIN ERROR                    A has no inverse.
```

Example 3

```
    A←3 3ρ1 2 3 1 3 5 2 5 9
    A
1 2 3
1 3 5
2 5 9
      ⌹ A
 2 ⁻3  1
 1  3 ⁻2
⁻1 ⁻1  1
```

APL solution to a system of linear equations

As pointed out in the previous section, a system of n linear equations with n unknowns can be expressed as a matrix equation $(A \cdot X) = B$. Also, if A^{-1} exists, the solution is given by $X = (A^{-1} \cdot B)$. In APL, this solution is given by

$$X \leftarrow (⌹A) + . \times B$$

or, even more simply, by

$$X \leftarrow B ⌹ A \qquad \text{(the matrix divide operation).}$$

Examples

1.
$$2x + 3y = 4$$
$$4x + 7y = 6$$

```
    A←2 2ρ2 3 4 7
    B←4 6

    (⌹A)+.×B
5 ⁻2
```
So, $x = 5$, $y = ⁻2$.

```
    B⌹A
5 ⁻2
```

```
    A+.×(5 ⁻2)
4 6
```
It checks.

2.
$$x + 2y + 3z = 1$$
$$x + 3y + 5z = 2$$
$$2x + 5y + 9z = 3$$

```
    A←3 3ρ1 2 3 1 3 5 2 5 9
    B←1 2 3

    (⌹A)+.×B
⁻1 1 0
```
So, $x = ⁻1$, $y = 1$, $z = 0$.

$$B \boxed{\div} A$$
‾1 1 0

$$A+.\times(‾1\ 1\ 0)$$
1 2 3   It checks.

3.

$$x+2y=5$$
$$3x+6y=4$$

$A\leftarrow 2\ 2\rho 1\ 2\ 3\ 6$
$B\leftarrow 5\ 4$

$(\boxed{\div}A)+.\times B$
DOMAIN ERROR   Since A is singular.

4.8 Applications

Let us now consider a couple of applications of systems of linear equations.

Example 1

Ajax manufacturing company produces three large products called A, B, and C. To produce each product requires time on three machines called I, II, and III. Each unit of A requires 2 hours on I, 3 hours on II, and 1 hour on III. Each unit of B requires 1 hour on I, 1 hour on II, and 4 hours on III. Each unit of C requires 3 hours on I, 1 hour on II, and 2 hours on III. On a busy day, machine I is available for 13 hours, machine II for 12 hours, and machine III for 10 hours. How many of each product can be produced on this day?

Let x represent the number of A's that can be produced on this day, y represent the number of B's, and z represent the number of C's. Then, the above paragraph can be summarized in the following system:

$$2x+3y+\ z=13$$
$$x+\ y+4z=12$$
$$3x+\ y+2z=10$$

$A\leftarrow 3\ 3\rho 2\ 3\ 1\ 1\ 1\ 4\ 3\ 1\ 2$ The matrix of coefficients.

$B\leftarrow 13\ 12\ 10$ The vector of right-hand values.

$B\boxed{\div}A$ Solving the system using the matrix
1 3 2   divide operation.

Thus, $x=1$, $y=3$, $z=2$. So, the company can produce 1 A, 3 B's, and 2 C's on this day.

Example 2

Last year, a man invested $60,000 in three operations: a hamburger stand, a miniature golf course, and a vegetable stand. He made a profit of 15 percent on the hamburger stand, 10 percent on the miniature golf course, and 20 percent on the vegetable stand. His total profit for the year was $9000. He invested twice as much in the hamburger stand as in the miniature golf course and the vegetable stand together. How much did he initially invest in each operation?

Let x be his initial investment in the hamburger stand, y his investment in the miniature golf course, and z his investment in the vegetable stand. Then, as a system of linear equations, we have

$$x + \quad y + \quad\quad z = 60000 \qquad \text{His total investment.}$$

$$0.15x + 0.10y + 0.20z = 9000 \qquad \text{His profit.}$$

$$x - \quad 2y - \quad 2z = 0 \qquad\quad \text{Since } x = 2 \cdot (y + z).$$

```
A←3 3ρ1 1 1 .15 .10 .20 1 ¯2 ¯2
B←60000 9000 0

B ⊞ A
40000 10000 10000
```

Thus, he invested $40,000 in the hamburger stand, $10,000 in the miniature golf course, and $10,000 in the vegetable stand.

EXERCISES

1. Find the inverses of the matrices in Exercise 1 of Section 4.6, using the APL operation $\boxed{\div}$.

2. Test the validity of the following statements at an APL terminal:
   (a) $(A \cdot B)^{-1} = B^{-1} \cdot A^{-1}$ $\quad$ $((\boxed{\div}(A + . \times B)) = ((\boxed{\div}B) + . \times (\boxed{\div}A))$ in APL)
   (b) $(A \cdot B)^{-1} = A^{-1} \cdot B^{-1}$ $\quad$ $((\boxed{\div}(A + . \times B)) = ((\boxed{\div}A) + . \times (\boxed{\div}B))$ in APL)
   (c) $(A^{-1})^{-1} = A$ $\quad$ $((\boxed{\div}(\boxed{\div}A)) = A$ in APL)
   Use some matrices of your own choosing for these tests.

3. Use the matrix divide operation $\boxed{\div}$ to solve the following systems of linear equations:
   (a) $x + 2y = 5$
   $\ 3x + 4y = 6$
   (b) $2x + y + 3z = 5$
   $\ 3x + y + 2z = 4$
   $\ x + 2y + 3z = 6$
   (c) $2x + y + 3z = 4$
   $\ 4x + 3y + z = 3$
   $\ 2x + y - z = 5$
   (d) $2x + y + 3z + 4w = 8$
   $\ 3x + y + 5z + 2w = 2$
   $\ 2x + y + 4z + 3w = 7$
   $\ 3x + y + 2z + 5w = 0$
   (e) $3x_1 + 2x_2 - x_3 + 4x_4 - x_5 = 7$
   $\ x_1 + 3x_2 - 2x_3 + x_4 + x_5 = 1$
   $\ 5x_1 - 3x_2 + x_3 + x_4 + 2x_5 = 4$
   $\ 2x_1 + x_2 - x_3 + 5x_4 + x_5 = 6$
   $\ x_1 - x_2 + 3x_3 + 2x_4 \quad\quad = {}^-1$

101

4. Set up and solve the following problems:

(a) A small store has $900 to spend on shirts and pants. Each shirt costs $4 and each pair of pants costs $10. They wish to buy twice as many shirts as pants. How many of each should they buy?

(b) A warehouse manager has a chance to buy and store and later sell three items called I, II, and III. Each I requires 1 square foot of storage space, each II requires 2 square feet, and each III requires 3 square feet. Each I costs $20 to buy and store, each II costs $50, and each III costs $90. Each I will bring a profit of $10, each II a profit of $30, and each III a profit of $50. If the warehouse manager has 1000 square feet of space available, $10,000 to spend, and would like a profit of $5000, how many I's, II's and III's should he buy?

(c) A carpenter builds bookcases, tables, and chairs. Each bookcase costs $20 to build, requires 12 square feet of lumber, and takes 5 hours. Each table costs $30 to build, requires 20 square feet of lumber, and takes 9 hours. Each chair costs $10 to build, requires 4 square feet of lumber, and takes 2 hours. How many of each can he build with $9500, 5800 square feet, and 2600 hours?

Determinants 5

5.1 Definition of a determinant

Associated with every square matrix is a number, called its determinant, symbolized by $\det(A)$ or by $|A|$. As we will see, determinants give some useful information about the matrix and can be used to invert matrices and solve systems of linear equations. In this chapter, we shall consider some ways in which to evaluate determinants as well as some of the applications of determinants.

Definition of determinant

The determinant of an $n \times n$ matrix A is the sum of all the $n!$ possible products of the form

$$(^-1)^k \cdot a_{1j_1} \cdot a_{2j_2} \cdot a_{3j_3} \dots a_{nj_n}$$

where no two column indices are the same and where k is the number of transpositions of the column indices needed to put them in natural order. [*Note*: In each of the above products, there is exactly one element from each row and column. Also, $n! = n \cdot (n-1) \cdot (n-2) \dots 3 \cdot 2 \cdot 1$, and $n!$ is read as n factorial.]

Let us apply this definition to 2×2 and 3×3 matrices.

Two-by-two determinants

If

$$A = \begin{bmatrix} a_{11} & a_{12} \\ a_{21} & a_{22} \end{bmatrix},$$

then

$$\det(A) = (^-1)^0 \cdot a_{11} \cdot a_{22} + (^-1)^1 \cdot a_{12} \cdot a_{21} = a_{11} \cdot a_{22} - a_{12} \cdot a_{21}.$$

Example

$$A = \begin{bmatrix} 1 & 2 \\ 3 & 4 \end{bmatrix}.$$

Then, $\det(A) = 1 \cdot 4 - 2 \cdot 3 = {}^-2$.

Three-by-three determinants

If

$$A = \begin{bmatrix} a_{11} & a_{12} & a_{13} \\ a_{21} & a_{22} & a_{23} \\ a_{31} & a_{32} & a_{33} \end{bmatrix},$$

then

$$\det(A) = ({}^-1)^0 \cdot a_{11} \cdot a_{22} \cdot a_{33} + ({}^-1)^2 \cdot a_{12} \cdot a_{23} \cdot a_{31} + ({}^-1)^2 \cdot a_{13} \cdot a_{21} \cdot a_{32}$$
$$+ ({}^-1)^1 \cdot a_{11} \cdot a_{23} \cdot a_{32} + ({}^-1)^1 \cdot a_{12} \cdot a_{21} \cdot a_{33} + ({}^-1)^1 \cdot a_{13} \cdot a_{22} \cdot a_{31}$$
$$= a_{11} \cdot a_{22} \cdot a_{33} + a_{12} \cdot a_{23} \cdot a_{31} + a_{13} \cdot a_{21} \cdot a_{32} - a_{11} \cdot a_{23} \cdot a_{32}$$
$$- a_{12} \cdot a_{21} \cdot a_{33} - a_{13} \cdot a_{22} \cdot a_{31}.$$

Example

If

$$A = \begin{bmatrix} 1 & 2 & 3 \\ 4 & 5 & 6 \\ 7 & 8 & 9 \end{bmatrix},$$

then

$$\det(A) = 1 \cdot 5 \cdot 9 + 2 \cdot 6 \cdot 7 + 3 \cdot 4 \cdot 8 - 1 \cdot 6 \cdot 8 - 2 \cdot 4 \cdot 9 - 3 \cdot 5 \cdot 7$$
$$= 45 + 84 + 96 - 48 - 72 - 105 = 225 - 225 = 0.$$

For $n > 3$, this definition becomes more difficult to apply. Therefore, we consider some techniques for using elementary row operations to evaluate determinants. First, however, we shall list some properties of determinants. These properties will be listed without proofs. However, the reader can easily verify them for two-by-two and three-by-three determinants using the formulas developed in the previous discussion.

Property 1.

If A is a triangular matrix (i.e., all elements above or all elements below the main diagonal are zeros), then $\det(A)$ is the product of the diagonal elements.

Property 2.

If A has a complete row (or column) of zeros, then $\det(A) = 0$.

Property 3.

If a row (or column) of a matrix A is multiplied by a nonzero constant c, then $\det(A)$ is multiplied by c.

104

Property 4.

If a multiple of one row (or column) is added to another row (or column), then the value of det(A) is unchanged.

Property 5.

If two rows (or columns) of A are interchanged, then det(A) is multiplied by $^-1$ (i.e., det(A) changes sign).

Using elementary row operations to evaluate determinants

By using elementary row operations, a matrix A can be reduced to a triangular matrix B. Then, by Property 1, det(B) can be found by just multiplying the diagonal elements of B. If B should have a row (or column) of all zeros, then det(B)=0. We must use Properties 3–5 to keep track of how det(B) is related to det(A). If to arrive at B, a row has been multiplied by a constant c, then det(B) must be multiplied by $1/c$ to compensate for this. If a multiple of a row has been added to another row, there has been no change in the determinant. If two rows have been interchanged then det(B) must be multiplied by $^-1$ to compensate for this. With these remarks in mind, let us evaluate some determinants using these properties.

Example 1

```
A←3 3ρ1 3 4 3 0 2 2 1 ¯1
A
```
```
1 3  4
3 0  2
2 1 ¯1
```

```
A[2;]←A[2;]+(¯3)×A[1;]
A
```
```
1  3   4
0 ¯9 ¯10          By Property 4, no change in det(A).
2  1  ¯1
```

```
A[3;]←A[3;]+(¯2)×A[1;]
A
```
```
1  3   4
0 ¯9 ¯10          By Property 4, no change in det(A).
0 ¯5  ¯9
```

```
A[3;]←(1 ÷ ¯5)×A[3;]
A
```
```
1  3   4          By Property 3, the value of det(A) is
0 ¯9 ¯10          multiplied by ¯1/5.
0  1 1.8
```

105

$A[2;] \leftarrow A[2;] + 9 \times A[3;]$
A

```
1 3  4
0 0 6.2
0 1 1.8
```

By Property 4, no change in det(A).

$A[2\ 3;] \leftarrow A[3\ 2;]$
A

```
1 3  4
0 1 1.8
0 0 6.2
```

By Property 5, det(A) is multiplied by $^-1$.

By Property 1, since the matrix is now triangular, the determinant of the resulting matrix is the product of its diagonal elements, 6.2. Compensating for the changes in det(A) noted above, we get

$$\det(A) = (^-5) \cdot (^-1) \cdot (6.2) = 31.$$

Example 2

$A \leftarrow 3\ 3\rho\iota\ 9$
A

```
1 2 3
4 5 6
7 8 9
```

$A[2;] \leftarrow A[2;] + (^-4) \times A[1;]$
$A[3;] \leftarrow A[3;] + (^-7) \times A[1;]$
A

```
1  2   3
0 ⁻3  ⁻6
0 ⁻6 ⁻12
```

By Property 4, no change in det(A).

$A[3;] \leftarrow A[3;] + (^-2) \times A[2;]$
A

```
1  2  3
0 ⁻3 ⁻6
0  0  0
```

By Property 4, no change in det(A).

Since the resulting matrix has a complete row of 0's, then by Property 2, det(A)=0.

EXERCISES

1. Using the formula developed from the definition of determinant, write a program to evaluate 2×2 determinants.

2. Repeat Exercise 1 for 3×3 determinants.

3. Use the programs you wrote in Exercises 1 and 2 to evaluate the following determinants:

(a) $\det\begin{bmatrix} 2 & 3 \\ 4 & 7 \end{bmatrix}$ (d) $\det\begin{bmatrix} 3 & 5 & 2 \\ 4 & 0 & 7 \\ 3 & 9 & 1 \end{bmatrix}$

(b) $\det\begin{bmatrix} 4 & {}^-2 \\ 6 & {}^-3 \end{bmatrix}$ (e) $\det\begin{bmatrix} {}^-1 & 4 & 2 \\ 7 & 2 & 0 \\ 0 & 3 & 5 \end{bmatrix}$

(c) $\det\begin{bmatrix} 6 & 0 \\ 3 & 2 \end{bmatrix}$ (f) $\det\begin{bmatrix} 3 & 2 & 1 \\ 1 & 2 & 3 \\ 2 & 3 & 1 \end{bmatrix}$

4. Use elementary row operations and Properties 1–5 to evaluate the following determinants:
   (a) Exercise 3(a) above
   (b) Exercise 3(d) above

(c) $\det\begin{bmatrix} 1 & 2 & 3 & 4 \\ 5 & 6 & 7 & 8 \\ 9 & 10 & 11 & 12 \\ 13 & 14 & 15 & 16 \end{bmatrix}$

(d) $\det\begin{bmatrix} 1 & 2 & {}^-1 & 4 \\ 2 & 4 & 3 & 5 \\ {}^-1 & {}^-2 & 6 & 0 \\ 5 & 0 & {}^-3 & {}^-1 \end{bmatrix}$

5. Use your programs in Exercises 1 and 2 and matrices of your choice to test the validity of the following law of determinants:

$$\det(A \cdot B) = \det(A) \times \det(B).$$

6. Use the law in Exercise 5 and the definition of inverse of a matrix to prove that $\det(A^{-1}) = 1 \div \det(A)$, provided that A is nonsingular. If A is singular, then $\det(A) = 0$.

5.2 A program for evaluating determinants

Rather than do all the work ourselves each time, why not let the computer perform the elementary row operations and reduce A to a triangular matrix B and compute $\det(A)$? The following program will accomplish this.

Program 5.1 DET

    $\nabla D \leftarrow DETM; K; N; P$

[1] $D \leftarrow 1$

[2] $START: N \leftarrow (\rho M)[1]$

[3] $\rightarrow (N=1)/LAST$

[4] $K \leftarrow 1$

[5] $INCREMENT: K \leftarrow K+1$

[6] $P \leftarrow 1$

[7] $CORRECT: \rightarrow (M[1;1] \neq 0)/EXECUTE$

[8] $\rightarrow ((\wedge/M[;1]=0)=1)/END$

[9] $M \leftarrow {}^-1 \phi [1]M$

[10] $P \leftarrow P \times (^-1)*(N-1)$

[11] $\rightarrow CORRECT$

[12] $EXECUTE: M[K;] \leftarrow M[K;] - ((\div M[1;1]) \times M[K;1]) \times M[1;]$

[13] $\rightarrow (K<N)/INCREMENT$

[14] $D \leftarrow D \times M[1;1] \times P$

[15] $M \leftarrow 1 \ 1 \downarrow M$

[16] $\rightarrow START$

[17] $LAST: D \leftarrow D \times M[1;1]$

[18] $\rightarrow 0$

[19] $END: D \leftarrow 0$
$$\nabla$$

Proceeding to line 7; if $M[1;1] \neq 0$, the computer branches to the line labeled *EXECUTE*. However, if $M[1;1]=0$, the computer goes to line 8, which checks to see if the first column consists entirely of zeros. If it does consist entirely of zeros, the computer branches to the line labeled *END*, and the value of D is 0, according to Property 2. If the column does not consist entirely of zeros, the computer proceeds to line 9. Here, the bottom row of M is moved to the top. Line 10 makes the proper sign adjustment for this. (This is equivalent to interchanging the bottom row with each other row in turn working from bottom to top. Thus, the sign of the determinant is changed $N-1$ times in accordance with Property 5.) Line 11 sends the computer back to the line labeled *CORRECT* to see if now $M[1;1] \neq 0$. This continues until $M[1;1] \neq 0$.

When $M[1;1] \neq 0$, the computer is sent to the line labeled *EXECUTE*. Here, each row from $K=2$ to $K=N$ is replaced by the expression in line 12. This expression essentially makes the first element in each row below the first row into a zero. Line 13 accomplishes the repetition of this process on each of these rows below the first. When $K=N$, the computer goes to line 14, where the previous value of D is multiplied by the element $M[1;1]$ and P, the sign adjustment due to interchanging rows.

In line 15, the first row and first column of *M* are dropped yielding a matrix with one less row and column. Then, the computer is sent back to *START* and the entire process is repeated on this smaller matrix.

The size of the matrix is thus reduced until it is a 1×1 matrix. In the process, the diagonal elements are multiplied and the sign changes accounted for (lines 10 and 14). The final result is equivalent to evaluating the determinant of a triangular matrix by multiplying its diagonal elements.

Since this program is quite long, it would be nice if determinant were a keyboard operation in APL. Unfortunately, at the present time, it is not. However, Dr. Iverson has told this author that this addition to the APL system is under consideration. Perhaps, in the near future, determinant will be a keyboard operation in APL.

Examples

```
      M←3 3ρ2 1 3 4 3 0 2 1 ¯1
      M
2 1  3
4 3  0
2 1 ¯1
      DET M
¯8
      M←4 4ρ1 2 ¯1 4 2 4 3 5 ¯1 ¯2 6 0 3 0 ¯3 ¯1
      M
 1  2 ¯1  4
 2  4  3  5
¯1 ¯2  6  0
 3  0 ¯3 ¯1
      DET M
¯210
      M←4 4ρι 16
      M
 1  2  3  4
 5  6  7  8
 9 10 11 12
13 14 15 16
      DET M
0
      M←3 3ρι 9
      M
1 2 3
4 5 6
7 8 9
```

DET M
1.705302566E⁻13

This answer is essentially zero, since E^-13 means $10*^-13 = 0.000000000001$, so that $1.705302566E^-13 = 0.0000000000001705302566$, which is essentially 0. Actually, if one were to compute the determinant of this matrix by hand, one would get 0.

EXERCISES

1. Use the program *DET* to evaluate the following determinants:

(a) $\det \begin{bmatrix} 2 & 3 \\ 4 & 7 \end{bmatrix}$

(b) $\det \begin{bmatrix} 3 & 5 & 2 \\ 4 & 0 & 7 \\ 3 & 9 & 1 \end{bmatrix}$

(c) $\det \begin{bmatrix} 1 & 2 & -2 & 3 \\ 5 & -3 & 1 & 0 \\ 3 & 0 & -4 & 8 \\ 2 & 5 & 4 & 0 \end{bmatrix}$

(d) $\det \begin{bmatrix} 1 & 2 & 3 & 4 & 5 \\ 6 & 7 & 8 & 9 & 10 \\ 11 & 12 & 13 & 14 & 15 \\ 16 & 17 & 18 & 19 & 20 \\ 21 & 22 & 23 & 24 & 25 \end{bmatrix}$

2. Perform a paper and pencil trace of the program *DET* using the matrix

$$M = \begin{bmatrix} 1 & 2 & 3 \\ 4 & 5 & 6 \\ 7 & 8 & 9 \end{bmatrix}.$$

3. Use the *TRACE* command at an APL terminal to do Exercise 2.

5.3 Cofactors

Another technique for evaluating determinants of $n \times n$ matrices involves reducing the problem to evaluating a linear combination of the determinants of certain $(n-1) \times (n-1)$ submatrices of the matrix. This method is referred to as the Method of cofactors. We will consider this method as well as some other uses of cofactors.

Definition of minor

The *minor* A_{ij} of an $n \times n$ matrix A is the $(n-1) \times (n-1)$ matrix obtained from A by deleting the *i*th row and *j*th column of A.

A program for computing the minor A_{ij} follows:

110

Program 5.2 MINOR

$\nabla AIJ \leftarrow A$ MINOR $V; Q$

[1] $Q \leftarrow (V[1] \neq \iota(\rho A)[1])/[1]A$

[2] $AIJ \leftarrow (V[2] \neq \iota(\rho A)[2])/Q$
∇

This program works as follows: $(V[1] \neq \iota(\rho A)[1])$ yields a vector of 1's (for true) in every position except the position corresponding to $V[1]$ where it yields a 0 (since $V[1] \neq V[1]$ is false). Then, $(V[1] \neq \iota(\rho A)[1])/[1]A$ compresses this vector 1 1 ... 1 0 1 ... 1 (0 is in the position $V[1]$) on [1]A (the rows of A).

The result is that the rows corresponding to the 1's are kept, and the row corresponding to the 0 is deleted. The resulting matrix is called Q. Then, $(V[2] \neq \iota(\rho A)[2])/Q$ operates similarly on the columns of A, yielding AIJ.

Examples

```
    A←3 3ρι 9
    A
1 2 3
4 5 6
7 8 9

    V←1 1
```

    A MINOR V   Delete the first row and first column of A.

```
5 6
8 9

    V←2 2
```

    A MINOR V   Delete the second row and second column.

```
1 3
7 9
```

Associated with any element a_{ij} of a matrix A is a number c_{ij}, called the cofactor of a_{ij}.

Definition of cofactor

The cofactor of the element a_{ij} of the matrix A is the number

$$c_{ij} = (^-1)^{i+j} \cdot \det(A_{ij}), \quad \text{where } A_{ij} \text{ is the minor.}$$

111

5 Determinants

Example

Let

$$A = \begin{bmatrix} 1 & 2 & 3 \\ 4 & 5 & 6 \\ 7 & 8 & 9 \end{bmatrix}.$$

The cofactor of

$$a_{12} = (^-1)^{1+2} \cdot \det \begin{bmatrix} 4 & 6 \\ 7 & 9 \end{bmatrix} = 6.$$

The cofactor of

$$a_{21} = (^-1)^{2+1} \cdot \det \begin{bmatrix} 2 & 3 \\ 8 & 9 \end{bmatrix} = 6.$$

The cofactor of

$$a_{33} = (^-1)^{3+3} \cdot \det \begin{bmatrix} 1 & 2 \\ 4 & 5 \end{bmatrix} = ^-3.$$

The following program computes the cofactor c_{ij} described above:

Program 5.3 COFACTOR

```
∇ CIJ←A COFACTOR V; SIGN
[1]    SIGN←⁻1 * V[1]+V[2]
[2]    CIJ←SIGN×DET A MINOR V
       ∇
```

Example

```
A←3 3ρι 9
A
1 2 3
4 5 6
7 8 9
       A COFACTOR 1 2
6
       A COFACTOR 2 1
6
       A COFACTOR 3 3
⁻3
```

It will be useful to us to have a program that will form a matrix obtained from a matrix *A* by replacing each element of *A* with the corresponding cofactor. Thus, it forms a matrix of all the cofactors of *A*.

112

Program 5.4 *COFACTORS*

    ∇ *CO*← *COFACTORS A*; *I*; *J*

[1]   *CO*←ι 0

[2]   *I*←1

[3]   *INCREASE*: *J*←0

[4]   *ITERATE*: *J*←*J*+1

[5]   *CO*←*CO*, *A COFACTOR I*, *J*   Each element is replaced by its cofactor.

[6]   →(*J*<(ρ*A*)[1])/*ITERATE*

[7]   *I*←*I*+1

[8]   →(*I*≤(ρ*A*)[1])/*INCREASE*

[9]   *CO*←(ρ*A*)ρ *CO*   Line 9 forms a matrix of order (ρ*A*)
        ∇ from the vector in line 5.

Example

    A←3 3$\rho\iota$ 9
    A
1 2 3
4 5 6
7 8 9

    COFACTORS A
¯3   6 ¯3
 6 ¯12 6
¯3   6 ¯3

The way in which these cofactors are used to evaluate determinants is explained in the following theorem.

The method of cofactors

For any $n \times n$ matrix A, det(A) can be computed as the sum of the products of the elements of any row (or column) of A with their cofactors. In other words, if

$$A = \begin{bmatrix} a_{11} & a_{12} & a_{13} & \cdots & a_{1n} \\ a_{21} & a_{22} & a_{23} & \cdots & a_{2n} \\ \vdots & \vdots & \vdots & & \vdots \\ a_{n1} & a_{n2} & a_{n3} & \cdots & a_{nn} \end{bmatrix},$$

then det(A) = $a_{i1} \cdot c_{i1} + a_{i2} \cdot c_{i2} + a_{i3} \cdot c_{i3} + \ldots + a_{in} \cdot c_{in}$ (expansion by the ith row), where c_{ij} is the cofactor corresponding to a_{ij}.

113

We will omit the proof of the method of cofactors. The following program uses the method of cofactors to evaluate a determinant with the first row used as the row of expansion.

Program 5.5 *D̲E̲T̲*

$\nabla\ D\leftarrow\underline{DET}\ M$

[1] $D\leftarrow M[1;]+.\times(COFACTORS\ M)[1;]$
     ∇

[*Note*: The name of this program is *D̲E̲T̲* , with the letters underlined to distinguish it from the program *DET*.]

This program forms the inner product of the elements in the first row of *M* with their cofactors, which form the first row of *COFACTORS*.

Examples

     $M\leftarrow 3\ 3\rho\iota\ 9$
     M
1 2 3
4 5 6
7 8 9

     $\underline{DET}\,M$
0

     $M\leftarrow 2\ 2\rho\iota\ 4$
     M
1 2
3 4

     $\underline{DET}\,M$
⁻2

     $M\leftarrow 4\ 4\rho\iota\ 16$
     M
 1 2 3 4
 5 6 7 8
 9 10 11 12
13 14 15 16

     $\underline{DET}\,M$
0

As an exercise, the student is asked to try multiplying the elements of some row by the cofactors of some other row and adding the results in some matrices. The result should always be zero, because this is equivalent to evaluating the determinant of a matrix with two rows the same. Why should this yield a result of 0?

114

EXERCISES

1. Let

$$M = \begin{bmatrix} 1 & 2 & 3 & 4 \\ 5 & 0 & 9 & ^-2 \\ 3 & 1 & ^-1 & 3 \\ 8 & 2 & 0 & 0 \end{bmatrix}$$

(a) Find the minor M_{23} of M and the corresponding cofactor c_{23}.
(b) Find the minor M_{41} of M and the corresponding cofactor c_{41}.
(c) Find the minor M_{13} of M and the corresponding cofactor c_{13}.
(d) Find the matrix of cofactors (use *COFACTORS*).
(e) Use *DET* to evaluate $\det(M)$.
(f) Use <u>*DET*</u> to evaluate $\det(M)$.

2. Write a program for evaluating 3×3 determinants using the method of cofactors expanding about the third row.

3. Use the method of cofactors to evaluate the determinant of the matrix

$$M = \begin{bmatrix} 3 & 5 & 2 \\ 4 & 0 & 7 \\ 3 & 9 & 1 \end{bmatrix}$$

(a) Using the program you wrote in Exercise 2.
(b) Expanding about the second row.
(c) Expanding about the second column.

4. Let M be the matrix in Exercise 3.
(a) Multiply the elements of row 1 by the cofactors of the corresponding elements of row two and add the results.
(b) Compare your result with the determinant of the matrix

$$\begin{bmatrix} 3 & 5 & 2 \\ 3 & 5 & 2 \\ 3 & 9 & 1 \end{bmatrix}.$$

(c) Can you explain why these results should be equal?

5. Trace the program *COFACTORS* on the matrix

$$M = \begin{bmatrix} 1 & 2 & ^-1 & 4 \\ 2 & 4 & 3 & 5 \\ ^-1 & ^-2 & 6 & 0 \\ 5 & 0 & ^-3 & 1 \end{bmatrix}.$$

6. Prove the method of cofactors on a general 3×3 determinant using expansion about the first row. [*Hint*: Compare your result with the formula for 3×3 determinants in Section 5.1.]

5.4 Adjoints and inverses

Related to the idea of cofactors of a matrix A is another matrix called the adjoint of A. This adjoint can be used to compute the inverse of A, as we shall see.

115

Definition of adjoint of a matrix

For any $n \times n$ matrix A, the *adjoint* of A, symbolized by Adj(A), is the transpose of the matrix obtained from A by replacing each element of A by its cofactor.

Thus, the adjoint of A is just the transpose of the matrix of cofactors. Since we already have a program *COFACTORS* for obtaining this matrix of cofactors, it is quite simple to write a program for the adjoint of A.

*Program 5.6 **ADJOINT***

```
∇ ADJ ← ADJOINT A
```

[1] ADJ ← ⍉ COFACTORS A   This program merely transposes
```
∇
```
COFACTORS.

Example 1

```
A ← 3 3ρ1 2 3 1 3 5 2 5 9
A
1 2 3
1 3 5
2 5 9
```

```
COFACTORS A
  2   1  ¯1
 ¯3   3  ¯1
  1  ¯2   1
```

```
ADJOINT A
  2  ¯3   1
  1   3  ¯2
 ¯1  ¯1   1
```

ADJOINT A is the transpose of COFACTORS A.

Example 2

```
A ← 3 3ρι 9
A
1 2 3
4 5 6
7 8 9
```

```
COFACTORS A
 ¯3    6   ¯3
  6  ¯12    6
 ¯3    6   ¯3
```

116

        ADJOINT A
‾3   6 ‾3
 6 ‾12 6
‾3   6 ‾3   Note that in this example, *COFAC-*
                TORS A = ADJOINT A.

Use of the adjoint in inverting a matrix

The following theorem can be found in most linear algebra texts:

Theorem 1

For any $n \times n$ matrix, $(A \cdot \text{Adj}(A)) = (\text{Adj}(A) \cdot A) = (\det(A) \cdot I)$, where I is the $n \times n$ identity matrix.

We will not rigorously prove this theorem here. However, we will verify it with the following examples:

Example 1

    A←3 3ρ1 2 6 2 1 0 3 2 1
    A
1 2 6
2 1 0
3 2 1

    DET A
3

    A+.× ADJOINT A
3 0 0
0 3 0
0 0 3

    (ADJOINT A)+.× A
3 0 0
0 3 0
0 0 3

      I←(ι3)∘.=(ι3)

    (DET A)× I
3 0 0
0 3 0
0 0 3

Example 2

```
A←3 3ρι 9
A
1 2 3
4 5 6
7 8 9

DET A
0

A+.×ADJOINT A
0 0 0
0 0 0
0 0 0

(ADJOINT A)+.×A
0 0 0
0 0 0
0 0 0

(DET A)×I
0 0 0
0 0 0
0 0 0
```

An immediate consequence of this theorem, obtained by multiplying both sides of each equation by $(1 \div DET\ A)$, is the following result concerning the inverse of A:

Theorem 2

A is nonsingular if and only if **DET** $A \neq 0$. *Moreover, if* **DET** $A \neq 0$, *then*

$$A^{-1} = (1 \div \det(A)) \cdot (\mathrm{Adj}(A)).$$

Thus, we have the following simple program for inverting a matrix.

Program 5.7 INVERT

```
    ∇INV←INVERT A
[1]   →((DET A)=0)/END
[2]   INV←(ADJOINT A)÷(DET A)
[3]   →0
[4]   END : 'A IS SINGULAR'
    ∇
```

If *DET A* is 0, the program goes to *END* and prints out the message: '*A IS SINGULAR*' in accordance with Theorem 2, above. Otherwise, it computes *ADJOINT A* and divides it by *DET A*, also in accordance with Theorem 2. Then, it branches to 0 to end the program.

Example 1

```
    A←3 3ρ1 2 3 1 3 5 2 5 9
    A
1 2 3
1 3 5
2 5 9
    INVERT A
 2 ¯3  1
 1  3 ¯2
¯1 ¯1  1
```

Example 2

```
    A←3 3ρι 9
    A
1 2 3
4 5 6
7 8 9
    INVERT A
A IS SINGULAR                    Since DET A is 0.
```

The above theorem gives an easy way to determine whether or not a matrix is nonsingular using its determinant.

EXERCISES

1. Let

$$A = \begin{bmatrix} 3 & 5 & 2 \\ 4 & 0 & 7 \\ 3 & 9 & 1 \end{bmatrix}.$$

Do the following with pencil and paper.
(a) Find *DET A*.
(b) Find *ADJOINT A*.
(c) Test the validity of Theorem 1 using A.
(d) Use Theorem 2 to find the inverse of A.

2. Redo Exercise 1 at an APL terminal.

3. Redo Exercise 1 with the following matrices at an APL terminal:

(a) $A = \begin{bmatrix} 4 & 6 \\ 7 & 9 \end{bmatrix}$ (b) $A = \begin{bmatrix} 2 & 1 & 3 & 4 \\ 3 & 1 & 5 & 2 \\ 2 & 1 & 4 & 3 \\ 3 & 1 & 2 & 5 \end{bmatrix}$

119

4. Given the system of linear equations:

$$2x + y + 3z = 5$$
$$3x + 2y + 2z = 4$$
$$x + y + 3z = 3$$

(a) Express this system as a matrix equation $(A + . \times X) = B$.
(b) Use the program *INVERT* to invert the matrix of coefficients, A.
(c) Use this matrix inverse and matrix multiplication to solve the system.

5. Use Theorem 1 to prove the result $(DET\ ADJOINT\ A) = (DET\ A)*(N-1)$ for any nonsingular $N \times N$ matrix A.

5.5 Cramer's rule

In Chapter 4, we showed that given a system of n linear equations with n unknowns, $A \cdot X = B$, the solutions can be found by the matrix multiplication $X = A^{-1} \cdot B$ (or $B \boxplus A$ in APL), provided A^{-1} exists (i.e., provided *DET A* is not 0). Another technique for solving such a system involves determinants. It is known as Cramer's rule.

Cramer's rule

If $\det(A) \neq 0$, and if $A \cdot X = B$ is a system of n linear equations with n unknowns, then the value of the ith unknown, x_i, can be found by evaluating the quotient of determinants

$$x_i = \frac{\det \begin{bmatrix} a_{11} & a_{12} & \cdots & a_{1(i-1)} & b_1 & a_{1(i+1)} & \cdots & a_{1n} \\ \vdots & \vdots & & \vdots & \vdots & \vdots & & \vdots \\ a_{n1} & a_{n2} & \cdots & a_{n(i-1)} & b_n & a_{n(i+1)} & \cdots & a_{nn} \end{bmatrix}}{\det A},$$

for $i = 1, 2, \ldots, n$. In other words, in the numerator, the ith column of A is replaced by B.

Actually, this method of solving a system of linear equations is more work than the $B \boxplus A$ method of the last chapter. However, it does illustrate an application of determinants. In addition, it has the advantage that it enables one to solve for each variable separately. Thus, if one only wants the value of x_i, here is a way of getting it without having to compute the values of the other unknowns.

Example

$$2x + 6y + 4z = -12$$
$$6x + 6y + 4z = 8$$
$$3x + 2y + 4z = 13$$

```
      A←3 3ρ2 6 4 6 6 4 3 2 4
      A
2 6 4                               The matrix of coefficients.
6 6 4
3 2 4

      DET A
⁻64

      B← ⁻12 8 13

      A1←A
      A1[;1]←B                      Notice that the column correspond-
                                    ing to x, the first variable, has been
                                    replaced by B.
      DET A1
⁻320

      X←(DET A1)÷(DET A)
      X
5                                   So, x = 5

      A2←A
      A2[;2]←B                      Notice that the column correspond-
                                    ing to y, the second variable, has
                                    been replaced by B.
      DET A2
320

      Y←(DET A2)÷(DET A)
      Y
⁻5                                  So, y = ⁻5.

      A3←A
      A3[;3]←B                      Notice that the column correspond-
                                    ing to z, the third variable, has been
                                    replaced by B.
      DET A3
⁻128

      Z←(DET A3)÷(DET A)
      Z
2                                   So, z = 2.
```

We now consider a program for Cramer's rule. This is a program in which the student and the computer interact. The program asks the student for some information which it then uses.

121

Program 5.8 CRAMERS (optional)

∇CRAMERS

[1]   'WHAT IS A?'

[2]   A←☐

[3]   D←DET A

[4]   →(D=0)/RATS

[5]   'WHAT IS B?'

[6]   B←☐

[7]   'WHAT IS I?'

[8]   'I IS THE NUMBER OF THE VARIABLE YOU ARE SEEKING'

[9]   I←☐

[10] A[;I]←B

[11] C←DET A

[12] XI←C÷D

[13] 'XI IS= '; XI

[14] →0

[15] RATS : 'THE SYSTEM IS DEPENDENT'
      ∇

In line 2, the student is asked to enter the matrix of coefficients, *A*. If *DET A* is 0, line 4 sends the computer to *RATS* (line 15). Then, the computer prints out the true statement that 'THE SYSTEM IS DEPENDENT' and the program is ended. If *DET A* is not 0, however, the student is asked to enter the vector of constants, *B*, and the number of the variable he is seeking, *I*. In line 10, the *i*th column of *A* is replaced by *B*. In line 12, *XI* is computed in accordance with Cramer's rule. The value of *XI* is then printed out in line 13. Line 14 ends the program.

Example 1

CRAMERS
WHAT IS A?
☐:

    A←4 4ρι 16

THE SYSTEM IS DEPENDENT

To run this program, simply type CRAMERS.

The matrix of coefficients for some system.
Since *DET A* is 0.

122

Example 2

$$2x + y + 3z = 1$$
$$4x + 3y = 2$$
$$2x + y - z = 3$$

     CRAMERS
WHAT IS A?
□:
     $A \leftarrow 3\ 3\rho 2\ 1\ 3\ 4\ 3\ 0\ 2\ 1\ ^-1$

WHAT IS B?
□:
     $B \leftarrow 1\ 2\ 3$
WHAT IS I?
I IS THE NUMBER OF THE VARIABLE YOU ARE SEEKING
□:
     $I \leftarrow 1$
XI IS = 2.75

     CRAMERS
WHAT IS A?
□:
     $A \leftarrow 3\ 3\rho 2\ 1\ 3\ 4\ 3\ 0\ 2\ 1\ ^-1$
WHAT IS B?
□:
     $B \leftarrow 1\ 2\ 3$
WHAT IS I?
I IS THE NUMBER OF THE VARIABLE YOU ARE SEEKING
□:
     $I \leftarrow 2$
XI IS = ⁻3

     CRAMERS
WHAT IS A?
□:
     $A \leftarrow 3\ 3\rho 2\ 1\ 3\ 4\ 3\ 0\ 2\ 1\ ^-1$
WHAT IS B?
□:
     $B \leftarrow 1\ 2\ 3$
WHAT IS I?
I IS THE NUMBER OF THE VARIABLE YOU ARE SEEKING
□:
     $I \leftarrow 3$
XI IS = ⁻.5

EXERCISES

1. Solve the following system using Cramer's rule but not the program *CRAMERS*:

$$2x + y + 3z = 5$$
$$3x + y + 2z = 4$$
$$x + 2y + 3z = 6$$

2. Use the program *CRAMERS* to do Exercise 1.

3. Solve the following system using Cramer's rule but not the program *CRAMERS*:

$$2x + y + 3z + 4w = 5$$
$$3x + y + 5z + 2w = 7$$
$$2x + y + 4z + 3w = 1$$
$$3x + y + 2z + 5w = 0$$

4. Use the program *CRAMERS* to do Exercise 3.

5. Write a program for solving for the ith unknown in a system of linear equations by Cramer's rule with a specific result XI and without the dialog in *CRAMERS*. [*Note*: Since *CRAMERS* is probably in your present workspace, you'd better use a name for your program other than *CRAMERS*.]

Functions and graphing $\mathbf{6}$

One of the most important and useful concepts in mathematics is that of a function. In this and the next chapter, we will study some of the more common and useful functions and point out some of their applications.

6.1 Definition of a function

A *function*, F, is a set of ordered pairs,

$$F = \{(x,y) \mid \text{some relationship between } x \text{ and } y\},$$

with the additional property that for every value of the first coordinate, x, there is a *unique* corresponding value of the second coordinate, y.

Domain and range of a function

The variable used for the first coordinate is called the *independent variable*. The set of all values of the independent variable is called the *domain* of the function. The variable used for the second coordinate is called the *dependent variable*. The set of all values of the dependent variable is called the *range* of the function.

Functional notation

The relationship between the independent variable (x in the function above) and the dependent variable (y in the function above) is often given in the form of a formula or a rule that tells the way in which the unique value of the dependent variable is assigned to each value of the independent variable. If the name of the function is F, then this function or rule is usually symbolized by the *functional notation*

$$y = F(x) \qquad (\text{read as "} y \text{ equals } F \text{ of } x \text{"}).$$

In fact, this formula or rule is often referred to as the function, rather than the resulting set of ordered pairs.

The *symbolism* $F(a)$ (read as "F of a") is used to designate the value in the range assigned by the function F to the value a in the domain.

Let us consider some examples of functions now:

Example 1

Consider the formula $y = F_1(x) = x^2$, where $0 \leqslant x \leqslant 5$ and x is an integer. This formula describes the following set of ordered pairs

$$F_1 = \{(0,0), (1,1), (2,4), (3,9), (4,16), (5,25)\}.$$

Technically speaking, according to our definition, this set of ordered pairs is the function F_1. However, it is more common to refer to the formula $y = F_1(x) = x^2$ as the function.

The domain of this function is the set $D = \{0, 1, 2, 3, 4, 5\}$. The range is the set $R = \{0, 1, 4, 9, 16, 25\}$. Using the functional notation, $F_1(3) = 9$, since $3^2 = 9$.

Example 2

$$w = F_2(z) = \sqrt{z} \ .$$

Consider the following uses of functional notation with this function F:

(a) $F_2(9) = 3$, since $\sqrt{9} = 3$. (Note that the answer is *not* $^{\pm}3$, since then we would not have a function, since to one value of the independent variable z there would correspond two values of the dependent variable w. To be a function, there can only be one unique value of the dependent variable to each value of the independent variable.)
(b) $F_2(0) = 0$, since $\sqrt{0} = 0$.
(c) $F_2(^-1)$ is not a real number, since $\sqrt{-1}$ is not a real number. In this text, *we will restrict ourselves to real numbers*. Therefore, $^-1$ is not in the domain of F_2.

In this function, the domain was not specified. *If the domain of a function is not specified, it will be understood to consist of all real values of the independent variable for which there corresponds a unique* real value of the *dependent variable*. The domain of F_2 is the set $D = \{z \mid z \geqslant 0 \text{ and } z \text{ is a real number}\}$. The range of F_2 is the set $R = \{w \mid w \geqslant 0 \text{ and } w \text{ is a real number}\}$.

Example 3

$p = F_3(q) = 1/q$. Then, $F_3(3) = 1/3$, $F_3(1/2) = 2$, $F_3(^-2) = ^-1/2$, but $F_3(0)$ is not defined, since $1/0$ is not defined.

The domain of $F_3 = \{q \mid q \text{ is real and } q \neq 0\}$, since q is the independent variable. The range of $F_3 = \{p \mid p \text{ is real and } p \neq 0\}$, since p is the dependent variable.

Example 4

$A = C(r) = \pi \cdot r^2$ is the function giving the area, A, of a circle as a function of its radius, r.

$C(2) = \pi \cdot 2^2 = 4\pi$, so that the area of a circle of radius 2 units is 4π square units.

The domain of C is $\{r | r > 0\}$. The range of C is $\{A | A > 0\}$.

Example 5

$s = H(t) = {}^{-}16 \cdot t^2 + 320 \cdot t$ is a function which gives the height, s, in feet, of a projectile fired from ground level with an initial velocity of 320 feet/second, where t is the time elapsed in seconds.

$H(2) = 576$ feet; so after 2 seconds, the projectile is at a height of 576 feet.

$H(10) = 1600$ feet; so after 10 seconds, the projectile is at a height of 1600 feet.

$H(20) = 0$; so after 20 seconds, the projectile is back at ground level. It was in flight for 20 seconds.

Since it takes as long for the projectile to rise as it does for it to fall (gravity being the only force acting on it), the projectile reaches its highest point, 1600 feet, in 10 seconds.

The domain for $H = \{t | 0 \leqslant t \leqslant 20 \text{ seconds}\}$. The range for $H = \{s | 0 \leqslant s \leqslant 1600 \text{ feet}\}$.

Functions in APL

Let us now consider programs for each of the functions in the previous examples.

Example 1

```
     ∇ Y←F1 X                    The name of the program is F1.

[1]   Y←X∗2
      ∇

      F1 3
 9                               So, F₁(3)=9.
```

So, $F_1(3) = 9$.

```
      F1 2 3 4 5 6
 4 9 16 25 36                    APL functions can be applied to
                                 vectors.
```

APL functions can be applied to vectors.

```
      A←3 3⍴1 2 3 4
      A
 1 2
 3 4
```

```
      F1  A
1   4
9  16
```
APL functions can even be applied to matrices.

Example 2

```
      ∇ W←F2 Z
[1]    W←Z*.5
      ∇
```
Since $\sqrt{Z} = Z^{1/2} = Z^{0.5}$.

```
      F2 9
3
```
Since $\sqrt{9} = 3$. Note, the computer only gives the positive square root.

```
      F2 0
0
```
Since $\sqrt{0} = 0$.

```
      F2 ¯1
DOMAIN ERROR
```
Since $\sqrt{\,^-1}$ is not a real number, so that $^-1$ is not in the domain of $F2$.

Example 3

```
      ∇ P←F3 Q
[1]    P←1÷Q
      ∇
```

```
      F3 3
0.3333333333
```
Since $1 \div 3 = 0.3333333333$, approximately.

```
      F3 1÷2
2
```

```
      F3 0
DOMAIN ERROR
```
Since $1 \div 0$ is not defined, so that 0 is not in the domain of $F3$.

Example 4

```
      ∇ AREA←CIRCLE RADIUS
```
The name of the program is *CIRCLE*.

```
[1]    AREA←(○1)×RADIUS*2
      ∇
```
In APL, π is given by ○1.

```
      CIRCLE 1
3.141592654
```
The area of a circle of radius 1 is π.

```
      CIRCLE 2
12.56637062
```
This is $4 \cdot \pi$.

Example 5

∇ *S*← *HEIGHT T*   The name of this program is *HEIGHT*.

[1]   *S*←(⁻16× *T*∗2)+(320× *T*)
       ∇

HEIGHT 2
576

HEIGHT 10
1600

HEIGHT 20
0

HEIGHT 30
⁻4800   Of course, this is not realistic, since the projectile would be 4800 feet below ground level.

In general, a simple program of the following form can be used to compute the value of the dependent variable, *y*, assigned by a function *FN* to a value of the independent variable, *x*.

Program 6.1 FN

∇ *Y*← *FN X*

[1]   *Y*←(   )   (Insert the formula for *Y* in terms of
       ∇ *X* here.)

We now consider a program to generate the set of ordered pairs in a function. We will use the general *FN* program just described for a particular function whose pairs we are computing. In order to compute ordered pairs for a different function, we will have to alter the *FN* program before running the program below.

Program 6.2 PAIRS

∇ *ORDERED*← *X PAIRS Y*

[1]   *ORDERED*←$\lozenge$(2,(ρ*X*))ρ*X, Y*∇

To run this program, enter a vector of values *X*, and type

   X PAIRS FN X

We now apply *PAIRS* to some of the previous examples.

Example 1

         ∇ Y←FN X   Creating a function in APL.

[1]   Y←X*2 $y = x^2$
       ∇

       X←0 1 2 3 4 5
       X PAIRS FN X
0 0
1 1
2 4
3 9
4 16
5 25

Example 2

         ∇ FN[1]   Request to alter line 1 of *FN*.

[1]   Y←X*.5 ∇ $y = \sqrt{x}$.
       X←0,ι9

       X PAIRS FN X
0 0
1 1
2 1.414213562
3 1.732050808
4 2
5 2.236067977
6 2.449489743
7 2.645751311
8 2.828427125
9 3

Example 3

It might be interesting to consider the list of heights at the end of each second in Example 5. This can be done as follows:

         ∇ FN[1]

[1]   Y←HEIGHT X∇   Altering *FN* to fit the height function
                                         defined earlier.

       X←0,ι20

       X PAIRS FN X
0 0
1 304
2 576

130

3 816
4 1024
5 1200
6 1344
7 1456
8 1536
9 1584
10 1600
11 1584
12 1536
13 1456
14 1344
15 1200
16 1024
17 816
18 576
19 304
20 0

This display of the ordered pairs of this function helps us to see several important characteristics of the flight of this projectile. For example:

1. The maximum height of the projectile is 1600 feet.
2. The total time in flight of the projectile is 20 seconds.
3. The projectile falls at the same rate at which it rose.

Example 4

As a final example, we might like to examine the behavior of the function $F_3(x) = 1/x$ near $x = 0$.

$\nabla FN[1]$

[1] $Y \leftarrow 1 \div X \nabla$ Change *FN* to fit this new function.

$X \leftarrow (\iota 10) \times .1$

X PAIRS FN X
0.1 10
0.2 5
0.3 3.333333333
0.4 2.5
0.5 2   So, the closer x is to 0, the larger y is.
0.6 1.666666667
0.7 1.428571429
0.8 1.25
0.9 1.111111111
1.0 1.0

$X \leftarrow (11 - (\iota 10)) \times {}^{-}.1$
$X \ PAIRS \ FN \ X$

¯1.0	¯1.0
¯0.9	¯1.111111111
¯0.8	¯1.25
¯0.7	¯1.428571429
¯0.6	¯1.666666667
¯0.5	¯2.0
¯0.4	¯2.5
¯0.3	¯3.333333333
¯0.2	¯5.0
¯0.1	¯10.0

So, the closer negative values of x are to 0, the more negative y is.

EXERCISES

1. For each of the following functions, evaluate $F(4)$, $F(0)$, and $F(^{-}2)$.
   (a) $y = F(x) = x^3$ (d) $y = F(x) = -x^2 + 3x - 2$
   (b) $y = F(x) = 2x + 3$ (e) $y = F(x) = 2/(x+1)$
   (c) $y = F(x) = \sqrt[3]{x}$

2. Write programs for each of the functions in Exercise 1, and use these programs to check your answers in Exercise 1.

3. If the domain of the functions in Exercise 1 is the set

$$D = \{^{-}4, ^{-}3, ^{-}2, ^{-}1, 0, 1, 2, 3, 4\},$$

then list the ordered pairs for each of the functions.

4. Use the program PAIRS to do Exercise 3.

5. With the domain stated in Exercise 3, state the ranges for the functions in Exercise 1.

6. If the domain is not stated for the functions in Exercise 1, then what are the implied domains understood to be?

7. The absolute value function is defined as follows:

$$y = A(x) = |x| = \sqrt{x^2} .$$

The result will always be a nonnegative number with the same magnitude as x. Write an APL program to evaluate the absolute value of any number x. Note that in APL, there is a built-in absolute value function, $|X$. Thus,

```
|5
5

|¯5
5

|0
0
```

8. Write a program for finding the area of a right triangle. (The formula for the area of a right triangle is $A = \frac{1}{2} \cdot B \cdot H$, where B is the length of the base, and H is the height.)

6.2 Graphing

The real line

The *real number system* is often described as the set of all numbers which can be represented by a point on a number line. Consider a horizontal line. Locate a point on it, called the *origin*, and associate with this point the number 0. Establish a unit length with a compass. Now, the points associated with the positive integers are located to the right of the origin. For example, the positive integer 3 is located 3 units to the right of the origin 0. The points associated with the negative integers are located to the left of the origin. For example, the negative integer ⁻3 is located 3 units to the left of the origin. Thus, there is a point on this number line corresponding to each integer (see Figure 6.1).

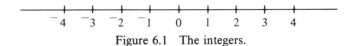

Figure 6.1   The integers.

A rational number is a number of the form p/q, where p and q are integers and $q \neq 0$. To locate the point corresponding to the rational number p/q, divide each unit length into q parts. Then, measure over p of the parts of length $1/q$, to the right if p is positive, left if p is negative. For example, 7/5 and ⁻3/5 are shown in Figure 6.2.

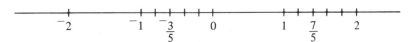

Figure 6.2   The rational numbers.

In ancient times, it was thought that the rational numbers completely filled up the number line. However, the Pythagorean theorem can be used to locate a number which is not rational, namely $\sqrt{2}$. To locate $\sqrt{2}$, construct the isosceles right triangle with leg of length 1 as in Figure 6.3. By the Pythagorean Theorem, the hypotenuse has length $\sqrt{1^2 + 1^2}$ $= \sqrt{2}$. Place a compass at 0 and open it the length of the hypotenuse. Then, swing an arc down to the number line. The point where the arc intercepts the number line represents $\sqrt{2}$.

It is not difficult to prove that $\sqrt{2}$ is irrational (not rational). However, we will not prove this here. There are an infinite number of other irrational numbers, such as $\sqrt{3}$ and π. All of these can be approximated on this number line, which we will refer to as the *real line*.

133

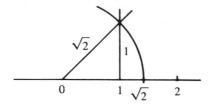

Figure 6.3 Constructing a real, irrational number.

The Cartesian plane

The French mathematician Rene Descartes established a one-to-one corre-spondence between ordered pairs of real numbers and points in the plane. Therefore, it is known as the *Cartesian plane*.

To construct the Cartesian plane, draw a horizontal real line, called the *x* axis. Perpendicular to this real line at the origin, draw a vertical real line, called the *y* axis. These two axes divide the plane into 4 regions called *quadrants*, as in Figure 6.4. The way in which to plot a point with coordinates (x,y) is illustrated by the points A, B, C, D, E, F, and G in the diagram. The first coordinate of a point is called its *x coordinate* or its *abscissa*. The second coordinate of the point is called its *y coordinate* or its *ordinate*.

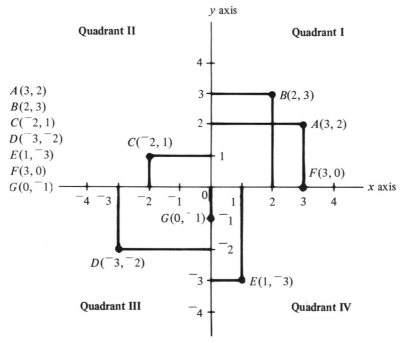

Figure 6.4 The Cartesian plane.

Definition of the graph of a function

The *graph* of a function $y = F(x)$ is the graph of all points corresponding to ordered pairs (x,y) satisfying the function (i.e., for which the function is a true statement.)

Example 1

$y = F_1(x) = x^2$, with domain $D = \{^-4, ^-3, ^-2, ^-1, 0, 1, 2, 3, 4\}$. The ordered pairs corresponding to this function are

$$F_1 = \{(^-4, 16), (^-3, 9), (^-2, 4), (^-1, 1), (0, 0), (1, 1), (2, 4), (3, 9), (4, 16)\}.$$

Thus, the graph of this function is as shown in Figure 6.5.

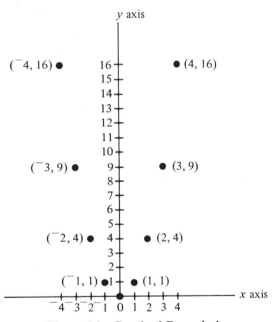

Figure 6.5   Graph of Example 1.

Example 2

$y = F_1(x) = x^2$, with domain $D = \{x | ^-4 \leqslant x \leqslant 4$ and x is a real number$\}$. In this case, there are an infinite number of ordered pairs for which the function is a true statement. It would be an endless task to plot all of these. Therefore, we will just plot enough of them to get a good idea of the pattern they trace out. Then, we will join these points by a smooth curve and hope that our curve will pass through the other points whose coordinates satisfy the function. Of course, this is only a reasonable guess at the graph of the function. More precise graphing techniques require more insight into the particular function being graphed, which we have not yet provided. The graph of Example 2 is shown in Figure 6.6. A graph with this shape is called a parabola.

135

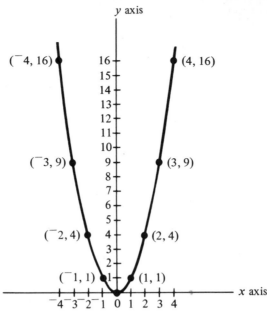

Figure 6.6   Graph of Example 2.

Example 3

$w = F_2(z) = \sqrt{z}$, with domain $D = \{z \mid 0 \leqslant z \leqslant 9\}$. In order to graph this function, we need some ordered pairs satisfying it to graph. We can use the program *PAIRS* to generate these pairs.

```
     ∇ FN[1]
[1]   Y←X*.5 ∇                          Altering FN to fit this function.
      X←0,ι9
      X PAIRS FN X
0 0
1 1
2 1.414213562
3 1.732050808
4 2
5 2.236067977
6 2.449499743
7 2.645751311
8 2.828427125
3 3
```

Thus, the graph is as shown in Figure 6.7. Note that since the dependent variable is *w*, the *y* axis has become the *w* axis, and that since the independent variable is now *z*, the *x* axis has become the *z* axis.

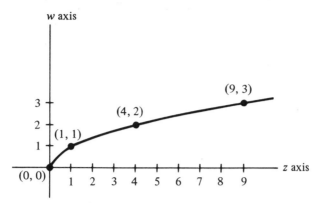

Figure 6.7   Graph of Example 3.

Example 4

$y = F_3(x) = 1/x$, where the domain is $D = \{x \mid {}^-1 \leqslant x \leqslant 1\}$.

∇ *FN*[1]

[1]   *Y* ← 1 ÷ *X*∇   Altering *FN* to fit this new function.
       X ← (ι10) × .1
       X PAIRS FN X

0.1 10.0
0.2 5.0
0.3 3.333333333
0.4 2.5
0.5 2.0
0.6 1.666666667
0.7 1.428571429
0.8 1.25
0.9 1.111111111
1.0 1.0

       X ← (11 − (ι10)) × ⁻.1
       X PAIRS FN X

⁻1.0 ⁻1.0
⁻0.9 ⁻1.111111111
⁻0.8 ⁻1.25
⁻0.7 ⁻1.428571429
⁻0.6 ⁻1.666666667
⁻0.5 ⁻2.0
⁻0.4 ⁻2.5
⁻0.3 ⁻3.333333333
⁻0.2 ⁻5.0
⁻0.1 ⁻10.0

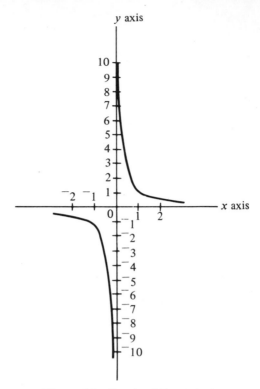

Figure 6.8   Graph of Example 4.

Thus, the graph is as shown in Figure 6.8. A graph with this shape is called a hyperbola. (Recall that $F_3(0)$ is not defined.)

The previous examples illustrate that APL and the program *PAIRS* can be used to quickly and easily generate a list of ordered pairs for a function $y = F(x)$. It is then an easy job to plot these points and join them, thereby graphing the function. However, it might also be of interest to the reader to consider an APL program for graphing the ordered pairs satisfying a function. The following program does this.

Program 6.3 **GRAPH**

    ∇ **GRAPH DOMAIN**

[1]   **PLANE←40 40ρ' '**   This creates a 40-by-40 blank matrix.

[2]   **PLANE[20;]←' − '**   Fill in the 20th row of *PLANE* with
                                          −'s.

[3]   **PLANE[20; 40]←'X'**   Labeling the 20th row with X for the
                                          x axis.

[4]	$PLANE[;20]\leftarrow`I'$	Fill in the 20th column with I's.
[5]	$PLANE[1;20]\leftarrow`Y'$	Labeling the 20th column with Y for the y axis.
[6]	$PLANE[20;20]\leftarrow`O'$	Labeling the origin.
[7]	$X\leftarrow DOMAIN[1]$	Starting X with the first element of $DOMAIN$.
[8]	$FCN: Y\leftarrow FN\ X$	The function being graphed, given by an external subprogram.
[9]	X,Y	Print out the pairs for the function.
[10]	$PLANE[20-Y;20+X]\leftarrow`*'$	Placing a $*$ at the point corresponding to the pair (x,y).
[11]	$X\leftarrow X+1$	Incrementing X.
[12]	$\rightarrow(X\leqslant\rho DOMAIN)/FCN$	Using up the entire $DOMAIN$.
[13]	$PLANE$	Print out the $PLANE$.
	∇	

To run this program, simply type $GRAPH$, after entering the subprogram FN for the function being graphed and the desired $DOMAIN$.

Example 1

Let us use this program to graph $y=F(x)=x^2$, where $^{-}4\leqslant x\leqslant 4$.

$\qquad\nabla\ FN[1]$

[1]	$Y\leftarrow X*2\nabla$	Altering FN to fit this function.

```
    GRAPH ⁻4 ⁻3 ⁻2 ⁻1 0 1 2 3 4
⁻4 16
⁻3  9
⁻2  4
⁻1  1
 0  0
 1  1
 2  4
 3  9
 4 16
```

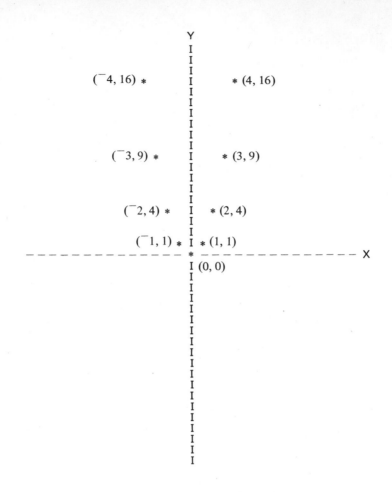

This program may also be used to graph other functions by merely altering the program *FN* to fit the new function. In addition, if it is possible for x or y to be either $< {}^-20$ or >20, then the size of the *PLANE* will also have to be altered.

Example 2

$y = F(x) = x^2 + 2x - 6$, with domain $D = \{x \mid {}^-6 \leqslant x \leqslant 4\}$.

    ∇ *FN*[1]

[1] $Y \leftarrow (X * 2) + (2 \times X) - 6 \nabla$ The new function.

    GRAPH ${}^-6$ ${}^-5$ ${}^-4$ ${}^-3$ ${}^-2$ ${}^-1$ 0 1 2 3 4

${}^-6$ 18
${}^-5$ 9
${}^-4$ 2

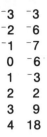

```
⁻3  ⁻3
⁻2  ⁻6
⁻1  ⁻7
 0  ⁻6
 1  ⁻3
 2   2
 3   9
 4  18
```

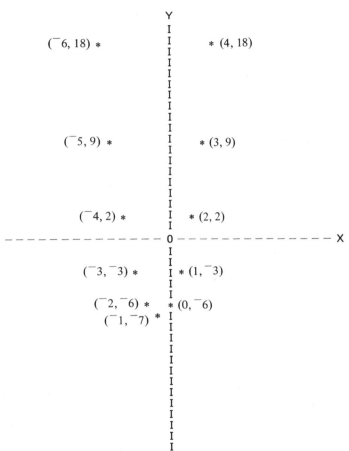

Another problem that may occur in using this program is that there may occur values of *x* or *y* that are not integers. In order to determine whether or not all of the *y* values are integers, one could print out the ordered pairs for the function using *PAIRS*. It would then be easier to just plot the corresponding points by hand then to modify the program *GRAPH*. Therefore, we omit this more complicated case.

EXERCISES

1. Plot the following points:
   (a) (5, 2)   (b) (2, 5)   (c) (¯4, 3)   (d) (3, ¯4)   (e) (¯2, ¯3)   (f) (¯3, 0)

2. Use the program *PAIRS* to print out a list of ordered pairs for the following functions with the specified domains. Then, graph these functions:
   (a) $y = x^2 - 3$, where $^-3 \leqslant x \leqslant 3$
   (b) $y = \sqrt{x} + 1$, where $0 \leqslant x \leqslant 8$
   (c) $y = 3x + 1$, where $^-2 \leqslant x \leqslant 2$
   (d) $y = x/(x+1)$, where $^-5 \leqslant x \leqslant 5$
   (e) $y = 3 \cdot x^2 - 2x + 1$, where $^-4 \leqslant x \leqslant 4$
   (f) $y = 2^x$, where $^-4 \leqslant x \leqslant 4$

3. Use the program *GRAPH* to graph the following functions with the specified domains:
   (a) $y = x^2 - x$, where $^-3 \leqslant x \leqslant 4$
   (b) $y = 2x + 3$, where $^-5 \leqslant x \leqslant 5$
   (c) $y = x^3 - 3x$, where $^-3 \leqslant x \leqslant 3$

6.3 Linear functions

A *linear function* is a function of the form $y = m \cdot x + b$, where m and b are constants.

Example 1

$y = 2x + 1$. To gain some insight into this linear function, let us examine some ordered pairs using the program *PAIRS*.

```
∇ FN[1]
```

[1]   Y←(2×X)+1 ∇   Revising *FN* for this new function.
      X←¯4 ¯3 ¯2 ¯1 0 1 2 3 4 5
      X PAIRS FN X

```
¯4 ¯7
¯3 ¯5
¯2 ¯3
¯1 ¯1
 0  1
 1  3
 2  5
 3  7
 4  9
 5 11
```

Notice that for each increase of 1 in x, y increases by 2, the coefficient of x in the linear function. Notice also, that when $x = 0$, $y = 1$, the constant term in the linear function.

Example 2

$y = {}^-3x + 4$. Again:

     ∇ *FN*[1]

[1]   *Y*←(‾3×*X*)+4 ∇

     X PAIRS FN X

‾4	16
‾3	13
‾2	10
‾1	7
0	4
1	1
2	‾2
3	‾5
4	‾8
5	‾11

Notice that for each increase of 1 in x, y decreases by 3 (or increases by ‾3), the coefficient of x in the linear function. Notice also, that when $x = 0$, then $y = 4$, the constant term in the linear function. These examples suggest that there is something significant about the m and b in the general linear function $y = m \cdot x + b$.

Slope

Let $y = m \cdot x + b$. Suppose that x increases by 1 to $x + 1$. The new value of y at this new value of x will be new $y = m \cdot (x + 1) + b = m \cdot x + m + b = (m \cdot x + b) + m$. Since m is constant, any increase of 1 in x increases y by this constant amount m. Thus, the graph of the linear function $y = m \cdot x + b$ is a straight line. The number m is called the *slope* of the line (see Figure 6.9).

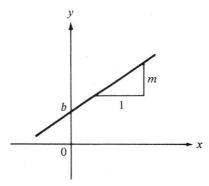

Figure 6.9 Graph of $y = m \cdot x + b$.

143

y intercept

When $x=0$, the value of y is $m\cdot 0+b=b$. Thus, the straight line whose equation is $y=m\cdot x+b$ intercepts the y axis (when $x=0$) at the point where $y=b$. This value, b, is called the y intercept of the line.

Graphing a line

A straight line is uniquely determined by two points. (That is, given two points, there is exactly one line that can pass through them.) Thus, when graphing a line, one only needs to plot two points and join them. Any two points that satisfy the equation of the line will do.

Example 1

$$y=2\cdot x+1$$

The two points $(0,1)$ and $(1,3)$ both satisfy this equation. Thus, plotting these two points and joining them, the graph is shown in Figure 6.10. Notice that this line rises as x increases.

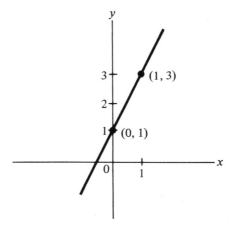

Figure 6.10   Graph of Example 1.

Example 2

$$y= {}^-3\cdot x+4$$

Plotting the two points $(0,4)$ and $(1,1)$ and joining them, the graph is as shown in Figure 6.11. Notice that this line falls as x increases.

Significance of the sign of the slope

Since the slope, m, of a line is the amount of increase in y for each increase of 1 in x, if m is positive, every increase of 1 in x causes the line to rise by m. Thus, lines with positive slope rise as x increases. If m is negative, then every increase of 1 in x causes the y value to increase by the negative

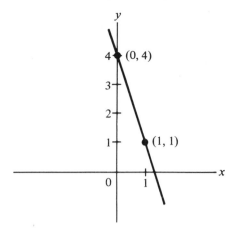

Figure 6.11   Graph of Example 2.

amount m. Thus, y actually decreases. Therefore, lines with negative slope fall as x increases.

The slope of a line is a measure of its steepness. The larger m is (in absolute value), the steeper the line. Parallel lines have the same slope.

Horizontal and vertical lines

On a *horizontal line*, y never increases. Thus, when x increases by 1, y remains the same. The slope of a horizontal line is 0. So, the equation of a horizontal line is of the form $y = 0 \cdot x + b = b$.

Example

Graph the equation $y = 2$. This is done in Figure 6.12.

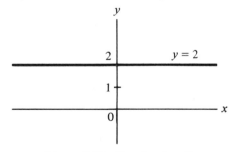

Figure 6.12   Graph of $y = 2$.

On a *vertical line*, x never increases. x always has the same value. Thus, a vertical line has an equation of the form $x = a$, for some constant a.

Example

Graph the equation $x = 2$. This is done in Figure 6.13. Since x never increases, a vertical line is often said to have no slope. (Some texts say that it has infinite slope.)

145

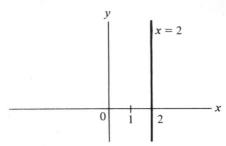

Figure 6.13 Graph of $x=2$.

Linear equations

In Chapter 4, we studied linear equations of the form $a_1 \cdot x + a_2 \cdot y = c$. If one solves such an equation for y, one gets

$$a_2 \cdot y = {}^- a_1 \cdot x + c, \quad \text{or } y = -\frac{a_1}{a_2} \cdot x + \frac{c}{a_2},$$

which is of the form $y = m \cdot x + b$, where $m = -a_1/a_2$ and $b = c/a_2$. Thus, the graph of such a linear equation is a straight line. It, therefore, only takes two points to graph a linear equation.

Example

Graph $3x + 2y = 6$. If $x=0$, then $y=3$, so that $(0,3)$ is a point on the line, called the *y intercept*. If $y=0$, then $x=2$, so that $(2,0)$ is a point on the line called the *x intercept*. Plotting these points and joining them yields the line shown in Figure 6.14.

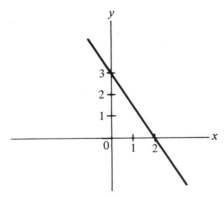

Figure 6.14 Graph of $3x + 2y = 6$.

Two-by-two systems of linear equations

In Section 4.2., we pointed out that a two-by-two system of linear equations has either one solution, no solutions, or an infinite number of solutions. Since a vector (x,y) must satisfy both equations simultaneously to be a solution of the system, then the point (x,y) must lie on the graph of

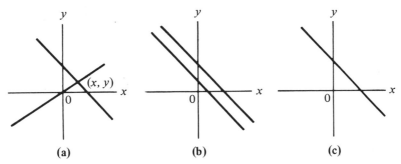

Figure 6.15 (a) Lines intersect: one solution, the point of intersection. (b) Parallel lines: no solutions. (c) Lines coincide, both equations represent the same line: an infinite number of solutions.

both lines. In other words, it must lie on the intersection of the lines, provided they intersect. Figure 6.15 explains the three possible conditions.

Examples

1. Find the point of intersection of the following lines, if they intersect (Figure 6.16):

$$2x - 3y = 5$$
$$x + 4y = 8$$

The method of Section 4.7 is used to solve the system.

```
A←2 2ρ2 ¯3 1 4
B←5 8

B ÷ A
4 1
```

Thus, these lines intersect at the point $(4, 1)$.

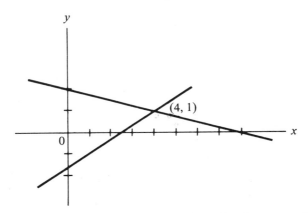

Figure 6.16 The method of Section 4.7 is used to solve the system (Example 1).

147

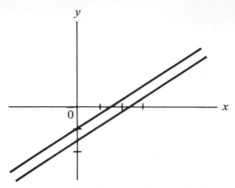

Figure 6.17   Graph of Example 2.

2. Find the point of intersection of the lines that follow, if they intersect (Figure 6.17):

$$2x - 3y = 5$$
$$4x - 6y = 7$$

$A \leftarrow 2\ 2\rho 2\ ^{-}3\ 4\ ^{-}6$
$B \leftarrow 5\ 7$

$B \boxplus A$
DOMAIN ERROR

This example must be in either case 2 or case 3. To see which case, let us solve the equations for y to put them in the form $y = m \cdot x + b$. From the first equation,

$$y = \frac{2}{3} \cdot x - \frac{5}{3}.$$

From the second equation,

$$y = \frac{2}{3} \cdot x - \frac{7}{6}.$$

Thus, both lines have the same slope, $m = 2/3$. However, they have different y-intercepts, b. Thus, they are parallel lines and do not intersect.

3. Find the point of intersection of the following lines, if they intersect (Figure 6.18):

$$2x - 3y = 5$$
$$4x - 6y = 10$$

$A \leftarrow 2\ 2\rho 2\ ^{-}3\ 4\ ^{-}6$
$B \leftarrow 5\ 10$

$B \boxplus A$
DOMAIN ERROR

148

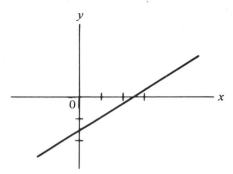

Figure 6.18 Graph of Example 3.

Solving these equations for y yields

$$y = \frac{2}{3} \cdot x - \frac{5}{3}$$

in both cases. Thus, both equations represent the same line. There are an infinite number of solutions to this system. Any point on the first line is automatically on the second line. (Notice that the second equation is a multiple of the first equation.)

Finding the slope of a line passing through two points

Let $P(x_1, y_1)$ and $Q(x_2, y_2)$ be two points. There is exactly one line through these two points. The slope of this line can be computed from the following equation:

$$m = \frac{\text{Rise}}{\text{Run}} = \frac{\text{Change in } y}{\text{Change in } x} = \frac{y_2 - y_1}{x_2 - x_1}.$$

To see why this is so, suppose the equation of the line is $y = m \cdot x + b$. Since (x_1, y_1) lies on this line, then $y_1 = m \cdot x_1 + b$. Since (x_2, y_2) lies on this line, then $y_2 = m \cdot x_2 + b$. Computing the difference quotient,

$$\frac{y_2 - y_1}{x_2 - x_1} = \frac{(m \cdot x_2 + b) - (m \cdot x_1 + b)}{x_2 - x_1} = \frac{m \cdot x_2 - m \cdot x_1}{x_2 - x_1} = \frac{m \cdot (x_2 - x_1)}{x_2 - x_1} = m.$$

Example

Find the slope of the line joining the points $(1, 3)$ and $(4, 9)$.

$$m = \frac{9 - 3}{4 - 1} = \frac{6}{3} = 2.$$

Program 6.4 SLOPE

$\nabla\, M \leftarrow P\ \textit{SLOPE}\ Q$ A simple program to compute slope given two points P and Q.

[1] $M \leftarrow (Q[2] - P[2]) \div (Q[1] - P[1])$

∇

Example

    $P \leftarrow 1\ 3$
    $Q \leftarrow 4\ 9$

    P SLOPE Q

2

Finding the equation of the line passing through two points

Given two points $P(x_1, y_1)$ and $Q(x_2, y_2)$, we can find the slope of the line through them, m. We know that the equation of the line through them has the form $y = m \cdot x + b$. Therefore, we only need to find b. To do this, use the fact that the coordinates of either of these points must satisfy the equation. Substitute the coordinates of either of these points into the equation and solve for b.

Example

Find the equation of the line joining the points $(1, 3)$ and $(4, 9)$. We found the slope of this line to be 2 (see above). Therefore, the equation looks like $y = 2x + b$. Substituting the coordinates $(1, 3)$ for x and y respectively, we get $3 = 2 \cdot 1 + b$. So, $b = 1$. If we had used the coordinates of the point $(4, 9)$, we would get $9 = 2 \cdot 4 + b$. So, $b = 1$. It doesn't matter which point is used to compute b. The equation of the line is $y = 2x + 1$.

Applications of linear functions

Anytime the relationship between two quantities is linear, then this relationship can be expressed by an equation of the form $y = m \cdot x + b$.

Example 1

Assume that the cost, C, of producing x items is a linear function of x. Assume also that the initial cost of getting ready to produce the items is $500. Assume also that the cost per item is $5. Find the formula for cost, C, in terms of the number of items produced, x.

Since this cost is a linear function of x, the formula must be of the form $C = m \cdot x + b$. The slope, m, is the amount of increase in C per increase of 1 in x. This was given to be $5. b is the cost when $x = 0$, which was given to be $500. Thus, the formula is $C = \$5x + \$500$. For example, the cost of producing 10 items is $C = \$5 \cdot 10 + \$500 = \$550$.

Example 2

Suppose the price of renting a wheel barrow from Rent-All Corporation is $5 plus an hourly rate of $3 per hour. Find a formula for the price, P, of renting a wheel barrow for x hours.

The formula is of the form $P=m \cdot x+b$, in which m is the increase in price per hour, which is \$3 per hour; b is the price when $x=0$, or the fixed price, which is \$5. Therefore, the formula is $P=\$3 \cdot x+5$. So, if a wheel barrow is needed for 5 hours, the price would be $P=\$3 \cdot 5+5=\$20$.

Example 3

The relationship between degrees Centigrade and degrees Fahrenheit is linear. Find this relationship.

0 degrees Centigrade corresponds to 32 degrees Fahrenheit. 100 degrees Centigrade corresponds to 212 degrees Fahrenheit. Thus, this problem entails finding the equation of the line passing through the two points $(0,32)$ and $(100,212)$. The slope of this line is given by

```
P←0 32
Q←100 212

P SLOPE Q
```
1.8

Let C denote degrees Centigrade and F denote degrees Fahrenheit. Then, the linear relationship has the form

$$F=1.8C+b.$$

Substituting in the coordinates of the point $(0,32)$, we get

$$32=1.8 \cdot 0+b=b.$$

Thus, the equation for this relationship is $F=1.8 \cdot C+32$. For example, 20 degrees Centigrade is equivalent to

$$F=1.8 \cdot 20+32=36+32=68 \text{ degrees Fahrenheit.}$$

EXERCISES

1. Graph the following lines:
(a) $y=2x+5$
(b) $y=-x+2$
(c) $2x+3y=12$
(d) $y={}^-1$
(e) $x=3$

2. Find the slope of each of the lines in Exercise 1.

3. Find the equations of the lines satisfying the following conditions:
(a) The slope is 5 and the y intercept is ${}^-1$.
(b) The slope is ${}^-2$ and the y intercept is 3.
(c) The slope is 0 and the y intercept is 5.
(d) The slope is 2 and it passes through the point $(3,1)$.
(e) The slope is ${}^-1$ and it passes through the point $(0,0)$.

4. Find the slopes of the lines passing through the following pairs of points:
(a) $(1,2)$ and $(3,7)$
(b) $({}^-1,2)$ and $(3,{}^-5)$
(c) $(3,0)$ and $(3,4)$
(d) $(2,1)$ and $(5,1)$

5. Find the equations of the lines in Exercise 4.

6. Find the intersections of the following pairs of linear equations, if they intersect:

(a) $2x+3y=6$
$\quad 3x+2y=4$
(b) $3x-y=5$
$\quad x+2y=3$
(c) $4x+y=-3$
$\quad x+3y=0$
(d) $2x+y=2$
$\quad 4x+2y=3$

7. Write a program to solve the linear equation $((A\times X)+B)=C$ (APL notation) for X (A, B, and C are constants).

8. The cost of renting a lawn mower from Ace Rental Co. is a linear function of the number of hours it is rented. The charge is $3 per hour, plus a fixed charge of $5. Find a formula for the charge for renting a lawn mower for x hours.

9. The number of dandelions in Mr. Jone's lawn is a linear function of the number of weeks from now. If right now he has 25 dandelions, and if he gets 10 new dandelions per week, find the function for the number of dandelions x weeks from now.

10. A machine costs $10,000 new. Each week, its value decreases by $10. Find a function for its value in x weeks.

11. The cost of producing 0 items is $100. The cost of producing 10 items is $250. Assuming that cost is a linear function of the number of items produced, find the cost function.

12. To rent a car costs $10, plus $.08 per mile. Find a formula for the cost of renting a car as a function of the number of miles.

6.4 Quadratic functions

A quadratic function is a function of the form $y=a\cdot x^2+b\cdot x+c$, where a, b, and c are constants and $a\neq0$.

The graph of a quadratic function is called a *parabola*. To help us gain some insight into quadratic functions and parabolas, let us graph the following two examples using the program *GRAPH* of Section 6.2.

Example 1

$$y=x^2-4x+4.$$

∇FN [1]

[1] $Y\leftarrow(X*2)+(-4\times X)+4\ \nabla$ The new function being graphed.

$\quad\ GRAPH\ -2\ -1\ 0\ 1\ 2\ 3\ 4\ 5\ 6$

-2 16   Command to execute the program
-1 9   *GRAPH.*
 0 4

```
1   1
2   0
3   1
4   4
5   9
6  16
```

The ordered pairs for this function with $^-2 \leqslant x \leqslant 6$.

The points are joined by hand to form the parabola. This parabola *opens upward*. The lowest point, called the *vertex*, is at the point $(2, 0)$. The line $x = 2$ is an *axis of symmetry*. That is, for each point to the left of this line, there is a corresponding point with the same height to the right of it.

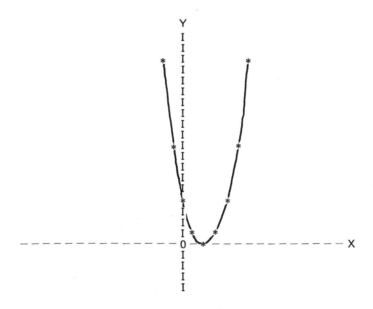

Example 2

$$y = -x^2 + 2x + 3.$$

$\nabla\, FN\, [1]$

[1] $Y \leftarrow (-(X*2)) + (2 + X) + 3\, \nabla$ The new function.

   $GRAPH\ ^-2\ ^-1\ 0\ 1\ 2\ 3\ 4$ This time, we'll only go to $x = 4$.

```
-2  -5
-1   0
 0   3
```

1	4
2	3
3	0
4	⁻5

This parabola *opens downward*. The highest point, called the *vertex*, is at the point (1,4). The line $x=1$ is the *axis of symmetry*. Again, the points are joined to form the parabola.

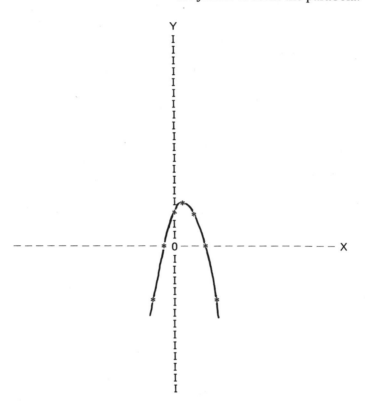

Useful characteristics of quadratic functions in graphing parabolas

Let us list some characteristics of quadratic functions which are useful in graphing the resulting parabolas.

Given a general quadratic function, $y = F(x) = a \cdot x^2 + b \cdot x + c,\ a \neq 0,$

1. If $a > 0$, the parabola opens upward. If $a < 0$, the parabola opens downward.
2. The x coordinate of the vertex is given by $x = -b/2 \cdot a$. The y coordinate is given by $F(-b/2 \cdot a)$.
3. The vertical line $x = -b/2 \cdot a$ is the axis of symmetry of the parabola.
4. When $x = 0$, $y = c$, so that the y intercept is at the point $(0, c)$.

5. The x intercepts (if there are any) are found by solving the quadratic equation $a \cdot x^2 + b \cdot x + c = 0$.

Let us now check out these characteristics with the above two examples.

Example 1

$$y = x^2 - 4x + 4.$$

Here, $a = 1$, $b = {}^-4$, and $c = 4$. Since $a > 0$, the parabola opens upward. Since $(-b/2 \cdot a) = -({}^-4/2 \cdot 1) = 2$, then the x coordinate of the vertex is $x = 2$. The y coordinate of the vertex is $F(2) = 2^2 - 4 \cdot 2 + 4 = 0$. Thus, the vertex is at the point $(2, 0)$. The axis of symmetry is $x = -b/2 \cdot a = 2$. The y intercept is the point $(0, 4)$. The x intercept is found by solving $x^2 - 4x + 4 = (x - 2)^2 = 0$. The only solution to this equation is $x = 2$, so that the only x intercept is at the point $(2, 0)$ (see the graph at the beginning of this section).

Example 2

$$y = F(x) = -x^2 + 2x + 3.$$

Here, $a = {}^-1$, $b = 2$, and $c = 3$. Since $a < 0$, the parabola opens downward. Since $-b/2 \cdot a = -2/{}^-2 = 1$, then the x coordinate of the vertex is $x = 1$. The y coordinate of the vertex is $F(1) = {}^-1 + 2 + 3 = 4$. Thus, the vertex is the point $(1, 4)$. The axis of symmetry is $x = (-b/2 \cdot a) = 1$. The y-intercept is the point $(0, 3)$. The x intercepts are found by solving $-x^2 + 2x + 3 = (x + 1) \cdot (x - 3) = 0$. The only solutions are $x = {}^-1$ and $x = 3$, so the x intercepts are the points $({}^-1, 0)$ and $(3, 0)$ (see the graph at the beginning of this section).

Quadratic equations and the quadratic formula

To get the x intercepts of the quadratic function $y = a \cdot x^2 + b \cdot x + c$, it is necessary to solve the equation $a \cdot x^2 + b \cdot x + c = 0$. Such an equation is called a *quadratic equation*. To solve such an equation, one can factor it into linear factors $(d \cdot x + e) \cdot (f \cdot x + g)$ and set each linear factor equal to 0 and solve for x (provided the function can be easily factored). Or, one can use the *quadratic formula*

$$x = \frac{-b \pm \sqrt{b^2 - 4 \cdot a \cdot c}}{2 \cdot a}.$$

We illustrated the factoring method in the two previous examples. In this text, we will emphasize the use of the quadratic formula.

The expression $b^2 - 4 \cdot a \cdot c$ is called the *discriminant*. There are three possible cases:

Case 1.

If $b^2 - 4 \cdot a \cdot c = 0$, there is one solution to the quadratic equation, namely $x = -b/2 \cdot a$.

Case 2.

If $b^2 - 4 \cdot a \cdot c > 0$, there are two solutions to the quadratic equation, namely

$$x = \frac{-b \pm \sqrt{b^2 - 4 \cdot a \cdot c}}{2 \cdot a}.$$

Case 3.

If $b^2 - 4 \cdot a \cdot c < 0$, then there are no real solutions, since $\sqrt{b^2 - 4 \cdot a \cdot c}$ would be the square root of a negative number, which is not a real number. (It is imaginary.)

Let us consider some examples of solving quadratic equations by the use of the quadratic formula.

Example 1

$$x^2 - 4x + 4 = 0.$$

Here, $a = 1$, $b = {}^-4$, and $c = 4$. Thus,

$$x = \frac{-b \pm \sqrt{b^2 - 4 \cdot a \cdot c}}{2 \cdot a} = \frac{4 \pm \sqrt{16 - 16}}{2} = \frac{4 \pm 0}{2} = 2.$$

Example 2

$$-x^2 + 2x + 3 = 0.$$

Here, $a = {}^-1$, $b = 2$, and $c = 3$. Thus,

$$x = \frac{-b \pm \sqrt{b^2 - 4 \cdot a \cdot c}}{2 \cdot a} = \frac{{}^-2 \pm \sqrt{4 + 12}}{{}^-2} = \frac{{}^-2 \pm \sqrt{16}}{{}^-2} = \frac{{}^-2 \pm 4}{{}^-2}.$$

So, we get the two solutions $x = ({}^-2 + 4)/{}^-2 = {}^-1$ and $x = ({}^-2 - 4)/{}^-2 = 3$.

Example 3

$$2x^2 - 5x + 1 = 0.$$

Here, $a = 2$, $b = {}^-5$, and $c = 1$. Thus, $x = (5 \pm \sqrt{25 - 8})/4 = (5 \pm \sqrt{17})/4$. So, there are two solutions, $x = (5 + \sqrt{17})/4$ and $x = (5 - \sqrt{17})/4$.

Example 4

$$x^2 - 9 = 0.$$

Here, $a = 1$, $b = 0$, and $c = {}^-9$. Thus, $x = (0 \pm \sqrt{0 + 36})/2 = \pm 6/2 = \pm 3$.

Example 5

$$3x^2 + x + 2 = 0.$$

Here, $a = 3$, $b = 1$, and $c = 2$. Thus,

$$x = \frac{-b \pm \sqrt{b^2 - 4 \cdot a \cdot c}}{2 \cdot a} = \frac{{}^-1 \pm \sqrt{1 - 24}}{6} = \frac{{}^-1 \pm \sqrt{{}^-23}}{6}.$$

However, $\sqrt{{}^-23}$ is not a real number. So, the solutions are imaginary.

156

We now consider a program for solving quadratic equations by the use of the quadratic formula.

Program 6.5 QUADRATIC

$\nabla X \leftarrow QUADRATIC\ COEFS: A; B; C; R; DISCRIMINANT$

[1] $A \leftarrow COEFS\ [1]$
[2] $B \leftarrow COEFS\ [2]$
[3] $C \leftarrow COEFS\ [3]$
[4] $DISCRIMINANT \leftarrow (B*2) - (4 \times A \times C)$

[5] $\rightarrow (DISCRIMINANT \geqslant 0)/ROOT$

[6]   '*THE SOLUTIONS ARE IMAGINARY*'
[7] $\rightarrow 0$
[8] $ROOT: R \leftarrow DISCRIMINANT * .5$

[9] $X \leftarrow ((-B) + R, -R) \div (2 \times A)$
       ∇

To illustrate the use of this program, we will redo the above examples using *QUADRATIC*.

Example 1

$$x^2 - 4x + 4 = 0.$$

QUADRATIC 1 ⁻4 4   To run the program, type
                               QUADRATIC, followed by the vector
2 2   of coefficients.

Example 2

$$-x^2 + 2x + 3 = 0.$$

QUADRATIC ⁻1 2 3
⁻1 3   The solutions are ⁻1 and 3.

Example 3

$$2x^2 - 5x + 1 = 0.$$

QUADRATIC 2 ⁻5 1
2.280776406 0.2192235936   These answers correspond to
                               $(5 + \sqrt{17})/4$ and $(5 - \sqrt{17})/4$
                               respectively.

157

Example 4

$$x^2 - 9 = 0.$$

QUADRATIC 1 0 ⁻9   Note that 0 coefficients must be included in the vector of coefficients.

⁻3 3

Example 5

$$3x^2 + x + 2 = 0.$$

QUADRATIC 3 1 2
THE SOLUTIONS ARE IMAGINARY

Since $b^2 - 4 \cdot a \cdot c$ is negative.

We conclude this section with a couple of applications of quadratic functions.

Applications of quadratic functions

Example 1

Suppose the cost of producing x items is given by the function

$$C = x^2 + 2x + \$2000.$$

Suppose also that these items sell at a price of $102 each, and that every item that is produced is sold.

(a) Find the number of items, x, that must be produced in order to maximize profit. Also, find this maximum profit.

Since the price is $102 each, and

$$\text{Revenue} = (\text{Price}) \cdot (\text{Number of items sold}),$$

then the revenue (in dollars) from the x items is given by $R = 102 \cdot x$. Now,

$$\text{Profit} = \text{Revenue} - \text{Cost} = R - C = 102 \cdot x - (x^2 + 2x + 2000)$$
$$= -x^2 + 100x - 2000 = F(x).$$

The graph of this function is a parabola that opens downward, since $a < 0$. Thus, the highest point (the point of maximum profit) will occur at the vertex. So,

$$x = \frac{-b}{2 \cdot a} = \frac{^-100}{^-2} = 50 \text{ items.}$$

The maximum profit is $F(50) = \$500$.

(b) Find the number of items, x, that must be produced to break even.

A company breaks even when its revenue equals its cost, or when its profits is 0. Thus, we need to solve the quadratic equation

$$-x^2 + 100 \cdot x - 2000 = 0.$$

QUADRATIC $^-$1 100 $^-$2000
27.63932023 72.36067978

Thus, the company breaks even when it produces approximately its 27th and its 72nd items.

Example 2

Suppose that to produce 500 widgets costs $2000. Thereafter, for every increase of 5 widgets produced, the cost is reduced by $0.02.

(a) Find a formula for cost.

Let $x=$ the number of increases of 5 widgets. Since the first 500 widgets cost $2000, then these widgets cost $4 each. Thereafter, the cost per widget decreases by $0.02x$, since for each increase of 1 in x (5 widgets), cost per widget decreases by 0.02. Now,

$$\text{Cost}=(\text{Number of widgets})\cdot(\text{Cost per widget})$$
$$=(500+5x)\cdot(4-0.02\cdot x)$$
$$=2000+20x-10x-0.1x^2$$
$$=\,^-0.1x^2+10x+2000=F(x).$$

(b) How many widgets yield maximum cost?

Since $a=\,^-0.1$ is negative, the graph of this function is a parabola that opens downward. Thus, the highest point (the point of maximum cost) is the vertex. So,

$$x=\frac{-b}{2\cdot a}=\frac{\,^-10}{\,^-0.2}=50.$$

Thus, maximum cost occurs when there are 50 increases of 5 widgets beyond the first 500. The total number of widgets is given by $500+5x$ $=500+5\cdot50=750$ widgets.

(c) What is this maximum cost?

$$F(50)=\,^-0.1\cdot(50)^2+10\cdot50+2000=\,^-250+500+2000$$
$$=\$2250.$$

We shall explore the concept of maximizing and minimizing functions in much more detail in a later chapter using differential calculus.

EXERCISES

1. Write a program for computing the x and y coordinates of the vertex of the parabola corresponding to the quadratic function

$$y=F(x)=a\cdot x^2+b\cdot x+c.$$

2. Compute both coordinates of the vertex of the following parabolas corresponding to each of the following quadratic functions:
(a) $y = F(x) = x^2 - 2$
(d) $y = F(x) = -x^2 + 8x - 6$
(b) $y = F(x) = x^2 - x - 2$
(e) $y = F(x) = 2x^2 - x - 3$
(c) $y = F(x) = -x^2 + 6x - 9$
(f) $y = F(x) = 3x^2 + 6x + 5$

3. Use the quadratic formula to find the roots of the quadratic functions in Exercise 2, provided the roots are real. [*Note*: A *root* of a quadratic function $y = a \cdot x^2 + b \cdot x + c$ is a solution to the corresponding quadratic equation.]

4. Redo Exercise 3 using the program *QUADRATIC*.

5. Use the characteristics of the parabolas in Exercise 2 that were mentioned in this section to graph these parabolas.

6. Redo Exercise 5 using the program *GRAPH*.

7. Suppose that the profit (in dollars) from the sale of x items is given by

$$P = F(x) = -x^2 + 500x - 40000.$$

(a) Find how many items must be sold to maximize profit.
(b) Find this maximum profit.
(c) Find how many items must be sold to break even.

8. If a manager of an apartment complex charges a monthly rent of $200, he will completely fill up his 80 apartments. For each increase of $10 in the monthly rent thereafter, 2 apartments will be empty.
(a) Find a quadratic function which expresses the monthly revenue in terms of the number of increases of $10 in monthly rent.
(b) How many such $10 increases in rent will lead to maximum revenue?
(c) What is the optimum rent? How many empty apartments result?

9. An object projected vertically upward from a height of 6 feet with an initial velocity of 128 feet per second has height at any time t given by the quadratic function

$$h = F(t) = {}^-16 \cdot t^2 + 128 \cdot t + 6,$$

where t is the number of seconds that have elapsed since it was projected vertically upward and h is the number of feet up after t seconds. Find the amount of time it takes for the object to reach its maximum height, and find the maximum height.

10. Suppose that the cost (in dollars) of producing x items is given by the formula

$$C = 0.01 \cdot x^2 + 2x.$$

Suppose also that these items sell for $22 each and that every item that is produced is sold.
(a) Find the number of items, x, that must be produced to maximize profit.
(b) Find this maximum profit.
(c) Find the number of items that must be produced to break even.

6.5 Polynomials

Linear functions and quadratic functions are special cases of a larger class of functions called polynomials which we now investigate.

Definition

Let n be a positive integer. Then, a *polynomial of degree n* is a function of the form

$$y = F(x) = a_n \cdot x^n + a_{n-1} \cdot x^{n-1} + \ldots + a_2 \cdot x^2 + a_1 \cdot x + a_0,$$

where, $a_0, a_1, \cdots, a_{n-1}, a_n$ are constant real numbers, and $a_n \neq 0$.

We have already studied polynomials of degree 1 (linear functions) and of degree 2 (quadratic functions). If $n > 2$, then, in general, it is more difficult to completely pin down the characteristics of polynomials of degree n. To do so requires calculus, as we will see in Chapter 8. However, with the aid of the computer, we can do a good job in graphing particular examples.

Domain of polynomials

Unless explicitly restricted, the domain of any polynomial is the entire set of real numbers. That is, any real number may be substituted into a polynomial $y = F(x)$ in place of x, resulting in a unique real answer, y.

We can use the program *PAIRS* to print out a set of ordered pairs for any polynomial between any two values of x. Using this set of ordered pairs, we can graph the polynomial. Let us consider a couple of examples now.

Example 1

$$y = F(x) = 2x^3 + 3x^2 - 12x - 10.$$

This is a polynomial of degree 3. In APL, this would be

```
      ∇FN [1]
[1]    Y←(2×X*3)+(3×X*2)+(¯12×X)−10 ∇
```

Altering *FN* for this new function.

```
      X←¯10 ¯9 ¯8 ¯7 ¯6 ¯5 ¯4 ¯3 ¯2 ¯1 0 1 2 3 4 5 6 7 8 9 10
      X PAIRS FN X
¯10 ¯1590
 ¯9 ¯1117
 ¯8  ¯746
 ¯7  ¯465
 ¯6  ¯262
 ¯5  ¯125
 ¯4   ¯42
 ¯3    ¯1
```

⁻2	10
⁻1	3
0	⁻10
1	⁻17
2	⁻6
3	35
4	118
5	255
6	458
7	739
8	1110
9	1583
10	2170

Using these pairs, we can graph the function (Figure 6.19). The arrowheads indicate the direction of the curve as indicated by the pairs. The roller coaster shape of this curve is typical of the graphs of polynomials of degree 3.

Example 2

$$y = F(x) = x^4 - x^3 - 7x^2 + x + 6.$$

This is a polynomial of degree 4.

∇FN [1]

[1] $Y \leftarrow (X*4) + (-X*3) + (^-7 \times X*2) + X + 6$ ∇

Altering *FN* for this new function.

$X \leftarrow ^-10\ ^-9\ ^-8\ ^-7\ ^-6\ ^-5\ ^-4\ ^-3\ ^-2\ ^-1\ 0\ 1\ 2\ 3\ 4\ 5\ 6\ 7\ 8\ 9\ 10$
X PAIRS FN X

⁻10	10296
⁻9	6720
⁻8	5158
⁻7	2400
⁻6	1260
⁻5	676
⁻4	210
⁻3	45
⁻2	0
⁻1	0
0	6
1	0
2	⁻12
3	0
4	90
5	336

Since $y = 0$ at both $x = ^-2$ and $x = ^-1$, it would be useful to know the value of $F(^-1.5)$. Thus,

FN ⁻1.5
⁻2.6125

162

6   840
7 1728
8 3150
9 5280
10 8316

Using these pairs, the graph looks as shown in Figure 6.20. The shape of this curve is typical of graphs of polynomials of degree 4.

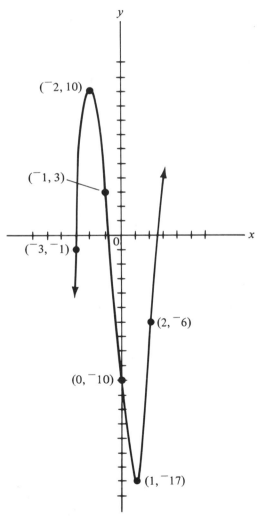

Figure 6.19   A third degree polynomial.

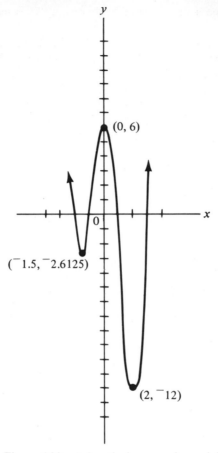

Figure 6.20   A fourth degree polynomial.

Polynomials are everywhere *continuous*. That is, there are no gaps in the graph of a polynomial, and it is possible to graph a polynomial without ever lifting one's pencil from the paper.

Roots of polynomials

A *root* of a polynomial

$$y = F(x) = a_n \cdot x^n + a_{n-1} \cdot x^{n-1} + \ldots + a_1 \cdot x + a_0$$

is a solution to the polynomial equation

$$a_n \cdot x^n + a_{n-1} \cdot x^{n-1} + \ldots + a_1 \cdot x + a_0 = 0.$$

In terms of the graph of a polynomial, a root is represented as an x intercept on the graph, since it is a value of x for which $y = 0$. Note that in terms of the graph of the polynomial of degree 3 in Example 1, there are 3 x intercepts. Therefore, there are three roots. In terms of the graph of the

164

polynomial of degree 4 in Example 2, there are 4 x intercepts. Therefore, there are 4 roots.

The *Fundamental theorem of algebra* states that a polynomial of degree n has n roots (not necessarily all distinct or all real). We have already investigated techniques for finding the roots of polynomials of degrees 1 and 2. There exist formulas for finding the roots of general polynomials of degree 3 and 4 also. However, these formulas are beyond the scope of this text. It has been proved (by the mathematicians Abel and Galois) that for polynomials of degree $\geqslant 5$, there are no such formulas for finding the roots. However, there are many techniques for approximating the roots of any particular polynomial to any desired degree of accuracy. These techniques are made much easier by the use of the computer. We will discuss one such technique here.

This technique is not very sophisticated, but it is quite easy to use if one has the computer available to do the computations.

A technique for approximating the real roots of a polynomial

Our technique is based on the fact that if a polynomial changes sign between a value $x=a$ and a value $x=b$, then there must exist a root between these two values. In terms of the graph, a polynomial cannot go from below the x axis to above it (or vice versa) without passing through an x intercept. Our main tool in this technique is the program *PAIRS*. Let us illustrate the method using the polynomial of Example 1.

Example

$$y = F(x) = 2x^3 + 3x^2 - 12x - 10.$$

$\nabla FN\,[1]$

[1] $Y \leftarrow (2 \times X * 3) + (3 \times X * 2) + (^-12 \times X) - 10\ \nabla$

Let us reprint the set of ordered pairs for this function.

$X \leftarrow ((\iota 10) - 11), 0, \iota 10$
$X\ PAIRS\ FN\ X$

¯10	¯1590
¯9	¯1117
¯8	¯746
¯7	¯465
¯6	¯262
¯5	¯125
¯4	¯42
¯3	¯1	Sign change here.
¯2	10

¯1	3
0	¯10
1	¯17
2	¯6
3	35
4	118
5	255
6	458
7	739
8	1110
9	1583
10	2170

Sign change here. (between ¯1 and 0)

Sign change here. (between 2 and 3)

Since the y value of the polynomial changes sign between $x = ¯3$ and $x = ¯2$, then there must be a root between these two values. Similarly, there must be a root between ¯1 and 0, and also between 2 and 3. To see in more detail where the sign changes between ¯3 and ¯2, let us use *PAIRS* to print out the ordered pairs between ¯3 and ¯2, incrementing by .1 each time.

$$X \leftarrow ¯3, \ ¯3 + (\iota 10) \times .1$$

X PAIRS FN X

¯3	¯1
¯2.9	1.252
¯2.8	3.216
¯2.7	4.904
¯2.6	6.328
¯2.5	7.5
¯2.4	8.432
¯2.3	9.136
¯2.2	9.624
¯2.1	9.908
¯2	10

Sign change here. (at ¯2.9 1.252)

Since the sign changes between ¯3 and ¯2.9, the root must be between these two numbers. To get closer to the root, let us use *PAIRS* to print out the ordered pairs between ¯3 and ¯2.9 incrementing by .01 each time.

$$X \leftarrow ¯3, \ ¯3 + (\iota 10) \times .01$$

X PAIRS FN X

¯3	¯1
¯2.99	¯0.761498
¯2.98	¯0.525984
¯2.97	¯0.293446
¯2.96	0.063872

Sign change here. (at ¯2.96 0.063872)

166

```
⁻2.95  0.16275
⁻2.94  0.386432
⁻2.93  0.607186
⁻2.92  0.825024
⁻2.91  1.039958
⁻2.90  1.252
```

Since the sign changes between ⁻2.96 and ⁻2.95, the root must be between these two numbers. To get even closer to the root, we could use *PAIRS* to print out the ordered pairs between ⁻2.96 and ⁻2.95, incrementing by 0.001 each time.

$$X \leftarrow {}^-2.96, \ {}^-2.96 + (\iota 10) \times .001$$

```
X PAIRS FN X
⁻2.96    ⁻.063872
⁻2.959  ⁻.041077158
⁻2.958  ⁻.018311824          Sign change here.
⁻2.957   .004424014
⁻2.956   .027130368
⁻2.955   .04980725
⁻2.954   .072454672
⁻2.953   .095072646
⁻2.952   .117661184
⁻2.951   .140220298
⁻2.95    .16275
```

Since the sign changes between ⁻2.958 and ⁻2.957, the root must be between these two numbers. We could get even closer by incrementing by 0.0001 in *PAIRS*. However, this is close enough to illustrate the technique. Since the y value at ⁻2.957 is closer to 0 than the y value at ⁻2.958, then to three decimal places, ⁻2.957 is the best approximation to the desired root.

Recall that there are also roots between ⁻1 and 0, and between 2 and 3. These could be approximated in the same way we approximated the root between ⁻3 and ⁻2.

Example 2

$$y = F(x) = x^4 - x^3 - 7x^2 + x + 6.$$

```
     ∇FN [1]

[1]  Y←(X∗4)+(−X∗3)+(⁻7×X∗2)+X+6 ∇

     X←((ι10)−11),0,ι10

     X PAIRS FN X
⁻10  10296
 ⁻9   6720
```

167

¯8	5158
¯7	2400
¯6	1260
¯5	676
¯4	210
¯3	45
¯2	0
¯1	0
0	6
1	0
2	¯12
3	0
4	90
5	336
6	840
7	1728
8	3150
9	5280
10	8316

The roots for this function are ¯2, ¯1, 1, and 3, since for these values, the y coordinates are all 0. No approximation is necessary. Lucky!!!

The above technique for approximating the real roots of a polynomial might seem quite tedious to the reader. However, with the aid of the computer, it can be accomplished very rapidly.

A program for approximating roots

Suppose that, due to a sign change in the y coordinates of a polynomial, we know that our polynomial has a root between a value $x = A$ and a value $x = B$. Then, the following program can be used to approximate this root correct to three decimal places.

Program 6.6 ROOT

$$\nabla R \leftarrow A \ ROOT \ B \ ; \ X$$

[1] $X \leftarrow A$

[2] $OLDY \leftarrow FN \ X$ Compute the previous y, called $OLDY$.

[3] $X \leftarrow X + .001$ Increment X by 0.001

[4] $NEWY \leftarrow FN \ X$ Compute the new y, called $NEWY$.

[5] $\rightarrow ((OLDY \times NEWY) > 0)/2$ If $(OLDY \times NEWY) > 0$ then $OLDY$ and $NEWY$ have the same sign, so we branch back to 2.

[6]   →((|OLDY)<(|NEWY))/END

> If the absolute value of *OLDY* is less than the absolute value of *NEWY*, branch to *END*.

[7]   R←X

> Otherwise, *R* is *X*, which corresponds to *NEWY*.

[8]   →0

> End the program with *R* being the *X* corresponding to *NEWY*.

[9]   END : R←X− .001
    ∇

> *R* is the value of *X* corresponding to *OLDY*.

This program keeps computing *OLDY* and *NEWY* until they have opposite signs (there has been a sign change). It then prints out the value of *X* corresponding to the smaller value of *y* in absolute value.

Before executing this program, you need a subprogram *FN* for the function whose root is being approximated. You also need to examine *PAIRS* for the function to see where the sign changes occur.

Example

   ∇FN [1]

[1]   Y←(2×X∗3)+(3×X∗2)+(⁻12×X)−10 ∇

> This is our previous Example 1.

We know that the roots are between ⁻3 and ⁻2, between ⁻1 and 0, and between 2 and 3. We can use *ROOT* to approximate these roots.

    ⁻3 *ROOT* ⁻2
⁻2.957
    ⁻1 *ROOT* 0
⁻0.762
    2 *ROOT* 3
2.219

> Thus, the roots are approximately ⁻2.957, ⁻0.762, and 2.219, correct to three decimal places.

EXERCISES

1. Use the program *PAIRS* to generate a table of ordered pairs for the following functions, with $^-10 \leqslant x \leqslant 10$:
   (a) $y = F(x) = x^3 + 2x - 5$
   (b) $y = F(x) = 2x^3 + 3x^2 - 4x - 10$
   (c) $y = F(x) = x^4 - 5x^2 + 6$
   (d) $y = F(x) = x^4 - x^3 + x - 2$
   (e) $y = F(x) = x^5 - 2x^4 + 3x - 5$

2. Using the table you generated in Exercise 1, graph the functions in Exercise 1.

3. Each of the functions in Exercise 1 has a root between 1 and 2. Use the technique explained in this section to approximate these roots correct to three decimal places.

4. Use the program *ROOT* to do Exercise 3.

5. Use the technique of this section to approximate the three roots of the polynomial $y = F(x) = x^3 - 4x^2 + 2x + 2$.

6. Use the program *ROOT* to do Exercise 5.

7. If the polynomial $y = F(x) = x^3 + A \cdot x^2 + B \cdot x + C$ passes through the three points $(0,0)$, $(1,2)$, and $(2,6)$, find A, B, and C.

8. The volume of a sphere of radius x is given by the function $V = \frac{4}{3} \cdot \pi \cdot x^3$. Write a program for finding the volume of a sphere.

6.6 Rational functions

Polynomials are special cases of rational functions, which we now briefly consider.

Rational functions

A *rational function* is a function of the form

$$y = F(x) = \frac{P(x)}{Q(x)},$$

where $P(x)$ and $Q(x)$ are polynomials. Included as rational functions are the polynomials, since they can be thought of as quotients of polynomials where $Q(x) = 1$, a polynomial of degree 0.

Domains of rational functions

Since polynomials are defined everywhere and since division by zero is not defined, the domain of a rational function consists of all real numbers except for the values of x for which $Q(x)$ is zero (i.e., except for the roots of $Q(x)$).

Roots of rational functions

The only way in which a quotient can be zero is for the numerator to be zero. Thus, the roots of a rational function are the same as the roots of the numerator $P(x)$.

Let us now consider the graphs of some rational functions. As before, we will use the program *PAIRS* to generate a set of ordered pairs.

Example 1

$$y = F(x) = \frac{4 - x}{x - 2}.$$

170

This function has a root at $x=4$, since $x=4$ is the only root of the numerator. The domain of this function consists of all real numbers except for $x=2$. At $x=2$, the denominator is zero, and division by zero is not defined. Let us now use *PAIRS* to give us some points to use in graphing this function.

∇FN [1]

[1] $Y \leftarrow (4-X) \div (X-2)$ ∇
       $X \leftarrow ((\iota 5)-6),0,\iota 5$
       X PAIRS FN X

```
 ¯5 ¯1.2857142857
 ¯4 ¯1.3333333333
 ¯3 ¯1.4
 ¯2 ¯1.5
 ¯1 ¯1.6666666667
  0 ¯2
  1 ¯3
  2 9.999999999E999
```

The value at which we have division by 0. $9.9999999999E999$ is essentially infinity.

```
  3 1
  4 0
```

The root.

```
  5 ¯.3333333333
```

Before graphing this function, let us examine some pairs in the vicinity of $x=2$ (the value which is not in the domain of $F(x)$.)

$X \leftarrow 1, 1+(\iota 10) \times .2$

X PAIRS FN X

```
1   ¯3
1.2 ¯3.5
1.4 ¯4.333333333
1.6 ¯6
1.8 ¯11
2   9.99999999E999
2.2 9
2.4 4
2.6 2.333333333
2.8 1.5
3   1
```

Using these tables, we can graph

$$y = F(x) = \frac{4-x}{x-2}.$$

It is shown in Figure 6.21.

171

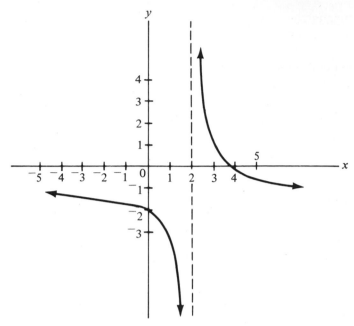

Figure 6.21   Graph of Example 1.

Vertical asymptotes

Notice that the closer x gets to 2, the closer the curve gets to the line $x=2$. This curve cannot ever touch the line $x=2$, since $x=2$ is not in the domain of the function. A line with the property that the curve approaches it continuously without ever touching it is called an *asymptote* of the function. The asymptotes of rational functions are the values of x which yield division by 0 (i.e., the roots of the denominator $Q(x)$).

Example 2

$$y = F(x) = \frac{4}{x-1}$$

In this example, the numerator, and therefore the rational function has no roots. This function is not defined at $x=1$, since at $x=1$, we would have division by 0. Thus, $x=1$ is a vertical asymptote for this function. To graph this function, we need some pairs to plot.

```
     ∇FN [1]

[1]   Y←4÷(X−1) ∇                    Our new function.

      X←((ι5)−6),0,ι5
      X PAIRS FN X
⁻5 ⁻.666666667
⁻4 ⁻.8
```

```
-3 -1
-2 -1.333333333
-1 -2
 0 -4
 1 9.99999999E999                    The vertical asymptote.
 2 4
 3 2
 4 1.333333333
 5 1
```

Let us also examine this function in the vicinity of the asymptote, $x = 1$.

```
        X←0,(ι10)×.2
        X PAIRS FN X
0    -4
0.2  -5
0.4  -6.66666667
0.6  -10
0.8  -20
1    9.99999999E999              The asymptote.
1.2  20
1.4  10
16   6.666666667
1.8  5
2    4
```

Thus, the graph is as shown in Figure 6.22.

Example 3

$$y = F(x) = \frac{x^2}{x^2 - 1}$$

By the previous discussions, $x = 1$ and $x = -1$ are vertical asymptotes, since for these values, the denominator is 0. The root is $x = 0$, since for this value, the numerator is 0.

```
        ∇FN [1]

[1]    Y←(X*2)÷((X*2)−1) ∇

        X←((ι5)−6),0,ι5
        X PAIRS FN X
-5 1.041666667
-4 1.066666667
-3 1.125
-2 1.333333333
-1 9.99999999E999                Since x = -1 is an asymptote.
 0 0                             This is our root.
```

173

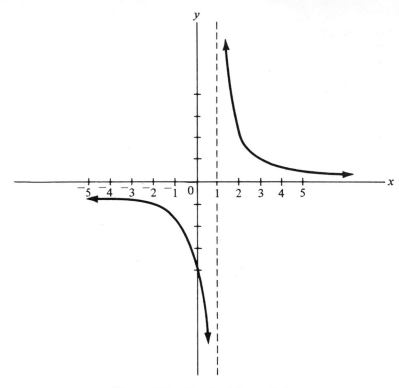

Figure 6.22   Graph of Example 2.

1 9.99999999E999   Since $x = 1$ is an asymptote.
2 1.3333333333
3 1.125
4 1.06666666667
5 1.041666667

We also want to examine this function in the vicinity of its asymptotes.

        X← ⁻2, ⁻2+(ι10)×.2, (ι10)×.2
        X PAIRS FN X
⁻2   1.333333333
⁻1.8 1.4464285715
⁻1.6 1.6410230769
⁻1.4 2.0416666667
⁻1.2 3.2727272727
⁻1   9.9999999E999 $x = ⁻1$ is an asymptote.
⁻0.8 ⁻1.7777777778
⁻0.6 ⁻.5625
⁻0.4 ⁻.1904785714
⁻0.2 ⁻.0416666667

0	0	The root.
0.2	⁻.0416666667	
0.4	⁻.04785714	
0.6	⁻.5625	
0.8	⁻1.7777777778	
1	9.99999999E999	$x = 1$ is an asymptote.
1.2	3.2727272727	
1.4	2.0416666667	
1.6	1.6410230769	
1.8	1.4464285715	
2	1.333333333	

Based on these tables of pairs, the graph is as shown in Figure 6.23.

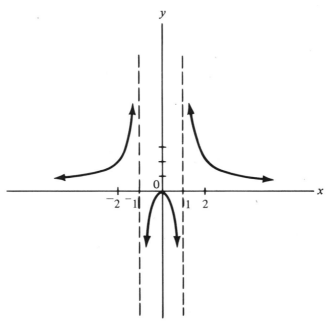

Figure 6.23   Graph of Example 3.

Applications

1. One important use of rational functions is in cost–benefit curves as the following:

   Let y be the cost, in thousands of dollars, of removing x percent of a certain pollutant from the air in a certain chemical factory. Suppose $y = 15 \cdot x / (100 - x)$. Then, the cost of removing 50 percent of this pollutant is

$$y = \frac{15 \cdot 50}{100 - 50} = \$15,000.$$

175

The cost of removing 90 percent of the pollutant is

$$y = \frac{15 \cdot 90}{100 - 90} = \$135{,}000.$$

The company would have to consider if it is really worth the expense of removing 90 percent of the pollutant or if some lower percent is permissible.

2. Suppose that a widget manufacturer is able to make x number one quality widgets and y number two quality widgets per day, where the relationship between x and y is given by the rational function

$$y = \frac{60 - 12x}{6 - x}.$$

   (a) If the manufacturer only produces number one quality widgets on a given day, then $y = 0$, so that $60 - 12x = 0$, or $x = 5$. Thus, he can produce a maximum of 5 number one quality widgets in a day.

   (b) If he only produces number two quality widgets, then $x = 0$, so $y = 60/6 = 10$ number two quality widgets.

EXERCISES

1. Find the roots and vertical asymptotes of the following rational functions:

   (a) $y = \dfrac{10}{x - 2}$

   (b) $y = \dfrac{5 - x}{x - 1}$

   (c) $y = \dfrac{x^2 - 1}{x^2 + 1}$

   (d) $y = \dfrac{1}{x^2 + 1}$

   (e) $y = \dfrac{x^2 + 1}{x^2 - x}$

2. Use the program *PAIRS* to print out a table of ordered pairs for the functions in Exercise 1, where $^-5 \leqslant x \leqslant 5$.

3. Use the program *PAIRS* to print out a table of ordered pairs in the immediate vicinity of the asymptotes to the functions in Exercise 1.

4. Use the information gathered above to graph the functions in Exercise 1.

5. Let y be the cost, in hundreds of dollars, of removing x percent of the impurities from the drinking water in a community. The function relating x and y is the rational function $y = 25 \cdot x / (100 - x)$.
   (a) Find the cost of removing 10 percent of the impurities.
   (b) Find the cost of removing 50 percent of the impurities.
   (c) Find the cost of removing 90 percent of the impurities.

6. Suppose it costs \$10 each to manufacture and distribute a gadget. If the manufacturer sells the gadgets for x dollars each, then the number he can sell is given by

$$n = \frac{100}{x - 10} + 5 \cdot (100 - x).$$

   How many can he sell at a price of \$20? \$30?

Exponential and logarithmic functions 7

All of the functions we have considered so far have been algebraic functions. An *algebraic function* is a function involving only the operations of addition, subtraction, multiplication, division, powers, and extraction of roots of expressions of the form $a \cdot x^n$, where a and n are real constants. Any function that is not algebraic is called a *transcendental function*. In this chapter, we will consider two important classes of transcendental functions, the exponential and logarithmic functions, as well as some applications of each.

7.1 Exponential functions

Definition

An *exponential function* is a function of the form $y = F(x) = a \cdot b^{k \cdot x}$, where a, b, and k are nonzero constants. We shall also require that b, called the *base* of the exponential function, be positive so that the function will be defined for all real powers. (For example, if $b = {}^-1$, then $b^{0.5} = \sqrt{b}$ would not be defined.)

The *domain of an exponential function* is the set of all real numbers, unless otherwise explicitly restricted. The following examples illustrate the general characteristics of an exponential function.

Example 1

Graph $y = F(x) = 2^x$.

We need a set of ordered pairs. Therefore, we will use the program **PAIRS**.

177

        ∇ FN[1]   Changing the function in FN to y =
[1]   Y←2 * XV   2ˣ.
        X←((ɩ5) – 6), 0, ɩ5
        X PAIRS FN X
¯5 .03125
¯4 .0625
¯3 .125
¯2 .25
¯1 .5
 0 1
 1 2
 2 4
 3 8
 4 16
 5 32

Thus, the graph is as shown in Figure 7.1.

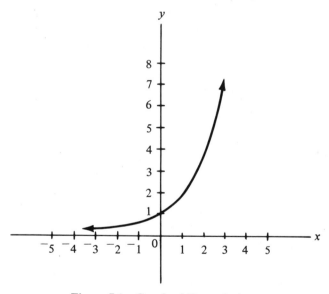

Figure 7.1   Graph of Example 1.

Example 2

Graph y = F(x) = 2⁻ˣ.

        ∇ FN[1]
[1]   Y←2 * – XV
        X←((ɩ5) – 6), 0, ɩ5
        X PAIRS FN X
¯5   32

```
⁻4   16
⁻3   8
⁻2   4
⁻1   2
 0   1
 1   .5
 2   .25
 3   .125
 4   .0625
 5   .03125
```

Thus, the graph is as shown in Figure 7.2.

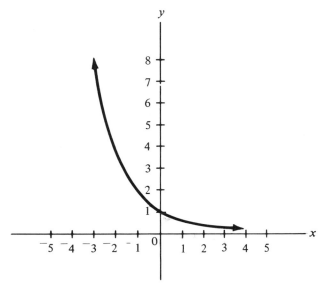

Figure 7.2   Graph of Example 2.

Since the domain of an exponential function includes all real numbers, we can use any real number for x. Consider the following examples:

Examples

2 ∗ .5 1.414213562	$2^{0.5}$ or $2^{1/2}$ or $\sqrt{2}$.
2 ∗ .4 1.319507911	$2^{0.4}$.
2 ∗ ⁻3 0.125	2^{-3} or $1/2^3$ or $1/8$.
2 ∗ ○1 8.824977827	2^{π}, since in APL, π is ○1.

```
    2*2*.5
2.665144143
```
$2^{2^{0.5}}$ or $2^{\sqrt{2}}$.

```
    8*1÷3
2
```
$8^{1/3}$ or $\sqrt[3]{8}$.

```
    ¯4*.5
DOMAIN ERROR
```
$(^-4)^{0.5}$ or $\sqrt{^-4}$, which is not a real number. This is why we have made our bases positive. We want to be able to include the entire real number system in the domain.

```
    0*0
1
```
This example might cause some mathematicians to cringe. Let us call this an APL curiousity. In APL, $b^0 = 1$ for all real b.

```
    2*3 4 5
8 16 32
```
This is $2^3, 2^4, 2^5$.

```
    3 4 5*2
9 16 25
```
This is $3^2, 4^2, 5^2$.

```
    3 4 5*3 2 1
27 16 5
```
This is $3^3, 4^2, 5^1$.

```
    4*ι4
4 16 64 256
```
This is $4^1, 4^2, 4^3, 4^4$.

```
    (ι4)*4
1 16 81 256
```
This is $1^4, 2^4, 3^4, 4^4$.

Negative and fractional exponents

A few words should be said about negative and fractional exponents for those who have not encountered them before. The following definitions explain these exponents:

$$b^{-n} = \frac{1}{b^n} \qquad b^{1/n} = \sqrt[n]{b}, \qquad b^{m/n} = \sqrt[n]{b^m} = \left(\sqrt[n]{b}\right)^m.$$

We have used these definitions in the previous examples. However, a couple more examples follow:

Examples

```
    4*¯2
0.0625
```
$4^{-2} = 1/4^2 = 1/16$.

```
    9*1÷2
3
```
$9^{1/2} = \sqrt{9}$.

```
      27 * 1 ÷ 3
3
```
$$27^{1/3} = \sqrt[3]{27}.$$

```
      27 * 2 ÷ 3
9
```
$$27^{2/3} = \sqrt[3]{27^2} = (\sqrt[3]{27})^2 = 3^2.$$

```
      2 * ⁻1
0.5
```
$$2^{-1} = 1/2^1 = 1/2.$$

```
      2 * .5
1.414213562
```
$$2^{0.5} = 2^{1/2} = \sqrt{2}.$$

The base e

Perhaps the most important exponential function in practical applications is the exponential function with base e. It is expressed as $y = F(x) = e^x$, where e is an irrational number approximately equal to 2.718281828. We will consider some of these applications in the next section. This exponential function is so important that it is often referred to as *the* exponential function.

The exponential function in APL

The importance of the exponential function with base e is further emphasized by the fact that it is a keyboard monadic function in APL. e^x in standard notation corresponds to $*X$ in APL.

The following examples illustrate the monadic use of $*$ in APL.

Examples

```
      *1
2.718281828
```
$$e^1 = e.$$

```
      *2
7.389056099
```
$$e^2.$$

```
      *3
20.08553692
```
$$e^3.$$

```
      *1 2 3
2.718281828  7.389056099  20.08553692
```
$$e^1, e^2, e^3.$$

```
      * ⁻1
0.3678794412
```
$$e^{-1}.$$

```
      * ⁻2
0.1353352832
```
$$e^{-2}.$$

```
      * .5
1.648721271
```
$$e^{0.5} = e^{1/2} = \sqrt{e}.$$

Graph of the exponential function

    ∇ *FN*[1]

[1]   *Y*←∗*X*∇ $y = e^x$ in APL.
    X←((ι5)−6),0,ι5
    X PAIRS FN X

⁻5	.006737946999
⁻4	.01831563889
⁻3	.04978706837
⁻2	.1353352832
⁻1	.3678694412
0	1
1	2.718281828
2	7.389056099
3	20.08553692
4	54.59815003
5	148.4131591

Thus, the graph of the exponential function is as shown in Figure 7.3.

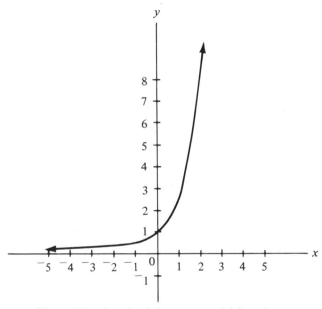

Figure 7.3   Graph of the exponential function.

EXERCISES

1. Use the program *PAIRS* to print out a table of ordered pairs for the following exponential functions with $^-5 \leqslant x \leqslant 5$, and use this table to graph the functions:
(a) $y = 4^x$
(b) $y = 3^{-x}$

(c) $y = e^{-x}$
(d) $y = 2^{x^2}$
(e) $y = \pi^x$ (Recall that in APL, π is ○1.)
(f) $y = 2^{|x|}$ (Recall that in APL, $|x|$ is $|X$.)

2. Evaluate the following with pencil and paper:
   (a) $25^{1/2}$ (e) $(1/2)^{-3}$
   (b) $16^{3/4}$ (f) $(\sqrt{2})^4$
   (c) 5^{-2} (g) $81^{1/4}$
   (d) e^0 (h) $125^{4/3}$

3. Check your answers to Exercise 2 at an APL terminal.

4. If the graph of an exponential function passes through the two points $(0,5)$ and $(1,20)$, find the function (i.e., find a and b, where $y = a \cdot b^x$).

5. Investigate the values of $(1 + (1/N))^N$, for N getting larger and larger. Some texts define e to be the limiting value of this expression as N approaches infinity.

7.2 Applications of exponential functions

In this section, we will consider some applications of exponential functions.

Application 1: Exponential growth

A quantity is growing exponentially if the amount of the quantity, y, present after x time intervals is given by a formula of the form $y = a \cdot b^x$, where a is the amount present when $x = 0$, and b is the rate of growth per time interval.

For example, suppose that the number of Japanese beetles on a golf course doubles each week during the summer. Suppose that on July 1, there are 100 Japanese beetles on the golf course. How many will there be on July 29?

$A \leftarrow 100$	When $x = 0$, there are 100 beetles present.
$B \leftarrow 2$	The rate of growth per week is 2.
$X \leftarrow 4$	There are 4 weeks between July 1 and July 29.
$Y \leftarrow A \times B * X$	$y = a \cdot b^x$ in conventional notation.
Y	Print out y.
1600	

So, there will be 1600 Japanese beetles in just 4 weeks.

Application 2: Compound interest

If an amount P, called the *principal*, is deposited in a savings bank at a yearly interest rate I for x years, and if the interest is compounded N times per year, then the total amount accumulated at the end of x years is given by

$$A = P \cdot \left(1 + \frac{I}{N}\right)^{N \cdot x}.$$

(a) Suppose that \$100 is deposited in a savings bank which has a yearly interest rate of 6 percent compounded quarterly, and is left there for 5 years. Find the amount accrued at the end of 5 years.

Since the interest rate of 6 percent is compounded quarterly, then each quarter, an interest rate of $(6/4)$ percent is given on the previous balance. Thus, the balances at the end of each quarter for the first year would be:

$$100 + 100 \cdot 0.015 = 100 \cdot (1 + 0.015)^1$$
$$\text{after 1 interest period,}$$

$$100 \cdot (1 + 0.015)^1 + 100 \cdot (1 + 0.015)^1 \cdot 0.015 = 100 \cdot (1 + 0.015)^2$$
$$\text{after 2 interest periods,}$$

$$100 \cdot (1 + 0.015)^2 + 100 \cdot (1 + 0.015)^2 \cdot 0.015 = 100 \cdot (1 + 0.015)^3$$
$$\text{after 3 interest periods,}$$

$$100 \cdot (1 + 0.015)^3 + 100 \cdot (1 + 0.015)^3 \cdot 0.015 = 100 \cdot (1 + 0.015)^4$$
$$\text{after 4 interest periods,}$$
$$\text{or one year.}$$

Continuing this process for 5 years, the total amount accumulated would be $A = 100 \cdot (1 + 0.015)^{20}$, as given by the formula. Thus, applying the formula to this problem yields the following result:

```
P←100
I←.06
N←4                          Quarterly means 4 times per year.
X←5

A←P×(1+I÷N)*N×X
A
134.6855007
```

Thus, the amount accrued is \$134.68.

(b) How much should be deposited in the above bank if one would like to have a total of $2000 at the end of 5 years? We now have the value of A, $2000 and would like to have the value of P. In a problem such as this, P is called the *present value*. Solving the above formula for P, we get the formula for present value

$$P = \frac{A}{(1+(I/N))^{N \cdot x}}.$$

$A \leftarrow 2000$

$P \leftarrow A \div (1 + I \div N) * N \times X$ Note that the values of I, N, and X have been specified previously.

P
1484.940836

Thus, $1484.94 should be deposited now if we want to have $2000 in the account at the end of 5 years.

It can be shown that as N, the number of times interest is compounded per year, gets larger and larger, the value of the quantity $(1+(I/N))^{N \cdot x}$ will get closer and closer to the quantity $e^{I \cdot x}$. This is very useful in doing problems where the interest is compounded continuously, where N increases without bound. The verification of this statement will be left as an exercise.

Application 3: Interest compounded continuously

If interest is *compounded continuously*, then the total amount in the account after x years if a principal P is deposited and the yearly interest rate is I is given by the formula

$$A = P \cdot e^{I \cdot x}.$$

Suppose $100 is deposited in a bank which has an interest rate of 6 percent compounded continuously. How much is this investment worth at the end of 10 years?

$P \leftarrow 100$
$I \leftarrow .06$
$X \leftarrow 10$

$A \leftarrow P \times (* I \times X)$
A
182.21188

Thus, the original $100 has become $182.21 in 10 years.

185

Application 4: Continuous growth

If a quantity grows continuously at a rate of K percent per time period, and if P is the amount of the quantity present after 0 time periods, then the total amount present after x time periods is given by the formula

$$A = P \cdot e^{K \cdot x}.$$

Suppose that the population of a country increases continuously at a yearly rate of 5 percent. Suppose that right now there are 1,000,000 people in the country. How many people will there be in 10 years? In 20 years?

 $P \leftarrow 1000000$

 $K \leftarrow .05$

 $X \leftarrow 10$

 $A \leftarrow P \times (*K \times X)$
 A
1648721.271

Thus, there will be 1,648,721 people in 10 years.

 $X \leftarrow 20$

 $A \leftarrow P \times (*K \times X)$
 A
2718281.828

Thus, there will be 2,718,281 people in 20 years.

Application 5: Continuous decay

If a quantity *decays* continuously at a rate of K percent per time period, and if P is the amount of the quantity present after 0 time periods (i.e., in the beginning), then the amount present after x time periods is given by the exponential function

$$A = P \cdot e^{-K \cdot x}.$$

This situation occurs in problems involving radioactive decay. For example, suppose that a radioactive substance decays at a rate of 5 percent per year. If a lump of this substance is 500 grams now, what will it be in 25 years?

 $P \leftarrow 500$
 $K \leftarrow .05$
 $X \leftarrow 25$

 $A \leftarrow P \times (* -K \times X)$
 A
143.2523984

Thus, there will be approximately 143.25 grams of this substance left from the original 500 grams in 25 years.

Further applications of exponential functions will be given in the exercises. However, the above examples should help to convince the reader of the importance of this class of functions.

EXERCISES

1. Write a program for the amount accumulated in x years if a principal P is deposited in a savings account at a bank which has an interest rate of 5 percent compounded semiannually.

2. Use your program in Exercise 1 to find the amount in the account if $400 is deposited in this bank for 15 years.

3. Repeat Exercises 1 and 2 if the bank compounds interest quarterly.

4. Repeat Exercises 1 and 2 if the bank compounds interest monthly.

5. Repeat Exercises 1 and 2 if the bank compounds interest continuously.

6. Find the amount of money that should be deposited in the bank of Exercise 1 if one would like to have $1000 in the account in 10 years.

7. If the population of a certain weed in a lawn triples every year, and if there are 50 such weeds in 1975, find the number of these weeds in the lawn in 1984, provided they are allowed to multiply.

8. If the size of a rash in a patient is cut in half every hour due to a wonder remedy, and if the rash covers 60 square inches of the patient's body at 1 o'clock, how much of the body will be covered by the rash at 5 o'clock, provided he applies the remedy as prescribed by his doctor?

9. Write a program for the continuous growth or decay function $A = P \cdot e^{K \cdot x}$.

10. If interest is compounded continuously at a rate of 5.25 percent per year,
    (a) Find the amount that will be in the account if $500 is deposited and left for 8 years.
    (b) Find the amount that should be deposited to yield a balance of $750 in 8 years.

11. Suppose that the number of bacteria in a culture increases continuously at a rate of 10 percent per hour. Suppose that at 10:00 AM there are 100 such bacteria present in the culture. Find the number of bacteria that will be present in the culture at 3 PM.

12. A radioactive substance disintegrates continuously at a rate of 8 percent per year. If there are 80 grams of this substance today, how much will there be 20 years from now?

13. An exponential function of the form $y = P + P \cdot e^{-K \cdot x}$ is often referred to as a *learning curve*, where P is the original production of the subject whose progress is being watched, K is a constant called the learning constant, x is the elapsed

187

time, and y is production of the subject at the end of x units of time. Suppose that a new employee on a production line can produce 200 pieces per day. Suppose that $K=0.1$. How many pieces should this employee produce after 10 days on the job?

14. Learning curves can also assume the form $y=P-P \cdot e^{-K \cdot x}$. Suppose the new employee scraps 25 pieces the first day. Using this learning curve, how many pieces will he scrap on the 10th day?

15. Describe the differences between the learning curves (if any) in Exercises 13 and 14.

16. In the text, it was stated that if N is large, the value of the quantity $(1+(I/N))^{N \cdot x}$ and the value of the quantity $e^{I \cdot x}$ will be approximately the same. Letting $I=0.06$, $x=5$, and $N=100$, verify this statement at an APL terminal.

7.3 Logarithmic functions

Closely related to the concept of exponential function is the concept of logarithmic function. In fact, the logarithmic function with base b is the inverse function of the exponential function with base b. Two functions which mean the same thing except that the roles of the independent and dependent variables are reversed and are called *inverse functions*.

Definition of logarithmic function

If $b>0$, but $b\neq1$, and if $x>0$, then $y=F(x)=\log_b x$ (read as "y is the logarithm to the base b of x") means the same as $x=b^y$. ($\log_1 x$ is not defined, since then we would have $x=1^y=1$, and the only value x could have would be 1. Thus, we don't allow a base of 1.)

Examples to Illustrate the Definition

$$\log_{10} 1000=3, \quad \text{since } 10^3=1000.$$
$$\log_2 32=5, \quad \text{since } 2^5=32.$$
$$\log_{10}.01=^-2, \quad \text{since } 10^{-2}=.01.$$
For any base b, $\log_b 1=0$, since $b^0=1$.
For any base b, $\log_b b=1$, since $b^1=b$.

Logarithms in APL

The APL notation for $\log_b x$ is B⊛X. [*Note:* ⊛ is an overstrike of * and ○.]

Examples

        10⊛1000
3 $\log_{10} 1000.$

2⊛32	
5	$\log_2 32.$
10⊛.01	$\log_{10} .01.$
‾2	
5⊛1	$\log_5 1.$
0	
5⊛5	$\log_5 5.$
1	
10⊛‾2	$\log_{10} {}^-2$ is not defined, since x is not
DOMAIN ERROR	> 0. There is no way to raise 10 to a power and get a negative.
10⊛0	$\log_{10} 0$ is not defined, since there is
DOMAIN ERROR	no way to raise 10 to a power and get 0.
10⊛.01 .1 10 100 1000	$\log_{10} 0.01, \log_{10} 0.1, \log_{10} 10,$
‾2 ‾1 1 2 3	$\log_{10} 100, \log_{10} 1000.$

Natural logarithms

In many practical applications of logarithms, the base is the number e. Logarithms with base e are called *natural logarithms*. $\log_e x$ is usually denoted as $\ln x$.

By our definition of logarithm with base e, $y = \ln x$ means the same as $x = e^y$.

Examples

$$\ln e = 1, \quad \text{since } e^1 = e.$$
$$\ln 1 = 0, \quad \text{since } e^0 = 1.$$

Natural logarithms in APL

The APL notation for $\ln x$ is ⊛ *X*. That is, if no base is indicated to the left of the logarithm symbol, ⊛ , the base is understood by the computer to be e.

Examples

⊛ *1	$\ln e = 1.$
1	
⊛ *2	$\ln e^2 = 2.$
2	

189

        ⊛ 1   ln 1 = 0.
0

        ⊛ 2   ln 2.
0.6931471806

        ⊛ 10   ln 10.
2.302585093

        ⊛ ⁻2   ln ⁻2 is not defined.
DOMAIN ERROR

        ⊛ 0   ln 0 is not defined.
DOMAIN ERROR

        ⊛ 1 2 10   ln 1, ln 2, ln 10
0 .6931471806 2.302585093

To get a feeling for logarithms as functions, we now graph a couple of logarithmic functions.

Example 1

Graph $y = \log_2 x$. We can use the program *PAIRS* to generate a set of ordered pairs for this function.

        ∇ FN[1]

[1]   Y←2⊛X∇
        X←ι10
        X PAIRS FN X
1   0
2   1
3   1.584962501
4   2
5   2.321928095
6   2.584962501
7   2.807354922
8   3
9   3.16992501
10 3.321928095

Let us also examine this function for values of x getting closer and closer to 0.

        X←(11−ι10)×.1
        X PAIRS FN X
1   0
0.9 ⁻.1520030934
0.8 ⁻.3219280949
0.7 ⁻.5145731728
0.6 ⁻.7369655942

0.5 ⁻1
0.4 ⁻1.321928095
0.3 ⁻1.736965594
0.2 ⁻2.321928095
0.1 ⁻3.321928095

Using these two tables of ordered pairs, the graph of $y = \log_2 x$ looks as shown in Figure 7.4.

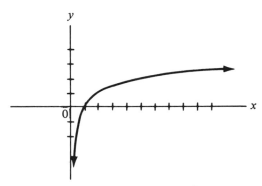

Figure 7.4   Graph of Example 1.

Example 2

Graph $y = \ln x$.

      ∇ *FN*[1]

[1]   *Y*←⊛ *X*∇   The APL natural log function.

       X←ι10
       X PAIRS FN X
1   0
2   .6931471806
3   1.098612289
4   1.386294361
5   1.609437912
6   1.791759469
7   1.945910149
8   2.079441542
9   2.197224577
10 2.302585093

Examining this function close to zero, we get:

       X←(11 − ι10)×.1
       X PAIRS FN X
1   0
0.9 ⁻.1053605157

191

0.8 ⁻.2231435513
0.7 ⁻.3566749439
0.6 ⁻.5108256238
0.5 ⁻.6931471806
0.4 ⁻.9162907319
0.3 ⁻1.203972804
0.2 ⁻1.609437912
0.1 ⁻2.302585093

Thus, based on these tables of ordered pairs, the graph of $y = \ln x$ is as shown in Figure 7.5.

These graphs are typical of the graphs of logarithmic functions.

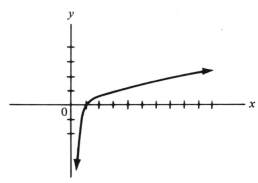

Figure 7.5   Graph of Example 2.

EXERCISES

1. Use the program *PAIRS* to generate tables of ordered pairs for the following functions where $1 \leqslant x \leqslant 10$, and where $0.1 \leqslant x \leqslant 1$. Then use these pairs to graph these functions:
   (a) $y = \log_3 x$ (b) $y = \log_{10} x$ (c) $y = \ln x^2$

2. Use the definition of $y = \log_b x$ to evaluate the following logarithms:
   (a) $\log_5 25$ (c) $\log_8 2$ (e) $\log_5 0.04$
   (b) $\log_3 81$ (d) $\log_4 4$ (f) $\ln e^{\sqrt{2}}$

3. Repeat Exercise 2 at an APL terminal.

4. Evaluate the following logarithms at an APL terminal:
   (a) $\log_{10} 2.35$ (b) $\log_{10} 235$ (c) $\log_{10} 23500$
   [Do you see any relationship among these logs?]

5. Evaluate the following logs at an APL terminal:
   (a) $\log_5 4$ (b) $\log_5 6$ (c) $\log_5 24$
   [Do you see any relationship between the answer to (c) and those of (a) and (b)?]

6. Evaluate the following natural logs at an APL terminal:
   (a) $\ln 4$ (b) $\ln 15$ (c) $\ln 60$ (d) $\ln 16$
   (e) How are (c) and (d) related to (a) and (b)?

192

7.4 Properties and applications of logarithms

Much of the usefulness of logarithms can be attributed to the following three properties.

Properties of logarithms

If M and N are positive real numbers, then

1. $\log_b(M \cdot N) = \log_b M + \log_b N$
2. $\log_b(M/N) = \log_b M - \log_b N$
3. $\log_b(M^k) = k \cdot \log_b M$

To illustrate the way in which these rules are derived, we shall prove Rule 2. The proofs of the other two rules are done similarly and are left as exercises.

Proof of Property 2:

Let $x = \log_b M$ and $y = \log_b N$. Then, by the definition of log to the base b, $M = b^x$ and $N = b^y$. We are interested in M/N. Thus,

$$\frac{M}{N} = \frac{b^x}{b^y} = b^{x-y},$$

since exponents are subtracted when you divide. Using the definition of logarithm to the base b on this expression, we get

$$x - y = \log_b\left(\frac{M}{N}\right).$$

However,

$$x - y = \log_b M - \log_b N.$$

Therefore,

$$\log_b\left(\frac{M}{N}\right) = \log_b M - \log_b N.$$

Application 1: Computations with logs

A decade or more ago, before the invention of the inexpensive pocket calculator or the readily accessible computer, one of the main applications of logarithms was to simplify arithmetical computations. Since our number system is based on base 10, logarithms to base 10, called *common logs*, were used.[1]

[1] John Napier (1550–1617) actually invented logarithms for this purpose. He also invented an early form of the slide rule, which is based upon logarithms.

Example 1: Multiplication

Property 1 can be used to convert a multiplication problem to an addition problem. Consider the problem of multiplying 32.4 by 41.8.

Let $M = 32.4$ and $N = 41.8$. We want to compute $M \cdot N$. By Property 1, $\log_{10}(M \cdot N) = \log_{10} M + \log_{10} N$. Let us use APL to find and add these logs. Actually logs to any base can be used. However, for historical reasons, we will use logs to base 10.

    M←32.4

    N←41.8

    10⊛M $\log_{10} M.$
1.51054501

    10⊛N $\log_{10} N.$
1.621176282

    (10⊛M)+(10⊛N) $\log_{10} M + \log_{10} N.$
3.131721292

Thus, we have now computed $\log_{10}(M \cdot N)$. However, we really want $M \cdot N$. Thus, we want the number whose log to the base 10 is 3.131721292. In other words, we want $10^{3.131721292}$. (This is often referred to as the *antilog* of 3.131721292.) Using APL, we get the following:

    10∗3.131721292
1354.32

Thus, $M \cdot N = 1354.32$. Let us check this using APL:

    M×N
1354.32   It checks!

This process was particularly useful when one wanted to multiply several numbers. It is much easier to add several logarithms and then compute the antilog than to multiply the numbers. However, if one has a calculator or a computer terminal handy, it is even easier to just multiply the numbers directly. Therefore, today, we would probably not use the above process to multiply numbers. We are presenting this process because it has historical interest and because it illustrates an application of the properties of logarithms. For these same reasons, let us also illustrate the way in which to use the rules of logs to simplify division, exponentiation, and extraction of roots.

Example 2: Division (optional)

Property 2 can be used to convert a division problem to a subtraction problem. Consider the problem of dividing M by N. By Property 2,

$$\log_{10}\left(\frac{M}{N}\right) = \log_{10} M - \log_{10} N.$$

Using APL,

    M

32.4

    N

41.8   Just to recall the values of *M* and *N*.

    (10⊛*M*)−(10⊛*N*) $\log_{10} M - \log_{10} N$.
¯.1106312716

This is the value of $\log_{10}(M/N)$. To get M/N, we need the number with this log. Thus,

    10∗¯.1106312716 $10^{-0.1106312716} = M/N$, since

                                  $\log_b(M/N) = {}^{-}0.1106312716$.

.7751196172   Thus, $M/N = 0.7751196172$.

    M ÷ *N*

.7751196172   It checks.

Example 3: Exponentiation (optional)

Property 3 can be used to convert a problem in raising a number to a power to a problem in multiplication. Consider the problem of computing $M^{0.4}$. By Property 3, $\log_{10}(M^{0.4}) = 0.4 \cdot \log_{10} M$. Using APL,

    .4×(10⊛*M*) $0.4 \cdot \log_{10} M$.
.6042180041

This is the value of $\log_{10}(M^{0.4})$. To get $M^{0.4}$, we need the antilog.

    10∗0.6042180041

4.019925496   Thus, $M^{0.4} = 4.019925496$.

    M ∗ .4

4.019925496   It checks.

Example 4: Root extraction (optional)

Since $M^{1/k} = \sqrt[k]{M}$, then the problem of extracting a root can be handled in the same way as the problem of exponentiation. Let us use logs and APL to compute $\sqrt[3]{M}$. We use the fact that $\sqrt[3]{M} = M^{1/3}$.

    M

32.4   Recalling the value of *M*.

    (1÷3)×(10⊛*M*) $\frac{1}{3} \cdot \log_{10} M = \log_{10}(M^{1/3})$.
.5035150034

    10∗0.5035150034   Getting the antilog.

3.187975708   Thus, $\sqrt[3]{M} = 3.187975708$.

    M ∗ 1÷3

3.187975708   It checks.

Application 2: Solving exponential equations

One very important application of logarithms is in solving exponential equations. An *exponential equation* is an equation of the form $a^x = b$, where a and b are constant known values. To solve it means to find the value of x. To do this, we could use logarithms to any base. We will use natural logarithms. The procedure is as follows:

$$a^x = b.$$

Take the natural log of both sides. Thus,

$$\ln a^x = \ln b.$$

Using Property 3,

$$x \cdot \ln a = \ln b, \quad \text{or} \quad x = \frac{\ln b}{\ln a}.$$

Example

Solve $3^x = 7$ for x. By the above discussion, $x = \ln 7 / \ln 3$. In APL,

```
X←(⍟7)÷(⍟3)
X
1.771243749
```

So, $x = 1.771243749$.

```
3*X
7
```

Thus, $3^x = 7$, and it checks.

Application 3: Radioactive half-life

Suppose that the amount of a radioactive substance present after x years is given by the exponential function

$$A = 500 \cdot e^{-0.05 \cdot x}.$$

The *half-life* of this substance is the amount of time, T, required for exactly half of this substance to disintegrate. The procedure for finding T is as follows:

$$A = 500 \cdot e^{-0.05 \cdot x}$$

At time $x = 0$, $A = 500$. To find the half-life, we need to find the value of x, which we are calling T, for which $A = 250$ (half of the initial amount). In other words, we need to solve the equation

$$250 = 500 \cdot e^{-0.05 \cdot T}$$

for T. Thus,

$$\frac{250}{500} = e^{-0.05 \cdot T}, \quad \text{or} \quad .5 = e^{-0.05 \cdot T}.$$

Taking the natural log of both sides yields

$$\ln 0.5 = \ln(e^{-0.05 \cdot T}) = -0.05 \cdot T \cdot \ln e \quad \text{(By property 3)}$$
$$= -0.05 \cdot T \quad \text{(Since } \ln e = 1\text{)}.$$

196

Thus,

$$T = \frac{\ln 0.5}{-0.05}.$$

Using APL, we get

   $T \leftarrow (\circledast.5) \div {}^-.05$
   T
13.86294361   Thus, the half-life of the substance is
                                     almost 14 years.

   $500 \times ({*}^-.05 \times T)$
250                                  $500 \cdot e^{-0.05 \cdot T} = 250$, so it checks.

Application 4: Continuous compound interest revisited

In Section 7.2., we pointed out that if interest is compounded continuously
for x years at a yearly interest rate I, then the amount, A, in the account
after x years if a principal P is deposited is given by the exponential
function

$$A = P \cdot e^{I \cdot x}.$$

At a 6 percent interest rate compounded continuously, how long does it
take for a deposit to double its initial value?
   Thus, we have

$$2 \cdot P = P \cdot e^{I \cdot x} = P \cdot e^{0.06 \cdot x}, \quad \text{or} \quad 2 = e^{0.06 \cdot x}.$$

Applying ln to both sides yields

$$\ln 2 = \ln e^{0.06 \cdot x} = 0.06 x \cdot \ln e = 0.06 \cdot x.$$

Thus, $x = (\ln 2)/0.06$. Solving this problem in APL, we get

   $X \leftarrow (\circledast 2) \div .06$
   X
11.55245301   Thus, the deposit doubles its value in
                                     about 11.55 years.

   ${*}.06 \times X$ $e^{0.06 \cdot x} = 2$, so it checks.
2

EXERCISES

1. Prove Properties 1 and 3 of logarithms.

2. Solve the exponential equation $5^x = 9$ for x.

3. Write an APL program to solve an exponential equation $A^X = B$ for x.

197

4. Evaluate $\ln 5$ and $\ln 8$ on an APL terminal. Use the properties of logarithms and the values you got for $\ln 5$ and $\ln 8$ to evaluate the following logarithms:
(a) $\ln 40$ (b) $\ln 1.6$ (c) $\ln 25$ (d) $\ln 200$ (e) $\ln \sqrt{5}$

5. Use common logs and the properties of logarithms to compute the following $(M = 58.6,\ N = 2.79,\ P = 341)$:
(a) $M \cdot N \cdot P$ (b) P/M (c) N^5 (d) $\sqrt{P}$

6. If the amount of a radioactive substance present after x years is given by $A = 80 \cdot e^{-0.08 \cdot x}$, find the half-life of the substance.

7. If a bank compounds interest continuously at a rate of 5 percent, how long does it take for the deposit to double in value?

8. In Exercise 7, how long does it take for the deposit to triple in value?

9. If the revenue from the sales of x items is given by the formula $y = 1000 \cdot \ln(x+1)$ dollars, find the revenue from the sale of 25 of these items.

10. Prove the following statement about logarithms:
$$\log_b(b^x) = x.$$

Differential calculus 8

Calculus—an essential tool of any mathematician, engineer, or scientist—is one of the most important branches of modern mathematics. In recent years, calculus has also become an important tool in such areas as business administration, economics, psychology, and sociology. In all of these areas, we are interested in instantaneous rates of change, and calculus is the tool for finding such rates of change. In an introductory text such as this, we cannot attempt to present a thorough coverage of calculus. In fact, in most colleges, calculus is offered in a 3- to 5-course sequence. It is our intention to introduce the student only to some of the more important concepts and applications of calculus. It is hoped that this will give the student some appreciation of this vital area of mathematics. Perhaps it might even inspire some students enough to take part or all of the calculus sequence.

There are two branches to calculus: differential and integral calculus. Using differential calculus, we can answer such questions as "How fast is an object moving at any instant?" and "How many items should be produced in order to maximize profit or minimize cost?" Using integral calculus, we can answer such questions as "Given the acceleration of a moving object at any time t, what is its velocity at time t?" and "What is the area under the normal probability curve between two specified values?" As we shall see, these two branches of calculus are very closely related to each other via the "Fundamental theorem of calculus." First, we will treat differential calculus.

8.1 The limit of a function

We begin our study of calculus by considering the idea of limit.

Definition of the limit of a function

Let $F(x)$ be a function and let c and L be real numbers. The symbolism

$$\lim_{x \to c} F(x) = L \quad \text{(read "the limit as } x \text{ approaches } c \text{ of } F(x) \text{ is } L\text{")}$$

means that as x gets closer and closer to c, $F(x)$ gets closer and closer to L. Or, in other words, we can make $F(x)$ as close to L as we desire by making x close enough to c. If no such real number L exists, we say that the limit does not exist.[1] [*Note*: In this definition, we are considering the values of $F(x)$ for x "close to" c. We are not considering the value of $F(x)$ at c, although in some instances, the value of the limit L will be $F(c)$. There are other instances where $F(c)$ does not exist, but the limit L does exist. In fact, the derivative is one of these instances.]

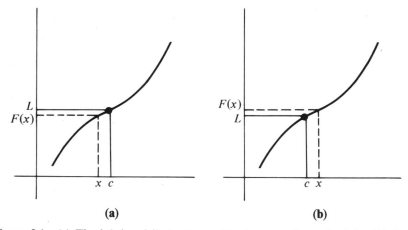

Figure 8.1   (a) The left-hand limit: As x gets close to c from the left, $F(x)$ gets close to L. (b) The right-hand limit: As x gets close to c from the right, $F(x)$ gets close to L.

One way of evaluating $\lim_{x \to c} F(x)$ is to examine the values of $F(x)$ for values of x getting closer and closer to c. We will examine $F(x)$ as x gets close to c from the left (the left-hand limit; Figure 8.1a), and as x gets close to c from the right (the right-hand limit; Figure 8.1b). These limits should be the same. Otherwise, we will say that the limit does not exist. In order to find these left-hand and right-hand limits, we can use the program *FN* as in the following examples.

---

[1]Our definition of limit is not the one given in most calculus texts. The words "close" and "closer" are usually made more precise using the so called "δ, ε" definition. However, the more intuitive definition will suffice for our purposes.

Example 1

Find $\lim_{x \to 2}(3x^2)$.

　　　∇FN [1]

[1]　　$Y \leftarrow 3 \times X * 2 \ \nabla$　　　　　　　　　　Altering *FN* to fit this function.

　　　FN 1.9 1.99 1.999 1.9999　　Four values as x gets closer and
　　　　　　　　　　　　　　　　closer to 2 from the left.

10.83 11.8803 11.98803 11.99880003

　　　FN 2.1 2.01 2.001 2.0001　　Four values as x gets closer and
　　　　　　　　　　　　　　　　closer to 2 from the right.

13.23 12.1203 12.012003 12.00120003

　　From these results, it can be seen that as x gets closer and closer to 2, from the left as well as from the right, $3x^2$ gets closer and closer to 12. Thus

$$\lim_{x \to 2}(3x^2) = 12.$$

Example 2

Find

$$\lim_{x \to 3}\left(\frac{2x+1}{x^2}\right).$$

　　　∇FN [1]

[1]　　$Y \leftarrow ((2 \times X)+1) \div (X * 2) \ \nabla$　　The new *FN*.

　　　FN 2.9 2.99 2.999 2.9999　　The left-hand limit.
.8085612366 .7807518932 .7780741852 .7778074085

　　　FN 3.1 3.01 3.001 3.0001　　The right-hand limit.
.749219563 .7748258849 .7774815926 .7777481493

　　From these results, we get

$$\lim_{x \to 3}\left(\frac{2x+1}{x^2}\right) = 0.778 = \frac{7}{9}.$$

　　Notice that in the two examples above, $\lim_{x \to c} F(x) = F(c)$. Functions with this property are said to be continuous at $x = c$. Not all functions are continuous at $x = c$. The following example illustrates this.

Example 3

Find

$$\lim_{x \to 1}\left(\frac{x^2-1}{x-1}\right).$$

201

In this function, $F(1)=0/0$, which is not defined. Thus, $\lim_{x\to c}F(x)\neq F(c)$. However, we can still evaluate this limit as follows:

∇FN [1]

[1] $Y\leftarrow((X*2)-1)\div(X-1)$ ∇

FN .9 .99 .999 .9999 .99999   The left-hand limit.
1.9 1.99 1.999 1.9999 1.99999

FN 1.1 1.01 1.001 1.0001 1.00001

The right-hand limit.

2.1 2.01 2.001 2.0001 2.00001

Thus,

$$\lim_{x\to1}\left(\frac{x^2-1}{x-1}\right)=2.$$

This example can also be solved by using a little algebra. Since

$$\frac{x^2-1}{x-1}=\frac{(x-1)(x+1)}{x-1}=x+1,$$

then

$$\lim_{x\to1}(x+1)=2.$$

Example 4

Find

$$\lim_{x\to1}\left(\frac{5}{x-1}\right).$$

∇FN [1]

[1] $Y\leftarrow5\div(X-1)$ ∇

FN .9 .99 .999 .9999   The left-hand limit.
⁻50. ⁻500. ⁻5000. ⁻50000.

FN 1.1 1.01 1.001 1.0001   The right-hand limit.
50. 500. 5000. 50000.

From these results, it seems that as x gets closer and closer to 1 from the left, $F(x)$ gets more and more negative, and as x gets closer and closer to 1 from the right, $F(x)$ gets larger and larger, with no apparent upper value. Thus, the limit in this example does not exist.

Example 5

Find

$$\lim_{h \to 0}\left(\frac{(2+h)^2-4}{h}\right).$$

(Note that the letter used in this limit problem is of no significance, since the answer is a number.

∇FN [1]

[1] $Y \leftarrow (((2+X)*2)-4) \div X \ \nabla$

FN ‾.1 ‾.01 ‾.001 ‾.0001 ‾.00001

The left-hand limit.

3.9 3.99 3.999 3.9999 3.999990003

FN .1 .01 .001 .0001 .00001

The right-hand limit.

4.1 4.01 4.001 4.0001 4.000009994

Thus, it appears that the limit is 4. This can also be done algebraically as follows:

$$\lim_{h \to 0}\frac{(2+h)^2-4}{h} = \lim_{h \to 0}\frac{4+4h+h^2-4}{h} = \lim_{h \to 0}\frac{h\cdot(4+h)}{h}$$
$$= \lim_{h \to 0}(4+h)=4.$$

Example 6

Find $\lim_{x \to 0}(1+x)^{1/x}$.

∇FN [1]

[1] $Y \leftarrow (1+X)*(1 \div X) \ \nabla$

FN ‾.1 ‾.01 ‾.001 ‾.0001 ‾.00001

The left-hand limit.

2.867971991 2.731999026 2.719642216 2.718417755 2.718295421

FN .1 .01 .001 .0001 .00001 The right-hand limit.

2.59374246 2.704813829 2.716923932 2.718145927 2.718268236

From these results, it appears that $\lim_{x \to 0}(1+x)^{1/x}$ is about 2.718. Actually, the value of this limit is usually taken to be the definition of the number *e*.

A limit program

In all of these examples, if one were to average the left-hand limit, given approximately by $F(c- 0.0001)$, and the right-hand limit, given approximately by $F(c+0.0001)$, he would have a good estimate of $\lim_{x \to c} F(x)$, provided this limit exists. The limit does not exist if the left-hand and right-hand limits differ substantially, as in Example 4.

Program 8.1   LIMIT

∇ *ESTIMATE* ← *LIMIT C*; *LEFT* ; *RIGHT*

[1]	*LEFT* ← *FN C* − .0001	The approximate left-hand limit.
[2]	*RIGHT* ← *FN C* + .0001	The approximate right-hand limit.
[3]	*ESTIMATE* ← (*LEFT* + *RIGHT*) ÷ 2	

The average of *LEFT* and *RIGHT*.

[4] → ((| *LEFT* − *RIGHT*) < .01)/0 If *LEFT* and *RIGHT* are at least as close as 0.01, then the limit will be assumed to exist. Thus, the program is ended with the *ESTIMATE* of the limit as on line 3. Otherwise, the program prints out the message on line 5.

[5] *ESTIMATE* ← ' THE LIMIT DOES NOT EXIST'

The values 0.0001 and 0.01 used in this program are arbitrary. The reader should feel free to experiment with other values.

∇

Let us redo the previous examples using this program *LIMIT*.

Example 1

Find $\lim_{x \to 2}(3x^2)$.

∇ *FN* [1]

[1]   *Y* ← 3 × *X* * 2 ∇

LIMIT 2

12

This answer, 12, is the average of 11.9988 and 12.0012, the approximations of the left and right-hand limits given in the previous solution to Example 1.

Example 2

Find

$$\lim_{x \to 3} \left(\frac{2x+1}{x^2} \right).$$

∇ *FN* [1]

[1]   *Y*←((2×*X*)+1)÷(*X*∗2) ∇

LIMIT 3
.7777777789

Notice that this estimate 0.77777777789 is closer to the real answer of 7/9 than either of the estimates of the left- and right-hand limits.

Example 3

Find

$$\lim_{x \to 1} \left(\frac{x^2-1}{x-1} \right).$$

∇ *FN* [1]

[1]   *Y*←((*X*∗2)−1)÷(*X*−1) ∇

LIMIT 1
2

The exact value of the limit.

Example 4

Find

$$\lim_{x \to 1} \left(\frac{5}{x-1} \right).$$

∇ *FN* [1]

[1]   *Y*←5÷(*X*−1) ∇

LIMIT 1
THE LIMIT DOES NOT EXIST

Since the left-hand estimate, ⁻50000, and the right-hand estimate, 50000, differ by more than 0.01.

Example 5

Find

$$\lim_{h \to 0} \left(\frac{(2+h)^2-4}{h} \right).$$

205

∇FN [1]

[1] $Y \leftarrow (((2+X)*2)-4) \div X \ \nabla$

$LIMIT$ 0

4

Example 6

Find $\lim_{x \to 0} (1+X)^{1/x}$.

∇FN [1]

[1] $Y \leftarrow (1+X)*(1 \div X) \ \nabla$

$LIMIT$ 0

2.718281841   The approximate value of e.

EXERCISES

1. Use the program *FN* and the computer to find the left- and right-hand limits to the following functions.

(a) $\lim_{x \to 1} (3x+2)$

(b) $\lim_{x \to 1} \left(\dfrac{2x^2-3}{3x+2} \right)$

(c) $\lim_{x \to 3} \left(\dfrac{x^2-9}{x-3} \right)$

(d) $\lim_{x \to 3} \left(\dfrac{x^2-2x-3}{x-3} \right)$

(e) $\lim_{x \to 1} \left(\dfrac{2}{x-1} \right)$

(f) $\lim_{x \to 1} \left(\dfrac{x^2+1}{x-1} \right)$

(g) $\lim_{x \to 0} \left(\dfrac{(1+x)^2-1}{x} \right)$

(h) $\lim_{x \to 0} (e^x)$

2. Use the program *LIMIT* to find the limits in Exercise 1, if these limits exist.

8.2 Slope of a curve and the definition of derivative at a point

In Section 6.3., we considered the concept of the slope of a line. Recall that the slope of a line joining two points $P(x_1,y_1)$ and $Q(x_2,y_2)$ is found by computing the difference quotient $m=(y_2-y_1)/(x_2-x_1)$. This slope is a measure of the steepness of the line. That is, the larger m is in absolute value, the steeper the line. If $m>0$, the line is a rising line. If $m<0$, the line is a falling line. If $m=0$, the line is horizontal. Finally, recall that the slope of a line is a constant. That is, the result is the same no matter which two points are used to compute the slope.

In this section, we would like to consider the idea of the slope of the curve given by any function $y=F(x)$ at a point $P(x,y)$ on it. Unlike straight lines, the slope of a curve will be different at each point on the curve.

206

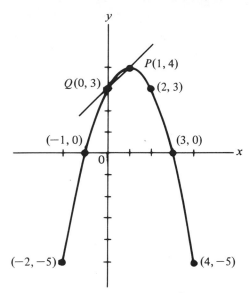

Figure 8.2   Graph of $y = -x^2 + 2x + 3$.

Example

Consider the quadratic function $y = F(x) = -x^2 + 2x + 3$ (Example 2, Section 6.4.) Let us try to define and compute the slope of this curve (shown in Figure 8.2) *at* the point $P(1,4)$.

Consider the line joining the point $Q(0,3)$ to the point $P(1,4)$. This is called a *secant line* of the curve. The slope of this line is $m = (4-3)/(1-0) = 1$.

Now, consider a point Q closer to P on the curve, say $Q(0.5,3.75)$. The slope of the secant line joining this point Q to P is $m = (4-3.75)/(1-0.5) = .5$.

Considering a point $Q(0.9,3.99)$ even closer to P, the slope of this secant line is $m = (4-3.99)/(1-0.9) = 0.1$.

In the table below, we have listed points Q which are approaching $P(1,4)$ from both the left and right, together with the slopes of the secant lines QP (see Figure 8.3).

x	y	m
0	3	1
0.5	3.75	0.5
0.9	3.99	0.1
0.99	3.999	0.01
1.01	3.999	⁻0.01
1.1	3.99	⁻0.1
1.5	3.75	⁻0.5
2	3	1

207

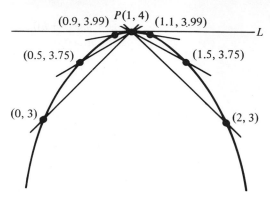

Figure 8.3   Possible secant lines PQ.

Notice that as Q gets closer and closer to P from the left, the slopes of the secant lines QP get closer and closer to 0. Also, as Q gets closer and closer to P from the right, the slopes of the secant lines QP get closer and closer to 0. Thus, it would seem reasonable to say that 0 is the slope of the curve at $P(1,4)$. Also, from the diagram, we can see that as Q approaches P along the curve, from the left or the right, the secant lines QP rotate into a limiting line L, called the *tangent line* to the curve *at P*.

Using this example as a model, we can make the following definition:

Definition of the slope of a curve at a point on it (the derivative)

The slope of a curve $y = F(x)$ at a point $P(x,y)$ on it is defined to be

$$\lim_{h \to 0} \frac{F(x+h) - F(x)}{h}.$$

This limit, if it exists, is called the *derivative of the function $y = F(x)$ at the point $P(x,y)$*, and is symbolized by $F'(x)$.

Note 1:

The derivative will exist for all examples in this text.

Note 2:

Other symbols for the derivative include $y', dy/dx, D_x F(x)$.

Geometric explanation of the derivative

The slope of the secant line QP is

$$m_{QP} = \frac{\text{Change in } y}{\text{Change in } x} = \frac{F(x+h) - F(x)}{h}.$$

As illustrated in Figure 8.4, the slope of the curve $y = F(x)$ at the point P is the limiting value of the slopes of the secant lines QP as Q approaches P along the curve. As Q approaches P along the curve, h approaches 0 and

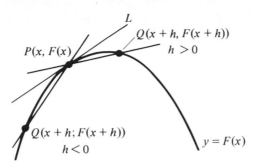

Figure 8.4 Possible secant lines QP as Q approaches P.

vice versa. Thus, the slope of the curve $y = F(x)$ at the point $P(x,y)$ is given by the limit of the difference quotient

$$\lim_{h \to 0} \frac{F(x+h) - F(x)}{h},$$

which is the *derivative* $F'(x)$ at P. Also, as Q approaches P along the curve, these secant lines rotate into the limiting line L, called the *tangent line* to the curve $y = F(x)$ at the point $P(x,y)$.

Examples

1. Find the slope of the curve $y = F(x) = -x^2 + 2x + 3$ at the point $P(2,3)$.
   We need to compute

   $$F'(2) = \lim_{h \to 0} \frac{F(2+h) - F(2)}{h}.$$

   Algebraically, this is done as follows:

   $$F'(2) = \lim_{h \to 0} \frac{-(2+h)^2 + 2(2+h) + 3 - (-2^2 + 2 \cdot 2 + 3)}{h}$$

   $$= \lim_{h \to 0} \frac{-(4 + 4h + h^2) + 4 + 2h + 3 + 4 - 4 - 3}{h}$$

   $$= \lim_{h \to 0} \frac{-2h - h^2}{h} = \lim_{h \to 0} \frac{h \cdot (-2 - h)}{h} = \lim_{h \to 0} (-2 - h) = {}^-2.$$

   So, the slope of this curve at the point $P(2,3)$ is ${}^-2$.

2. Find the slope of the curve $y = F(x) = -x^2 + 2x + 3$ at the point $(0,3)$.
   We need

   $$F'(0) = \lim_{h \to 0} \frac{F(0+h) - F(0)}{h} = \lim_{h \to 0} \frac{-h^2 + 2h + 3 - 3}{h}$$

   $$= \lim_{h \to 0} \frac{h \cdot (-h + 2)}{h} = \lim_{h \to 0} (-h + 2) = 2.$$

   The desired slope is 2.

8 Differential calculus

We will find it useful to have a program for computing derivatives at a point which yields a specific result D. The following program computes the difference quotient $(F(x+h)-F(x))/h$ for $h=0.0000000001$. This gives a fairly good estimate of $F'(x)$ in most cases.

Program 8.2 DERIVATIVE

    ∇D←DERIVATIVE X; H

[1]   H←.0000000001   This value for H is arbitrary. The
[2]   D←((FN X+H)−(FN X))÷H   reader is urged to experiment with
      ∇   other values for H. If H is too small,
                                               the computer *will always give* 1, *since*
                                               $(0÷0)=1$ in APL.

In the next section, we will consider some rules for computing derivatives. Let us apply the program *DERIVATIVE* to some examples.

Example 1

Given the function $y=F(x)=-x^2+2x+3$, find the derivatives $F'(0)$, $F'(1)$, and $F'(2)$.

    ∇FN [1]

[1]   Y←(−X*2)+(2×X)+3 ∇   We need a subprogram for our function.

      DERIVATIVE 0
1.999999999   The real answer is 2.

      DERIVATIVE 1
0

      DERIVATIVE 2
⁻2.0000000001   The real answer is ⁻2.

Example 2

Find the equation of the line tangent to the curve $y=F(x)=-x^2+2x+3$ at the point $(0,3)$.

The equation of a line is of the form $y=m\cdot x+b$. The slope of this line is $F'(0)=2$. So, the line looks like $y=2x+b$. However, when $x=0$, $y=3$, so that $b=3$. Therefore, the equation of the tangent line is $y=2x+3$.

Example 3

Given the function $y=F(x)=3x+2$, find the derivatives $F'(-1)$, $F'(0)$, $F'(1)$, and $F'(0)$.

    ∇FN [1]

[1]   Y←(3×X)+2 ∇   Altering *FN* for the new function.

210

DERIVATIVE ⁻1 0 1 3
3 3 3 3

In APL, we can do all of these derivatives at once.

Notice that the result is always 3. This is reasonable, since this function is a linear function with slope $m = 3$, and the derivative at any point gives the slope of the curve of the function at that point.

EXERCISES

1. Consider the function $y = F(x) = x^2 - 4x + 4$ and the point $P(1,1)$.
   (a) Find the slopes of the secant lines joining the points Q with the following x coordinates to the point P:
   $$x = 0; \ x = 0.5; \ x = 0.9; \ x = 0.99; \ x = 0.999.$$
   (b) Repeat Part (a) with the following x coordinates:
   $$x = 2; \ x = 1.5; \ x = 1.1; \ x = 1.01; \ x = 1.001.$$
   (c) Based on the results you get in Parts (a) and (b), what is a good estimate of the slope of this curve at the point P?
   (d) Use the program DERIVATIVE to find the slope of this curve at P.

2. Find the slopes of the following curves at the indicated points:
   (a) $y = F(x) = -x^2 + 6x - 9$ at the point $(2, ⁻1)$. (Use the computer.)
   (b) $y = F(x) = 3x^2 + 6x + 5$ at the point $(⁻1, 2)$.
   (c) $y = F(x) = 5x + 1$ at the point $(1, 6)$.
   (d) $y = F(x) = 2x^3 + 3x^2 - 12x - 10$ at the point $(2, ⁻6)$.

3. Find the equations of the lines tangent to the curves in Exercise 2 at the points indicated.

4. Using the definition of derivative and the necessary algebra, find the indicated derivatives of the following functions:
   (a) $y = F(x) = 2x + 5$; find $F'(1)$.
   (b) $y = F(x) = 2x^2 - x - 3$; find $F'(2)$.
   (c) $y = F(x) = x^3 + 2x - 5$; find $F'(0)$.

5. Use the program DERIVATIVE to check your answers to Exercise 4.

8.3 Differentiating polynomials

The derivative function

If x is a variable, then the *derivative of a function* $y = F(x)$ is a new function symbolized and defined as follows:

$$y' = F'(x) = \frac{dy}{dx} = \lim_{h \to 0} \frac{F(x+h) - F(x)}{h}.$$

The process of computing this derivative function is known as *differentiation*. The function $y = F(x)$ is said to be *differentiable* at any value of x for which this derivative exists.

In the previous section, we saw that for any particular point $P(x,y)$ on the graph of $y = F(x)$, the derivative $F'(x)$ can be used to find the slope of the curve at the point P. We will consider some more applications of derivatives in the next section. However, first we will consider some rules which will enable us to find the derivative functions for polynomials.

Rule 1

If $y = F(x) = k$, where k is a constant, then $F'(x) = 0$ (i.e., the derivative of any constant is 0).

This rule should be obvious, since the graph of $y = k$ is a horizontal line and the slope of a horizontal line is 0. Using the definition of derivative, this rule can be proven as follows:

$$F'(x) = \lim_{h \to 0} \frac{F(x+h) - F(x)}{h} = \lim_{h \to 0} \frac{k - k}{h} = \lim_{h \to 0} 0 = 0.$$

Examples

1. If $y = F(x) = 5$, then $F'(x) = 0$.
2. If $y = F(x) = \pi$, then $F'(x) = 0$.

Rule 2

If $y = F(x) = x^n$, where n is any constant real number, then $F'(x) = n \cdot x^{n-1}$.

A general proof of this rule is beyond the scope of this text. However, we shall verify this rule for the following particular case to help convince the reader of the validity of this rule:

Let $y = F(x) = x^3$.

$$F'(x) = \lim_{h \to 0} \frac{F(x+h) - F(x)}{h} = \lim_{h \to 0} \frac{(x+h)^3 - x^3}{h}$$

$$= \lim_{h \to 0} \frac{x^3 + 3x^2 \cdot h + 3x \cdot h^2 + h^3 - x^3}{h}$$

$$= \lim_{h \to 0} \frac{h \cdot (3x^2 + 3x \cdot h + h^2)}{h} = \lim_{h \to 0} (3x^2 + 3x \cdot h + h^2)$$

$$= 3x^2 + 3x \cdot 0 + 0^2 = 3x^2.$$

Thus, this rule works in this particular case.

Examples

1. If $y = F(x) = x^5$, then $F'(x) = 5x^4$.
2. If $y = F(x) = x = x^1$, then $F'(x) = 1 \cdot x^0 = 1$.
3. If $y = F(x) = x^\pi$, then $F'(x) = \pi \cdot x^{\pi - 1}$.
4. If $y = F(x) = \sqrt{x} = x^{1/2}$, then $F'(x) = \frac{1}{2} \cdot x^{-1/2} = 1/2\sqrt{x}$.
5. If $y = F(x) = 1/x^3 = x^{-3}$, then $F'(x) = -3 \cdot x^{-4} = -3/x^4$.

Rule 3

If $y = F(x) = k \cdot G(x)$, where k is constant and $G(x)$ is differentiable, then $F'(x) = k \cdot G'(x)$.

This is derived as follows:

$$F'(x) = \lim_{h \to 0} \frac{F(x+h) - F(x)}{h}$$

$$= \lim_{h \to 0} \frac{k \cdot G(x+h) - k \cdot G(x)}{h}$$

$$= k \cdot \lim_{h \to 0} \frac{G(x+h) - G(x)}{h}$$

$$= k \cdot G'(x).$$

Examples

1. If $y = F(x) = 3x$, then $F'(x) = 3 \cdot 1 = 3$.
2. If $y = F(x) = 2x^3$, then $F'(x) = 2 \cdot 3x^2 = 6x^2$.
3. If $y = F(x) = 4\sqrt{x} = 4x^{1/2}$, then $F'(x) = 4 \cdot \frac{1}{2} \cdot x^{-1/2} = 2/\sqrt{x}$.

Rule 4

If $y = F(x) = G(x) + H(x)$, where $G(x)$ and $H(x)$ are differentiable, then $F'(x) = G'(x) + H'(x)$. Also, if $F(x) = G(x) - H(x)$, then $F'(x) = G'(x) - H'(x)$.

This is proven as follows:

$$F'(x) = \lim_{h \to 0} \frac{F(x+h) - F(x)}{h}$$

$$= \lim_{h \to 0} \frac{(G(x+h) + H(x+h)) - (G(x) + H(x))}{h}$$

$$= \lim_{h \to 0} \left[\frac{G(x+h) - G(x)}{h} + \frac{H(x+h) - H(x)}{h} \right] = G'(x) + H'(x).$$

Using these four rules, we can now differentiate any polynomial.

Example 1

Let $y = F(x) = 2x^3 + 3x^2 - 12x - 10$. Find the slope of the tangent line to the graph of this function at the point $P(2, ^-6)$.

The derivative function is

$$F'(x) = 6x^2 + 6x - 12 - 0$$

$$= 6x^2 + 6x - 12.$$

We need to evaluate the derivative of this function at the point P. Thus, we get $F'(2) = 6 \cdot 2^2 + 6 \cdot 2 - 12 = 24$.

Let us use the program *DERIVATIVE* as a check.

∇FN [1]

[1] $Y \leftarrow (2 \times X * 3) + (3 \times X * 2) + (^{-}12 \times X) - 10$ ∇

DERIVATIVE 2

23.9999999999   This is almost 24.

Example 2

Find the equation of the line tangent to the curve $y = F(x) = -x^2 + 2x + 3$ at the point $P(2,3)$.

The derivative of this function is the function $y' = F'(x) = ^{-}2x + 2$. At the point P, $F'(2) = ^{-}2 \cdot 2 + 2 = ^{-}2$. Thus, the slope of this tangent line is $^{-}2$. The equation of a line has the form $y = m \cdot x + b$. Since $m = ^{-}2$, we have $y = ^{-}2x + b$. Since $(2,3)$ is a point on this tangent line, we have $3 = ^{-}2 \cdot 2 + b$, or $b = 7$. Thus, the equation of this tangent line is $y = ^{-}2x + 7$.

Example 3

Find the slope of the curve $y = F(x) = 4\sqrt{x}$ at the point $(4, 8)$.

$F'(x) = 4 \cdot \frac{1}{2} \cdot x^{-1/2} = 2/\sqrt{x}$. Thus, the desired slope is $F'(4) = 2/\sqrt{4} = 1$. We can use the program *DERIVATIVE* to check this answer as follows:

∇FN [1]

[1] $Y \leftarrow 4 \times X * .5$ ∇ Altering *FN* to fit $y = 4\sqrt{x} = 4x^{1/2}$.

DERIVATIVE 4

1   It checks.

Example 4

Find the points at which the tangent lines to the curve of $y = F(x) = -x^2 + 2x + 3$ are horizontal.

Since the slope of a horizontal line is 0, we need to find the point or points at which $F'(x) = 0$. Since $F'(x) = ^{-}2x + 2$, then $F'(x) = 0$ when $x = 1$. Substituting this back into the original function, we get $y = F(1) = 4$. Thus, the desired point is $(1, 4)$.

A program for differentiating polynomials

The following program yields the vector of coefficients of the derivative of a polynomial. Since the four rules for differentiating polynomials are so basic and easy to use, this program is optional. It does illustrate a way to get the computer to perform these rules.

Program 8.3 DIFF (optional)

$\nabla COEFFS \leftarrow DIFF\ POLYNOMIAL\ ;\ N;\ EXPONENTS$

[1] $N \leftarrow \rho$ *POLYNOMIAL*   *N* is the number of coefficients in the polynomial.

[2] *EXPONENTS* $\leftarrow N - \iota N$ The vector of exponents in the polynomial.

[3] *COEFFS* $\leftarrow {}^-1 \downarrow$ *POLYNOMIAL* $\times$ *EXPONENTS*

∇ These are the coefficients of the derivative of the polynomial.

Examples

1. Use the program *DIFF* to differentiate the polynomial

$$F(x) = 3x^5 + 2x^4 + 3x^3 + 5x^2 + 10x + 1.$$

         DIFF 3 2 3 5 10 1   The coefficients of the polynomial.
15 8 9 10 10   The coefficients of the derivative.

Remember that the exponents are all reduced by 1 in the derivative of a polynomial in accordance with Rule 2. Thus, the derivative of this polynomial is $F'(x) = 15x^4 + 8x^3 + 9x^2 + 10x + 10$.

2. If $F(x) = x^6 + x^4 + 8x^2 - 3x + 2$, use *DIFF* to find $F'(x)$.

         DIFF 1 0 1 0 8 ${}^-3$ 2   Notice that the zero coefficients must be included in *POLYNOMIAL*.

6 0 4 0 16 ${}^-3$

Thus, $F'(x) = 6x^5 + 0x^4 + 4x^3 + 0x^2 + 16x - 3 = 6x^5 + 4x^3 + 16x - 3.$

A program for finding the derivative of a polynomial at a particular value

The following program yields the derivative of a polynomial at a particular value of x.

Program 8.4 POLY (optional)

         ∇ *VALUE* $\leftarrow X$ *POLY POLYNOMIAL* ; *N*; *EXPS*; *COEFFS*

[1] $N \leftarrow (\rho$ *POLYNOMIAL*$) - 1$ The degree of *POLYNOMIAL* is 1 less than the number of coefficients.

[2] *EXPS* $\leftarrow N - \iota N$ The exponents of the derivative.

[3] *COEFFS* $\leftarrow$ *DIFF POLYNOMIAL*

                                          The coefficients of the derivative given by the program *DIFF*.

[4]   *VALUE ← + / COEFFS × X * EXPS*
   ∇
                              This is equivalent to plugging the
                              value of x into the derivative poly-
                              nomial.

Examples

1. If $F(x)=3x^5+2x^4+3x^3+5x^2+10x+1$, find the value of the derivative
   when $x=1$ (i.e., find $F'(1)$).

   1 *POLY* 3 2 3 5 10 1
52

   Thus, $F'(1)=52$.
2. Use *POLY* to find $F'(2)$ for the function of Example 1.

   2 *POLY* 3 2 3 5 10 1
370

   Thus, $F'(2)=370$.
3. If $F(x)=x^6+x^4+8x^2-3x+2$, use *POLY* to find $F'(1)$.

   1 *POLY* 1 0 1 0 8 ‾3 2
23

   Thus, $F'(1)=23$.

EXERCISES

1. Use the four rules of this section to find the derivative functions for the
   following functions:
   (a) $y=F(x)=3x+5$
   (b) $y=F(x)=3x^3-x^2+4x+1$
   (c) $y=F(x)=x^4+2x^3-4x^2+5x+3$
   (d) $y=F(x)=6\sqrt{x}-(2/x^4)$
   (e) $y=F(x)=2x^{3/2}-3x^{-2}+5$

2. Use the program *DIFF* to check your answers to Exercise 1, Parts (a)–(c).

3. Use the four rules of this section to find the derivatives of the following
   functions at the points indicated:
   (a) $y=4x+3$, $(e,4e+3)$
   (b) $y=-x^2+5x-1$, $(2,5)$
   (c) $y=x^3-2x^2+x+1$, $(1,1)$
   (d) $y=3x^4-x$, $(1,2)$
   (e) $y=\dfrac{4}{\sqrt{x}}$, $(4,1)$

4. Use the program *POLY* to check your answers to Exercise 3, Parts (a)–(d).

5. Find the equations of the tangent lines to the following curves at the points indicated:
(a) $y = F(x) = 3x^2 - 4x + 1$, $(1, 0)$
(b) $y = F(x) = 4/x^2$, $(2, 1)$

6. Find the points at which the slope of the following curve is 0:
$$y = F(x) = 2x^3 - 3x^2 - 12x + 9.$$

8.4 Applications of derivatives

Let us now consider a few applications of derivatives.

Application 1: Slopes of curves

As we have already seen, the derivative of a function $y = F(x)$ can be used to find the slope of the curve of the function at any point on the curve.

Application 2: Increasing, decreasing

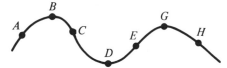

Figure 8.5   Curve of a function that both increases and decreases.

A function $y = F(x)$ is *increasing* at a point P if as x increases, y also increases as we proceed past P along the curve. (For example, the above curve in Figure 8.5 is increasing at the points A and E.) At a point where the curve is increasing, the tangent line has positive slope, so that the derivative $F'(x)$ is positive at such a point.

A function is decreasing at a point Q if as x increases, y decreases as we proceed past Q along the curve. (For example, the curve in Figure 8.5 is decreasing at the points C and H.) At a point where the curve is decreasing, the tangent line has negative slope, so that the derivative $F'(x)$ is negative at such a point.

Notice also that (at least in the case of polynomials) at the points where a curve changes from increasing to decreasing (or vice versa), the derivative will be zero (the tangent lines will be horizontal). At such a point, the derivative (the slope) changes from + to − or from − to + (see points B, D, and G above).

Example 1

Suppose that the profit (in dollars) from the manufacture and sale of x items is given by the function $P = F(x) = -x^2 + 100x - 2000$.

(a) Is profit increasing or decreasing when $x = 40$?
$$\frac{dP}{dx} = F'(x) = {}^{-}2x + 100.$$

217

So, $F'(40)=20$, which is positive. Thus, profit is increasing when $x=40$.

(b) Is profit increasing or decreasing when $x=60$? $F'(60)=\ ^-20$, which is negative. Thus, profit is decreasing.

(c) At what value of x does profit change from increasing to decreasing? We need to solve the equation $F'(x)=\ ^-2x+100=0$. Thus, $x=50$. This is the value of x for which we have maximum profit.

Example 2

Find the set of all values of x for which the following function is increasing and the set of all values of x for which it is decreasing.

$$y = F(x) = 2x^3 - 3x^2 - 12x + 10$$
$$y' = F'(x) = 6x^2 - 6x - 12.$$

Since the derivative $F'(x)=0$ at the points where the curve changes from increasing to decreasing (or vice versa), we first solve the equation $F'(x)= 6x^2 - 6x - 12 = 0$. Let us use the program *QUADRATIC* (Chapter 6) to solve this equation for x.

QUADRATIC 6 ¯6 ¯12
2 ¯1

Thus, the derivative changes sign when $x=2$ and when $x=\ ^-1$. We need only to examine the sign of $F'(x)$ on either side of these values of x. To do this, we will use a sign chart for the derivative $F'(x)$ (see Figure 8.6).

Figure 8.6   Sign chart for $F'(x)$.

The derivative is positive to the left of $x=\ ^-1$, since $F'(^-2)=24$ and the derivative doesn't ever change sign to the left of $x=\ ^-1$.[2] The derivative is negative between $x=\ ^-1$ and $x=2$, since $F'(0)=\ ^-12$ and the derivative doesn't ever change sign between $x=\ ^-1$ and $x=2$. The derivative is positive to the right of $x=2$, since $F'(3)=24$, and the derivative never changes sign to the right of $x=2$. Using this sign chart, we see that the function is increasing in the set $\{x|x<\ ^-1 \text{ or } x>2\}$. It is decreasing in the set $\{x|^-1<x<2\}$.

Application 3: Velocity

Suppose that the position of a moving object at time t is given by the function $s=F(t)$. The average rate of change of position, s, per unit of time, t, as t goes from some time t to a later time $t+h$ is given by the

---

[2]In order for the derivative to change sign, it would have to become 0. We have found that the only times the derivative becomes 0 is at $x=\ ^-1$ and $x=2$.

following difference quotient:

$$V_{ave} = \frac{\text{Change in position}}{\text{Change in time}} = \frac{F(t+h) - F(t)}{h}.$$

This is called the *average velocity*, V_{ave}, of the object between times t and $t+h$. The *instantaneous rate of change of position*, s, per unit of time t, *at* some time t is given by

$$V = \lim_{h \to 0} \frac{F(t+h) - F(t)}{h} = \lim_{h \to 0} V_{ave} = F'(t).$$

This is called the *instantaneous velocity at* time t (or simply the *velocity* at time t.)

Example 1

Suppose that the distance of an object from a starting point after t seconds is given by $s = F(t) = 3t^2 + 2t$ feet.

(a) Find the initial velocity of the object.
   The initial velocity is the velocity at time $t = 0$. $F'(t) = 6t + 2$, so $F'(0) = 2$ feet per second.
(b) Find the velocity at the end of 1 second.
   $F'(1) = 8$ feet/second.
(c) Find the velocity at the end of the 5th second.
   $F'(5) = 32$ feet/second.
(d) Find the average velocity from the first to the 5th second.

$$V_{ave} = \frac{F(5) - F(1)}{5 - 1} = \frac{85 - 5}{4} = 20 \text{ feet/second.}$$

Example 2

Suppose that $s = F(t) = {}^-16t^2 + 320t$ is the function which gives the height s, in feet, of a projectile fired vertically upward from ground level with an initial velocity of 320 feet/second, where t is the time elapsed in seconds. (This was Example 5 of Section 6.1.)

(a) Find the velocity at the end of the 5th second.
   $F'(t) = {}^-32t + 320$. So, $F'(5) = 160$ feet/second.
(b) Find the velocity at the end of the 15th second.
   $F'(15) = {}^-160$ feet/second. This velocity is negative since s is decreasing. The projectile is falling back to earth.
(c) Find the velocity at the end of the 10th second.
   $F'(10) = 0$ feet/second. This is the time at which the projectile reaches its highest point.

219

Application 4: Acceleration

The average rate of change of velocity, V, per unit of time between times t and $t+h$ is called the *average acceleration* of the object from time t to time $t+h$. This is given by the following difference quotient:

$$A_{ave} = \frac{\text{Change in velocity}}{\text{Change in time}} = \frac{V(t+h)-V(t)}{h} = \frac{F'(t+h)-F'(t)}{h},$$

since $F'(t)$ gives the velocity at time t. The *instantaneous acceleration* at time t is the instantaneous rate of change in velocity at time t and is given by

$$A = \lim_{h \to 0} A_{ave} = \lim_{h \to 0} \frac{F'(t+h)-F'(t)}{h} = V'(t) = \frac{dV}{dt} = F''(t).$$

We will refer to the instantaneous acceleration at time t as simply the *acceleration* at time t. Acceleration is the derivative of velocity, which is the derivative of position. Thus, acceleration is the derivative of the derivative of position. Such a derivative of a derivative is known as the *second derivative* of the original function. Thus, acceleration is the second derivative of the position function.

Example 1

Find the acceleration at any time t for the particle of Example 1 above. Since $s = F(t) = 3t^2 + 2t$ and $V = F'(t) = 6t + 2$, then $A = F''(t) = 6$ feet/second/second (usually denoted as feet/second^2). Thus, every second the velocity increases by 6 feet/second.

Example 2

Find the acceleration for the projectile of Example 2 above.
$s = F(t) = {}^-16t^2 + 320t$ and $V = F'(t) = {}^-32\ t + 320$, so $A = F''(t) = {}^-32$ feet/second^2. This is called the acceleration due to gravity. Since A is negative, the velocity is decreasing by 32 feet/second each second.

Application 5: Instantaneous rates of change in general

The previous two applications are illustrations of the use of the derivative for finding instantaneous rates of change of a function with respect to time.

Let $y = F(x)$ be *any* function. The *average rate of change* of y per unit change in x as x goes from x to $x+h$ is given by the difference quotient

$$\frac{\text{Change in } y}{\text{Change in } x} = \frac{F(x+h)-F(x)}{h}.$$

The instantaneous rate of change of y per unit change in x when x has a given value x is given by

$$\frac{dy}{dx} = F'(x) = \lim_{h \to 0} \frac{F(x+h) - F(x)}{h}.$$

Example 1

The volume of a sphere is given by the function $V = \frac{4}{3}\pi x^3$, where x is the radius. If a spherical balloon is being inflated with air, how fast is the volume increasing per unit increase in radius when the radius is 6 inches?

$$\frac{dV}{dx} = \frac{4}{3}\pi \cdot 3x^2 = 4\pi x^2.$$

So, when $x = 6$, $dV/dx = 144\pi$ cubic inches/inch.

Example 2

A water reservoir is being drained in such a way that the amount of water (in gallons) in the reservoir after t hours is given by the function

$$W = F(t) = 500t^2 - 50{,}000t + 1{,}250{,}000.$$

(a) How fast is the water running out initially?

$$\frac{dW}{dt} = F'(t) = 1000t - 50{,}000.$$

Initially, $t = 0$, so that $F'(0) = {}^-50{,}000$ gallons/hour.
(b) How fast is it running out when $t = 10$ hours?

$$F'(10) = {}^-40{,}000 \quad \text{gallons per hour.}$$

(c) How much water is in the reservoir when it is running out at the rate of 20,000 gallons per hour?

$$F'(t) = 1000t - 50{,}000 = {}^-20{,}000,$$

so that $t = 30$ hours. The amount of water in the reservoir when $t = 30$ hours is $W = F(30) = 200{,}000$ gallons.

Application 6: Economic analysis

Suppose that the total cost of producing x items is given by $C = F(x)$. Then, the instantaneous rate of change in cost per item produced at the level of production x is given by the derivative $dC/dx = F'(x)$. This is called *the marginal cost function*.

Similarly, if the revenue from the sale of x items is given by $R = G(x)$, then the instantaneous rate of change in revenue per item sold is given by the derivative $dR/dx = G'(x)$. This is called the *marginal revenue function*.

221

Example

Suppose that the cost (in dollars) of producing x items is given by the function

$$C = F(x) = x^2 + 2x + 2000.$$

(See Example 1 of applications of quadratic functions in Section 6.4.) Suppose also that the items sell at a price of $102 each, and that every item that is produced is sold.

(a) Find the marginal cost when the 20th item is produced.

$$\frac{dC}{dx} = F'(x) = 2x + 2.$$

So, $F'(20) = 22$, or cost is changing at a rate of $22 per item when the 20th item is produced.

(b) Find the marginal revenue function.

Since the price is $102 per item, the revenue is $R = G(x) = 102x$. Therefore, the marginal revenue is given by $dR/dx = G'(x) = 102$. Thus, revenue is changing at a rate of $102 per item sold. This is obvious, since this is the price.

(c) Find the number of items that must be produced and sold, x, in order for the marginal revenue to equal the marginal cost.

$$\text{M.R.} = \text{M.C.}$$
$$102 = 2x + 2$$
$$x = 50 \text{ items.}$$

Recall that this was the number of items that must be produced and sold in order to maximize profit. It is a fundamental law of economics that profit is maximized when marginal revenue equals marginal cost.

EXERCISES

It is optional to use any of the programs *DERIVATIVE*, *DIFF*, or *POLY* to help in doing any of the following exercises.

1. Find all values of x for which the following function is increasing and all values of x for which it is decreasing:

$$y = F(x) = x^2 - 4x + 3.$$

2. Repeat Exercise 1 with the function $y = F(x) = x^3 - 6x^2 + 9x + 6$.

3. Suppose that for a given company the profit from the sale and distribution of x items is given by the function

$$P = F(x) = 10x - 0.02x^2 - 0.0001x^3.$$

(a) Is the profit increasing or decreasing when $x = 100$?
(b) Is the profit increasing or decreasing when $x = 200$?

(c) At what values of x does profit change from increasing to decreasing or vice versa?

4. The area of a circle is given by the formula $A = \pi x^2$, where x is the radius. Find the rate of change in area per unit change in radius when $x = 5$ inches.

5. Find the equation of the tangent line to the curve $y = F(x) = 8\sqrt{x}$ at the point $(4, 16)$.

6. Suppose that the position of an object is given by the function

$$s = F(t) = 4t^2 + 12t + 9 \text{ feet.}$$

   (a) Find the average velocity of the object during the first 5 seconds.
   (b) Find the initial velocity.
   (c) Find the instantaneous velocity at the end of the 5th second.
   (d) Find the acceleration at any time t.

7. For a freely falling body dropped from a height of 1000 feet, the height at the end of t seconds is given by

$$H = F(t) = 1000 - 16t^2 \text{ feet.}$$

   (a) Find the velocity of the object at the end of 5 seconds.
   (b) Find the velocity of the object at the end of 10 seconds.
   (c) Find the acceleration at any time t.

8. The relation between sales and advertising cost, x, for a product is given by the function $S = 400x^2 - 50x$. How fast is sales changing per dollar of advertising cost when $x = \$5000$?

9. The cost (in dollars) of making x items is given by $C = 10 + 20 \cdot \sqrt{x}$.
   (a) What is the marginal cost when $x = 100$?
   (b) How fast is cost changing per item produced when $x = 25$?
   (c) What is x when the marginal cost is \$1 per unit produced?

10. The cost of producing x items is given by $C = F(x) = 4x - 0.0005x^2$ and the revenue from the sale of x of these items is given by $R = G(x) = x^2 - 8x$. Find x when marginal cost equals marginal revenue.

11. A fire is spreading along a river bank in such a way the $S = 2t - 0.5t^2$ gives the distance in miles from a starting point after t hours. When does it stop spreading?

12. According to Newton's law of universal gravitation, the force exerted by the earth on a space object is given by $F = -k/x^2$, where x is the distance of the object from the center of the earth in miles. If $k = 1000$, find the rate of change in F per unit change in x when $x = 10,000$ miles.

8.5 More rules of differentiation (optional)

In this section, we present some more rules of differentiation which will enable us to differentiate more complicated functions, including all of the functions we have studied so far in this text. Since these rules may be beyond the interest of many readers who want only a brief introduction to calculus and its applications, this section is optional. Although these rules may be proved using the definition of derivative, it is beyond the intent of this text to prove these rules, and we will omit the proofs. We will merely state the rules and present examples of their use.

Rule 5: Chain rule

If $y = F(u)$, where $u = G(x)$, then

$$\frac{dy}{dx} = \frac{dy}{du} \cdot \frac{du}{dx} = F'(u) \cdot G'(x).$$

[*Note*: We could think of y as a function of x, since $y = F(G(x))$.]

Example

Let $y = \sqrt{x^2 + 1}$; find dy/dx.
Let $u = x^2 + 1$. Then, $y = u^{1/2}$. By the chain rule,

$$\frac{dy}{dx} = \frac{dy}{du} \cdot \frac{du}{dx} = \frac{1}{2} \cdot u^{-1/2} \cdot 2x = \frac{2x}{\sqrt{u}} = \frac{2x}{\sqrt{x^2 + 1}}.$$

If we apply this chain rule to a power function, we get the following rule.

Rule 6: Power rule

If $y = G(x)^n$, then

$$\frac{dy}{dx} = n \cdot G(x)^{n-1} \cdot G'(x).$$

Example

$y = (3x^2 + 6x)^5$. Thus,

$$\frac{dy}{dx} = 5 \cdot (3x^2 + 6x)^4 \cdot (6x + 6).$$

Rule 7: Product rule

If $y = F(x) \cdot G(x)$, then

$$\frac{dy}{dx} = F(x) \cdot G'(x) + G(x) \cdot F'(x).$$

Examples

1. Let $y = x^2 \cdot x^3 = x^5$. Of course, $dy/dx = 5x^4$. We shall apply the product rule to this example to convince the reader that the derivative of a product is not the product of the derivatives. Let $F(x) = x^2$ and $G(x) = x^3$. Then, by the product rule,

$$\frac{dy}{dx} = (x^2) \cdot (3x^2) + (x^3) \cdot (2x) = 3x^4 + 2x^4 = 5x^4.$$

If one merely multiplied the derivatives of the functions $F(x)$ and $G(x)$, he would get $(2x) \cdot (3x^2) = 6x^3$, which would not be correct.

2. Let $y = (x^2 + 3x + 2) \cdot (x^3 + 3)$. By the product rule,

$$\frac{dy}{dx} = (x^2 + 3x + 2) \cdot (3x^2) + (x^3 + 3) \cdot (2x + 3) = 5x^4 + 12x^3 + 6x^2 + 6x + 9.$$

3. Let $y = x\sqrt{x^2 + 1}$. Using the product rule and the power rule, we get

$$\frac{dy}{dx} = x \cdot \frac{1}{2}(x^2 + 1)^{-1/2} \cdot 2x + \sqrt{x^2 + 1} = \frac{x^2}{\sqrt{x^2 + 1}} + \sqrt{x^2 + 1} .$$

Rule 8: Quotient rule

If $y = F(x)/G(x)$, then

$$\frac{dy}{dx} = \frac{G(x) \cdot F'(x) - F(x) \cdot G'(x)}{(G(x))^2}.$$

Examples

To convince the reader that the derivative of a quotient is not the quotient of derivatives, consider the following simple example:

1. Let $y = x^5/x^2 = x^3$. Obviously, $dy/dx = 3x^2$. Using the quotient rule, we get

$$\frac{dy}{dx} = \frac{(x^2) \cdot (5x^4) - (x^5) \cdot (2x)}{(x^2)^2} = \frac{5x^6 - 2x^6}{x^4} = 3x^2.$$

2. Let $y = 3x/(x^2 + 2)$. By the quotient rule,

$$\frac{dy}{dx} = \frac{(x^2 + 2) \cdot 3 - (3x) \cdot (2x)}{(x^2 + 2)^2} = \frac{6 - 3x^2}{(x^2 + 2)^2}.$$

Rule 9: Exponential functions

If $y = a^u$, where u is a function of x, then

$$\frac{dy}{dx} = (\ln a) \cdot (a^u) \cdot \left(\frac{du}{dx}\right).$$

Example 1: The exponential function

Since $\ln e = 1$, then if $y = e^u$, where u is a function of x, then

$$\frac{dy}{dx} = (e^u) \cdot \left(\frac{du}{dx}\right).$$

Further examples

1. Let $y = 2^{x^2}$. Then,

$$\frac{dy}{dx} = (\ln 2) \cdot (2^{x^2}) \cdot (2x) = (0.6931471806) \cdot (2^{x^2}) \cdot (2x).$$

2. Let $y = e^{x^2}$, then $dy/dx = (e^{x^2}) \cdot (2x)$.
3. Let $y = x^2 \cdot e^x$. Using the product rule,

$$\frac{dy}{dx} = (x^2) \cdot (e^x) + (e^x) \cdot (2x).$$

Rule 10: Logarithmic functions

If $y = \log_b u$, where u is a function of x, then

$$\frac{dy}{dx} = \left(\frac{1}{\ln b}\right) \cdot \left(\frac{1}{u}\right) \cdot \left(\frac{du}{dx}\right).$$

Example 1: Natural logs

Since $\ln e = 1$, then if $y = \ln u$, where u is a function of x, then

$$\frac{dy}{dx} = \left(\frac{1}{u}\right) \cdot \left(\frac{du}{dx}\right).$$

Natural logs are used more often than any other logs in calculus because of their simpler derivatives.

Further examples

1. Let $y = \log_{10} x^2$. Then,

$$\frac{dy}{dx} = \left(\frac{1}{\ln 10}\right) \cdot \left(\frac{1}{x^2}\right) \cdot (2x) = \left(\frac{1}{\ln 10}\right) \cdot \left(\frac{2}{x}\right).$$

2. Let $y = \ln(x^2)$. Then,

$$\frac{dy}{dx} = \left(\frac{1}{x^2}\right) \cdot (2x) = \frac{2}{x}.$$

EXERCISES

1. Find the derivative functions for the following:

(a) $y = \dfrac{x^3 + 1}{4x}$
(b) $y = (3x^2 + 5) \cdot (2x^5 + 7)$
(c) $y = (x^2 + 2x) \cdot \ln(3x)$
(d) $y = 10^{x^3 + 5x}$
(e) $y = e^{x^3 + 5x}$

(f) $y = \dfrac{\ln x}{e^x}$
(g) $y = \log_{10}(x^2 + 5x + 1)$
(h) $y = \ln(x^2 + 5x + 1)$
(i) $y = (x^2 + 5x + 1)^4$
(j) $y = (\ln x)^5$

2. Find the derivatives of the following functions at $x = 2$:

(a) $y = (2x^3 - 6x^2 + 1) \cdot (x^2 + 2x)$
(b) $y = \dfrac{2x^3 - 6x^2 + 1}{x^2 + 2x}$

(c) $y = (2x^3 - 6x^2 + 1)^5$
(d) $y = e^{x^2 - 2x}$
(e) $y = \ln(x^2 - 2x)$

3. Use the program *DERIVATIVE* to check your answers to Exercise 2.

4. Find the equation of the line tangent to the curve $y = \ln x$ at $(1, 0)$.

5. If the sales of a new product is given by the function $S = 20e^{0.01x^2}$, where x is the number of days the product has been on the market, find the rate of increase in sales per day on the market on the 10th day.

8.6 Theory of maxima, minima

Derivatives are useful in finding the maximum or minimum value of a function.

Definitions

A function $y = F(x)$ has a *relative maximum* at a point $P(a, F(a))$ if $F(a) \geqslant F(x)$ for all points $(x, F(x))$ on the graph near P. In other words, a relative maximum occurs at a highest point in the immediate vicinity. In Figure 8.7, the function has a relative maxima at the points A, C, and E.

A function $y = F(x)$ has an *absolute maximum* at a point $P(a, F(a))$ if $F(a) \geqslant F(x)$ for all x in the domain of F. The absolute maximum occurs at the point C in Figure 8.7.

A function $y = F(x)$ has a *relative minimum* at a point $P(a, F(a))$ if $F(a) \leqslant F(x)$ for all points $(x, F(x))$ on the graph near P. In other words, a relative minimum occurs at a lowest point in the immediate vicinity. In Figure 8.7, the function has relative minima at the points B and D.

A function has an *absolute minimum* at a point $P(a, F(a))$ if $F(a) \leqslant F(x)$ for all x in the domain of F. The absolute minimum occurs at the point D in Figure 8.7.

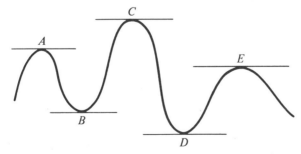

Figure 8.7 A function with both relative and absolute maxima and minima.

Critical values

The derivative is quite helpful in locating the *relative extrema* (relative maxima and minima) for a function $y = F(x)$ with domain an open interval $\{x \mid a < x < b\}$. Notice from Figure 8.7 that at the relative maxima and minima, the curve changes from increasing to decreasing or from decreasing to increasing and that the tangent lines are horizontal at these relative

227

extrema. Recall that the derivative of a function at a point P gives the slope of the tangent line at P and that the slope of a horizontal line is 0. Thus, it appears that the relative maxima and minima occur at points where the derivative $F'(x)$ is 0. Such points are called *critical points* and the values of x for which $F'(x)$ is 0 are called *critical values*.[3] When looking for relative extrema, therefore, one first locates the critical values.

The first derivative test for relative extrema

Let $y = F(x)$ and let $x = c$ be a critical value for F (i.e., a value for which $F'(c) = 0$). Then,

1. If the derivative $F'(x)$ changes sign from + to − as we pass by (moving from left to right) the point $(c, F(c))$, then the function changes from increasing to decreasing at this point. Therefore, the function has a relative maximum at this point.
2. If the derivative $F'(x)$ changes sign from − to + as we pass by (moving from left to right) the point $(c, F(c))$, then the function changes from decreasing to increasing at this point. Therefore, the function has a relative minimum at this point.
3. If the derivative does not change sign as we pass by the point $(c, F(c))$, then we have neither a relative maximum nor a relative minimum at this point.

Example 1

This last case can be illustrated by the following example:

$$y = F(x) = x^3$$

$F'(x) = 3x^2 = 0$ when $x = 0$. Thus, the critical point is $(0,0)$. However, $F'(x) = 3x^2$ never changes sign. Thus, this function has neither a relative maximum nor minimum at $(0,0)$, or at any other point for that matter. The graph of this function (Figure 8.8) bears this out.

Example 2

Find any relative maxima or relative minima for the function $y = F(x) = x^2 - 4x + 4$. (The graph of this function appeared as Example 1 in Section 6.4.)

First, we need to locate any critical values. $F'(x) = 2x - 4 = 0$ when $x = 2$. Thus, $x = 2$ is the only critical value. We now have to examine the sign change of $F'(x)$ as we pass by the value $x = 2$. Therefore, we make a sign chart for the derivative, examining the sign of the derivative to the left and right of $x = 2$ (see Figure 8.9). The derivative is negative for all x to the left of 2 and positive for all x to the right of 2. Thus, the function has a

[3]It is also possible to have a relative maximum or minimum at a point where $F'(x)$ does not exist. However, we shall omit such cases.

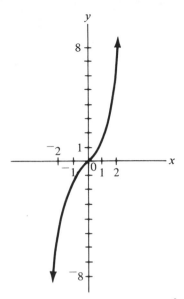

Figure 8.8   Graph of $y = F(x) = x^3$.

relative minimum at $x=2$. The value of this function at this relative minimum is $F(2)=2^2-4\cdot2+4=0$. Recall that the vertex of the parabola for this function occured at this minimum point $(2,0)$.

Figure 8.9   Sign chart for $F'(x)$ about $x=2$.

Example 3

Find any relative extrema for the function $y = F(x) = -x^2 + 2x + 3$. (See Example 2 of Section 6.4 and Figure 8.10a.)

Locating any critical values, we note that $F'(x) = {}^-2x + 2 = 0$ when $x = 1$. Examining the sign chart for $F'(x)$ about this critical value $x=1$ (Figure 8.10b), we discover that there is a relative maximum at $x=1$, $y = F(1) = 4$. This was the vertex of the parabola for this function also.

Example 4

Prove that the vertex of the parabola given by the quadratic function $y = F(x) = a\cdot x^2 + b\cdot x + c$ occurs at the point where $x = -b/2\cdot a$.

Since the vertex is a relative maximum or minimum, it must occur at a critical value of $y = F(x)$. However, $F'(x) = 2a\cdot x + b = 0$ when $x = -b/2\cdot a$.

229

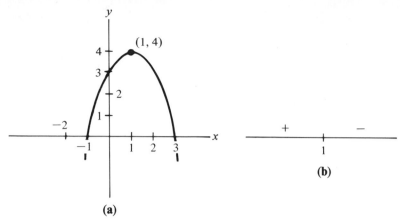

(a)

Figure 8.10   (a) Graph of $F(x)=y=-x^2+2x+3$. (b) Sign chart for $F'(x)$ around $x=1$.

Example 5

Find any relative extrema for the function $y=F(x)=2x^3+3x^2-12x-10$. (See Example 1 of Section 6.5. and Figure 8.11a.)

Locating the critical values, we note that

$$F'(x)=6x^2+6x-12=6(x+2)(x-1)=0$$

when $x=-2$ and $x=1$. The sign chart is shown in Figure 8.11b. Since these are the only values at which the derivative changes sign, we can use any value we wish to determine the signs of $F'(x)$ to the left and right of these values. Thus, a relative maximum occurs at the point $(-2, F(-2))=(-2,10)$ and a relative minimum occurs at $(1, F(1))=(1,-17)$.

Endpoint extrema

If a function is only defined on a closed interval $\{x \mid a \leqslant x \leqslant b\}$, then it is possible that the maximum or minimum value might occur at one of the endpoints of this interval, a or b. A function defined on such a closed interval must have an absolute maximum and an absolute minimum value. If either of these occurs in the interior of the interval, then it would occur at a critical value and would be found as in the previous examples. However, it is important to check for possible endpoint extrema also.

230

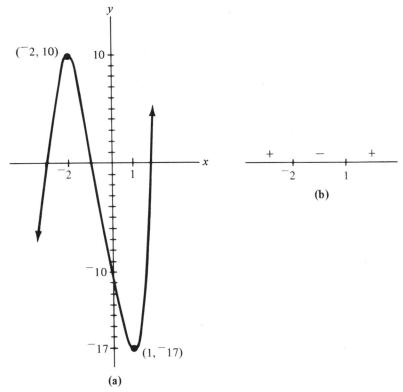

Figure 8.11   (a) Graph of $y = F(x) = 2x^3 + 3x^2 - 12x - 10$. (b) Sign chart for $F'(x)$ $= 0$ near $x = {}^-2$ and $x = 1$.

Example 6

Find the absolute maximum and absolute minimum values of the function $y = F(x) = -x^2 + 2x + 3$ in the restricted domain $\{x \mid 0 \leqslant x \leqslant 4\}$ (see Figure 8.12a).

As before, in Example 3, we get the critical value: $F'(x) = {}^-2x + 2 = 0$, so that $x = 1$ and $y = F(1) = 4$. The sign chart for $F'(x)$ about this critical value $x = 1$ and in this restricted domain is as shown in Figure 8.12b. Thus, the function is always increasing on the interval $\{x \mid 0 \leqslant x < 1\}$ and always decreasing on the interval $\{x \mid 1 < x \leqslant 4\}$. Thus, the absolute maximum point is at $(1, 4)$. Since the function must have an absolute minimum value, it must occur at one of the endpoints, 0 or 4, of the domain. At $x = 0$, $F(0) = 3$. At $x = 4$, $F(4) = {}^-5$. Therefore, the absolute minimum point for this function in the restricted domain is the point $(4, {}^-5)$. The graph of this function follows:

231

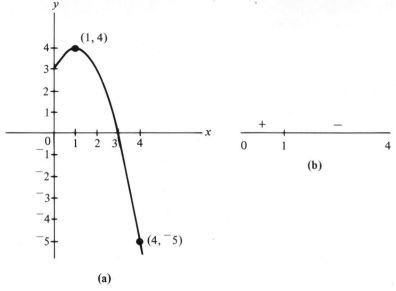

(a)

Figure 8.12 (a) Graph of $y = F(x) = {}^-x^2 + 2x + 3$, $\{x|0 \leqslant x \leqslant 4\}$. (b) Sign chart of $F'(x)$ around $x = 1$.

Example 7

Find the absolute maximum and absolute minimum values of the function $y = F(x) = 2x^3 + 3x^2 - 12x - 10$ in the restricted domain $\{x|0 \leqslant x \leqslant 5\}$.

$F'(x) = 6x^2 + 6x - 12 = 0$ yields solutions $x = {}^-2$ and $x = 1$. Since only $x = 1$ is in the restricted domain of $F(x)$, we are only interested in it. The sign chart for $F'(x)$ in this domain is as shown in Figure 8.13. This function is always decreasing on the interval $\{x|0 \leqslant x < 1\}$ and always increasing on the interval $\{x|1 < x \leqslant 5\}$. Thus, the absolute minimum point is $(1, F(1)) = (1, {}^-17)$. The absolute maximum must occur at one of the endpoints, 0 or 5, of the domain. At $x = 0$, $F(0) = {}^-10$. At $x = 5$, $F(5) = 255$. Thus, the absolute maximum occurs when $x = 5$ and $y = F(5) = 255$.

Figure 8.13   Sign chart for $F'(x) = 6x^2 + 6x - 12$, $5 \geqslant x \geqslant 0$.

In the next section, we shall consider some applied maxima, minima problems.

EXERCISES

1. Find any relative maxima, minima for the following functions:
   (a) $y = F(x) = -x^2 + 8x - 6$
   (b) $y = F(x) = 2x^2 - 5x - 3$
   (c) $y = F(x) = x^3 - 6x^2 + 9x + 6$
   (d) $y = F(x) = 2x^3 + 3x^2 - 4x - 10$
   (e) $y = F(x) = x^5 - 2x^4 + 3x - 5$

2. Find the absolute maximum and minimum values for the functions in Exercise 1 in the restricted domain $\{x|1 \leqslant x \leqslant 5\}$.

8.7 Applied maxima, minima

We now consider some applied maxima, minima problems.

Example 1

Suppose that the cost (in dollars) of producing x items is given by the function

$$C = x^2 + 2x + 2000.$$

Suppose also that these items sell for \$102 each and that every item that is produced is sold. Find the number of items, x, that must be produced and sold in order to maximize profit, and find this maximum profit.

Since revenue is the product of the price and the number of items sold, then revenue is given by $R = 102x$. Now, Profit = Revenue − Cost. Therefore, profit is given by

$$P = R - C = 102x - (x^2 + 2x + 2000) = -x^2 + 100x - 2000.$$

Thus, $dP/dx = {}^{-}2x + 100 = 0$ when $x = 50$. So, the critical value is $x = 50$ (see Figure 8.14). Thus, the maximum profit occurs when $x = 50$. This maximum profit is $F(50) = \$500$.

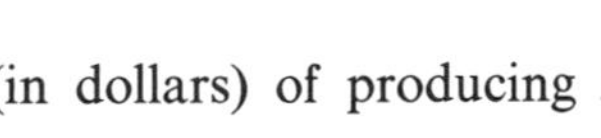

Figure 8.14 Sign chart for $F'(x)$ about $x = 50$.

Alternate approach to profit maximization

To find the maximum profit above, we solved the equation $dP/dx = 0$. Since $P = R - C$, then

$$\frac{dP}{dx} = \frac{dR}{dx} - \frac{dC}{dx}.$$

So, when $dP/dx = 0$, then $dR/dx = dC/dx$, or marginal revenue equals marginal cost. This rule, that profit is maximized when marginal revenue equals marginal cost is a fundamental rule in economics.

Example 2

A manager of an apple orchard consisting of 50 apple trees is trying to decide when to pick his apple crop. If he picks it now, his average apple tree will yield 80 pounds of apples which he can sell for \$.50 per pound. However, for each week he waits to pick the apples, the average yield per tree will increase by 10 pounds per tree, while the price will decrease by \$.03 per pound. How many weeks should he wait to pick his apples if he would like to maximize his revenue?

Let $x=$ the number of weeks he should wait to pick the apples.

Revenue $= R =$ (number of pounds picked)·(price per pound)

$$= 50 \cdot (80 + 10x) \cdot (50 - 3x)$$

$$= 20{,}000 + 13{,}000x - 1500x^2.$$

So,

$$\frac{dR}{dx} = 13{,}000 - 3000x = 0$$

when

$$x = \frac{13{,}000}{3000} = 4\frac{1}{3}.$$

Thus, to maximize his revenue, he should wait about 4 weeks and 2 days to harvest his apples (see Figure 8.15).

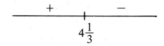

Figure 8.15   Sign chart for dR/dx about $x=4\frac{1}{3}$.

Example 3

A man has 100 feet of fencing which he wishes to use to fence in a rectangular yard for his dog. He will put the dog's yard against an existing fence, and therefore, only needs to fence in three sides. Find the dimensions of the yard of maximum area and find this maximum area.

In a problem such as this, a sketch (such as in Figure 8.16) is very helpful. Since he has 100 feet of fencing to do the job, then $x + 2y = 100$. We wish to maximize the area of the rectangle. Thus, we need a formula for this area. Accordingly, the area $= A = x \cdot y$. However, this area formula has two variables and doesn't use the fact that he has only 100 feet of fencing. However, from the equation $x + 2y = 100$, we get $x = 100 - 2y$. If we replace x in the area formula by this expression, we get

$$A = x \cdot y = (100 - 2y) \cdot y = 100y - 2y^2.$$

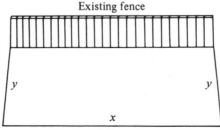

Figure 8.16   Illustration of Example 3.

Figure 8.17   Sign chart for dA/dy about $y=25$.

Now, $dA/dy=100-4y=0$ when $y=25$ feet. Thus, the maximum area occurs when $y=25$ feet (see Figure 8.17). At this value of y, $x=100-2y=50$ feet. So, $A=x\cdot y=50\cdot 25=1250$ square feet is the maximum area for the dog's yard.

Example 4

A rectangular shaped in-ground swimming pool with a square bottom is to hold 4000 cubic feet of water. Find the dimensions of the swimming pool of minimum surface area satisfying these restrictions.

Again, a sketch, as in Figure 8.18, will be helpful in visualizing the problem. Since the bottom of the pool is square, it is x feet on each side. The depth of the pool is labeled by y. The volume of the pool is given to be 4000 cubic feet. The volume of a rectangular box is given by length·width·depth. Thus, we have the following equation for the volume of the pool:

$$V=x\cdot x\cdot y=x^2\cdot y=4000.$$

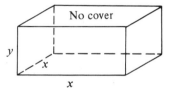

Figure 8.18   Illustration for Example 4.

We wish to minimize the surface area (perhaps to build the pool of least cost). Therefore, we need a formula for the surface area of the pool. This is given by

$$A=\text{Area of bottom}+\text{Area of 4 sides}=x^2+4xy.$$

Here again, we have two variables in our area formula and we have not used the fact that the volume of the pool must be 4000 cubic feet. Solving the volume formula for y, we get $y=4000/x^2$. Replacing y in the area formula by this value yields the following area formula:

$$A=x^2+4x\cdot\left(\frac{4000}{x^2}\right)=x^2+\frac{16{,}000}{x}=x^2+16{,}000x^{-1}.$$

So,

$$\frac{dA}{dx}=2x-16{,}000x^{-2}=2x-\frac{16{,}000}{x^2}=0.$$

Thus, $2x^3=16{,}000$ or $x=20$ feet.

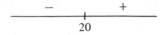

Figure 8.19 Sign chart for dA/dx about $x=20$ feet.

Thus, we have minimum surface area when $x=20$ feet (see Figure 8.19) and $y=4000/2=4000/400=10$ feet. The surface area is

$$A=x^2+4xy=400+800=1200 \text{ square feet.}$$

EXERCISES

1. The cost (in dollars) of producing x items is given by $C=50x+100$. The revenue from these x items is given by $R=100x-0.01x^2$. Find the number of items that must be produced in order to maximize profit and find this maximum profit.

2. If a farmer harvests his potatoes right now, he will get 200 bushels at a price of $2 per bushel. If he waits, he will get an increase of 25 bushels per week, while the price will drop by $.10 per bushel. How many weeks should he wait to harvest his potato crop in order to maximize his profit?

3. The area of a rectangular field is to be 1350 square feet. The field is to be fenced on all 4 sides with another fence running down the middle. Find the dimensions which require the least amount of fence.

4. A rectangular field is to be enclosed by a fence. One side of the fence is to be along a road and requires a stronger fence than the other three sides. The fence along the road costs $5 per foot. The fence along the other three sides costs $3 per foot. Find the dimensions of the field of maximum area that can be enclosed with $2500.

5. A rectangular box with a square base is to be made to hold 64 cubic inches. Find the dimensions of the box of minimum surface area.

6. A rectangular box with no top and square base is to be constructed to store apples. Material for the bottom costs $2 per square foot, while material for the sides costs $1 per square foot. Find the dimensions for the box of maximum volume that can be so constructed for $100.

7. A rectangular box is to be made from a piece of cardboard 12 inches long and 8 inches wide by cutting out a square from each corner and turning up the sides. Find the volume of the box of maximum volume that can be so made.

8. A manager of an apartment complex charges a monthly rent of $200 and completely fills up his 80 apartments. However, for each increase of $10 in rent thereafter, 2 apartments become empty. Find the rent which will yield maximum revenue. How many empty apartments will there be? What is the maximum revenue?

9. A beer company wishes to have a beer can that will hold 30 cubic inches of beer. A beer can is, of course, a right circular cylinder. Find the dimensions of

the beer can of minimum surface area. [*Note:* The volume of a right circular cylinder is given by $V=\pi \cdot r^2 \cdot h$ and the surface area by $A=2\pi \cdot r \cdot h+2\pi \cdot r^2$, where r is the radius of the base and h is the height.]

10. Suppose that the number of bacteria present in a certain culture in t days is given by the exponential function
$$B=F(t)=60t\cdot e^{-0.3t}.$$
Find the number of days in which the number of bacteria will be greatest.

8.8 Curve sketching using derivatives

Recall that the derivative $F'(x)$ of a function $y=F(x)$ at a point $P(x,y)$ gives the slope of the tangent line at that point. If $F'(x)>0$ at P, then the curve is increasing at P. If $F'(x)<0$ at P, then the curve is decreasing at P. We now want to consider what information about the graph of the function is given by the second derivative.

Second derivatives

Let $y=F(x)$. The first derivative is a new function, symbolized by $y'=dy/dx=F'(x)$. The first derivative can be differentiated yielding a new function called the *second derivative* of $y=F(x)$ and symbolized by $y''=d^2y/dx^2=F''(x)$.

Example 1
$$y=F(x)=x^3+3x^2-5x+2$$
$$y'=F'(x)=3x^2+6x-5 \qquad \text{The first derivative}$$
$$y''=F''(x)=6x+6 \qquad \text{The second derivative.}$$

Example 2
$$y=G(x)=\sqrt{x}=x^{1/2}$$
$$\frac{dy}{dx}=\tfrac{1}{2}\cdot x^{-1/2}=\frac{1}{2\sqrt{x}} \qquad \text{The first derivative.}$$
$$\frac{d^2y}{dx^2}=-\tfrac{1}{4}\cdot x^{-3/2}=\frac{-1}{4\sqrt{x^3}} \qquad \text{The second derivative.}$$

Geometric interpretation of the second derivative at a point

A curve is *concave downward* (opens downward) at a point P if the curve lies below the tangent line at P. It is *concave upward* (opens upward) if it lies above its tangent line at P.

From Figure 8.20, one can see that if a curve is concave downward at a point P, then the slope of the tangent line decreases as we pass by P from left to right along the curve (see points A, B, and C in the sketch). Since the slope of the tangent line is given by the first derivative $F'(x)$, then $F'(x)$ is a decreasing function as we pass by the point P from left to right

237

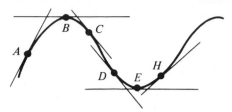

Figure 8.20   Curve with concave downward and concave upward sections.

along the curve. Thus, the derivative of $F'(x)$ is negative at P. In other words, $F''(x)$ is negative at P. Thus, at any point $P(x,y)$ where the curve is concave downward, $F''(x)<0$.

By a similar argument, at any point $P(x,y)$ where the curve is concave upward (points D, E, and H in the sketch), $F''(x)>0$.

A point at which the curve changes its concavity from concave downward to concave upward or vice versa is called an *inflection point*, I, of the curve. At an inflection point, $F''(x)$ changes sign.

Using the first and second derivatives of a function $y=F(x)$, we can sketch the curve for this function.

Example 1

$$y=F(x)=2x^3-3x^2-12x+10$$
$$y'=F'(x)=6x^2-6x-12$$
$$y''=F''(x)=12x-6.$$

(a) Find any relative maxima and relative minima.

We need to solve the quadratic equation $F'(x)=6x^2-6x-12=0$ in order to get the critical values. To do this, we can use the program QUADRATIC from Chapter 6.

QUADRATIC 6 ⁻6 ⁻12
2 ⁻1

So, $(^-1,F(^-1))=(^-1,17)$ is a relative maximum point and $(2,F(2))=(2,^-10)$ is a relative minimum point (see Figure 8.21).

(b) Where is the function increasing and where is it decreasing?

Since $F'(x)$ is positive in the set $\{x|x<^-1\}\cup\{x|x>2\}$, then the function is increasing for all x in this set. Since $F'(x)$ is negative for all x in the set $\{x|^-1<x<2\}$, then the function is decreasing for all x in this set.

(c) Find any inflection points for this function.

To find inflection points, we need to find where the second derivative $F''(x)$ changes sign, if ever. Thus, we need to solve $F''(x)=12x-6=0$, which yields $x=\frac{1}{2}$. To see if $F''(x)$ actually changes sign at $x=\frac{1}{2}$, we make a sign chart for $F''(x)$ about $x=\frac{1}{2}$ (Figure 8.22). Thus, the sign of $F''(x)$ changes from $-$ to $+$ at $x=\frac{1}{2}$, and the point $(\frac{1}{2},F(\frac{1}{2}))=(\frac{1}{2},4\frac{1}{4})$ is an inflection point.

(d) Find all values of x for which the curve is concave upward and concave downward.

Since $F''(x)<0$ for $x<\frac{1}{2}$, the curve is concave downward in the set $\{x|x<\frac{1}{2}\}$. Since $F''(x)>0$ for $x>\frac{1}{2}$, the curve is concave upward in the set $\{x|x>\frac{1}{2}\}$.

(e) Using the information gathered in Parts (a)–(d), sketch the graph of the function $y=F(x)=2x^3-3x^2-12x+10$.

This is done in Figure 8.23.

Figure 8.21   Sign chart for $F'(x)$ about $x=^-1$ and $x=2$.

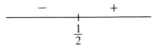

Figure 8.22   Sign chart for $F''(x)$ around $x=\frac{1}{2}$.

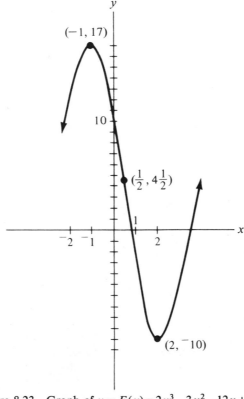

Figure 8.23   Graph of $y=F(x)=2x^3-3x^2-12x+10$.

Example 2

Repeat the above procedure for the function $y = F(x) = x^2 - 4x + 2$.

$$y' = F'(x) = 2x - 4 \quad \text{and} \quad y'' = F''(x) = 2.$$

(a) Maxima, minima: $F'(x) = 2x - 4 = 0$ when $x = 2$. So, $(2, F(2)) = (2, {}^-2)$ is a relative minimum (see Figure 8.24).

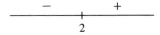

Figure 8.24   Sign chart for $F'(x) = 2x - 4$ around $x = 2$.

(b) Increasing, decreasing: From the sign chart for $F'(x)$ about $x = 2$ in Figure 8.24, the curve is decreasing in the set $\{x \mid x < 2\}$, and increasing in the set $\{x \mid x > 2\}$.
(c), (d) Inflection points and concavity: Since $F''(x) = 2 > 0$ for all x, the curve has no inflection points and is always concave upward (see Figure 8.25).

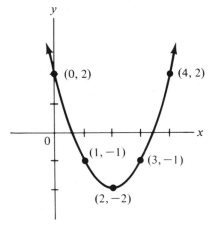

Figure 8.25   Graph of $y = F(x) = x^2 - 4x + 2$.

Example 3

$$y = F(x) = \ln x \qquad\qquad \text{By definition, } x > 0.$$

$$y' = F'(x) = \frac{1}{x} > 0 \quad \text{for all } x > 0.$$

$$y'' = F''(x) = \frac{{}^-1}{x^2} < 0 \quad \text{for all } x.$$

Thus, this curve has no relative maxima or minima, since $F'(x)$ is never 0. Since $F'(x) > 0$ for all $x > 0$, then the function is always rising (increasing). Also, the curve has no inflection points, since $F''(x)$ is never 0. Since $F''(x) < 0$ for all $x > 0$, then the curve is always concave

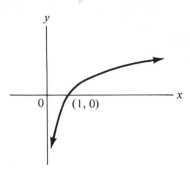

Figure 8.26   Graph of $y=F(x)=\ln x$. Note it is always increasing and concave downward.

downward. Since $\ln 1=0$, then the curve must pass through the point $(1,0)$. Thus, the shape of the curve must be as shown in Figure 8.26.

EXERCISES

In the following problems, find the relative maxima, relative minima, and the inflection points. Also, find the sets of values for x for which the functions are increasing, decreasing, concave upward, and concave downward. Finally, use the information you have gathered from the two derivatives to sketch the graphs of the functions.

1. $y=F(x)=^-3x^2+6x+1$

2. $y=F(x)=x^3-3x^2+4$

3. $y=F(x)=2x^3-3x^2-72x+10$

4. $y=F(x)=4x^3+2x^2-3x+5$

5. $y=F(x)=x\cdot\ln x$

6. $y=F(x)=e^{-x^2/2}$

241

9 Integral calculus

As in our study of differential calculus, we shall consider only a few major ideas of integral calculus. We hope to give the student an appreciation for the concepts of antiderivative and integral as well as a few applications of the concepts. In addition, we shall consider the "Fundamental theorem of calculus," which deals with the relationship between differential and integral calculus.

9.1 Antidifferentiation

We begin by considering the inverse process to that of differentiation, called *antidifferentiation*.

Definition of antiderivative (or indefinite integral)

An *antiderivative of a function* $F(x)$ is a function $G(x)$ having the property that $G'(x) = F(x)$.

(Another name for antiderivative is an *indefinite integral*. We prefer the name antiderivative because it describes the operation better.)

The symbol for an antiderivative of $F(x)$ with respect to x is

$$\int F(x)\,dx$$

read as "antiderivative of $F(x)$ with respect to x."

According to our definition, $\int F(x)\,dx = G(x)$ provided that $G'(x) = F(x)$.

Example

$$\int 5x^4\,dx = x^5, \quad \text{since } \frac{d}{dx}(x^5) = 5x^4.$$

However, this answer is not unique, since $\int 5x^4\,dx = x^5 + 3$ is also a true statement, since

$$\frac{d}{dx}(x^5 + 3) = 5x^4.$$

Also,

$$\int 5x^4\,dx = x^5 + 7$$

is true since

$$\frac{d}{dx}(x^5 + 7) = 5x^4.$$

In fact, the statement

$$\int 5x^4\,dx = x^5 + c$$

is true for any constant real number c, since the derivative of any constant is 0, so that

$$\frac{d}{dx}(x^5 + c) = 5x^4.$$

Therefore, we write $\int F(x)\,dx = G(x) + c$, where c is any arbitrary constant real number, called the constant of antidifferentiation, provided that $G'(x) = F(x)$.

We shall defer any rules or formulas for antidifferentiation until the next section in order that the reader be given the opportunity to really work with the definition of antiderivative. Right now, we would prefer that he make an educated guess at an antiderivative and then check his answer by differentiating it.

Further examples

1.

$$\int (3x^2 + 4x + 5)\,dx = x^3 + 2x^2 + 5x + c,$$

since

$$\frac{d}{dx}(x^3 + 2x^2 + 5x + c) = 3x^2 + 4x + 5.$$

243

2.

$$\int (^-32t+160)\,dt = ^-16t^2+160t+c,$$

since

$$\frac{d}{dt}(^-16t^2+160t+c)=^-32t+160.$$

3.

$$\int\left(6\sqrt{x}-\frac{4}{x^3}\right)dx=\int(6x^{1/2}-4x^{-3})\,dx=4x^{3/2}+2x^{-2}+c,$$

since

$$\frac{d}{dx}(4x^{3/2}+2x^{-2}+c)=6x^{1/2}-4x^{-3}=6\sqrt{x}-\frac{4}{x^3}.$$

4.

$$\int\frac{2}{x}\,dx=2\cdot\ln x+c,$$

provided $x>0$, since

$$\frac{d}{dx}(2\cdot\ln x+c)=\frac{2}{x}.$$

5.

$$\int e^{2x}\,dx=\tfrac{1}{2}e^{2x}+c,$$

since

$$\frac{d}{dx}\left(\tfrac{1}{2}e^{2x}+c\right)=e^{2x}.$$

Some applications of antidifferentiation

Since antidifferentiation is just the inverse process to differentiation, the applications of antidifferentiation are just the inverse to the applications of differentiation.

Example 1

Find the equation $y=F(x)$ for the curve passing through the point $(1,3)$ if the slope of this curve at any point on it is given by $m=2x+1$.

Since the slope of a curve at any point (x,y) is found by computing the derivative of the equation for the curve, the equation for the curve is found by computing the antiderivative of the slope function. Thus,

$$y=\int(2x+1)\,dx=x^2+x+c.$$

Since the point $(1,3)$ lies on the curve, $(1,3)$ must satisfy the equation for the curve. Thus, $3 = 1^2 + 1 + c$, or $c = 1$. Thus the equation of the curve is

$$y = x^2 + x + 1.$$

Example 2

Suppose that the marginal cost of producing the xth item is given by M.C. $= 10/\sqrt{x}$ and that the fixed cost is \$10. Find the total cost function $C = F(x)$.

Since the derivative of the total cost function is the marginal cost function, the antiderivative of the marginal cost function is the total cost function. Thus,

$$C = \int \frac{10}{\sqrt{x}} \, dx = \int 10x^{-1/2} \, dx = 20x^{1/2} + k = 20\sqrt{x} + k,$$

since

$$\frac{d}{dx}(20x^{1/2} + k) = 10x^{-1/2} = \frac{10}{\sqrt{x}}.$$

Also, since the fixed cost is \$10, then when $x = 0$, $C = 10$, so that $k = 10$. Therefore, the total cost function (in dollars) is

$$C = 20\sqrt{x} + 10.$$

Example 3

If the velocity of a falling object at any time t seconds after it first started falling is given by $v = {}^-32t + 160$, and if the initial height is 1500 feet, then find the formula for the height at any time t.

Since the derivative of the height formula yields the velocity formula, then the antiderivative of the velocity formula will yield the height formula. Thus,

$$s = \int ({}^-32t + 160) \, dt = {}^-16t^2 + 160t + c.$$

Also, since the initial height is 1500 feet, then when $t = 0$, $s = 1500$. Putting these values into the formula for s yields $c = 1500$. Thus, the height formula is

$$s = {}^-16t^2 + 160t + 1500.$$

Example 4

The instantaneous rate of change in a quantity P per unit change in a quantity q is given by $dP/dq = 10e^q + 5$. If $P = 20$ when $q = 0$, find the function relating P to q.

Since the derivative of the function $P=F(q)$ yields the instantaneous rate of change function dP/dq, then the antiderivative of the instantaneous rate of change function will yield the original function. Thus,

$$P= \int (10e^q +5) \, dq = 10e^q +5q+c,$$

since

$$\frac{d}{dq}(10e^q +5q+c) = 10e^q +5.$$

Also, since when $P=20$, $q=0$, then $20=10e^0+5\cdot 0+c$, so that $c=10$. Thus, the formula for P in terms of q is

$$P= 10e^q +5q+10.$$

EXERCISES

1. Use the definition of antiderivative to find the following antiderivatives:

(a) $\int (6x^2 -8x+2) \, dx$ (d) $\int \frac{6}{\sqrt{x}} \, dx$

(b) $\int (x^3 +x^2 +x+1) \, dx$ (e) $\int \frac{4}{x} \, dx$

(c) $\int (x^{3/2} -2x^{-3}) \, dx$ (f) $\int e^{-5x} \, dx$

2. Find the function whose derivative is $4x^3+6x$, if it passes through the point $(1,5)$.

3. If the marginal cost of producing x items is given by M.C.$=4x-200$, and if the fixed cost is \$100, then find the total cost function.

4. An object is moving in a straight line in such a way the its velocity at any time t is given by $v=6t^2-24t+12$. Find its distance from the starting point at any time t.

5. An object dropped from a height of 500 feet has for its velocity at any time t the function $v=\,^-20-32t$. Find a formula for its height at any time t.

6. The slope of the tangent line to a curve at any point $P(x,y)$ is given by $m=3e^{3x}$. This curve passes through the origin. Find the equation for the curve.

7. The instantaneous rate of change of the area of a certain geometric figure is given by $dA/dx=2\pi x$. Find the formula for A. What kind of a geometric figure is this?

9.2 Some formulas for antidifferentiation

Now that the reader has some feeling for the concept of the antiderivative, we present some formulas for antiderivatives. All of these formulas can be proved by differentiating the answers according to the definition of antiderivative.[1]

[1] These rules don't cover all cases. This is just a sample of a table of integral formulas. In general, antidifferentiation is a more difficult process than differentiation.

Rule 1

$$\int k\,dx = kx + c, \quad \text{where } k \text{ is a constant real number.}$$

Rule 2

$$\int x^n\,dx = \frac{1}{n+1}x^{n+1} + c, \quad \text{provided } n \neq {}^-1.$$

Rule 3

$$\int x^{-1}\,dx = \int \frac{1}{x}\,dx = \ln x + c, \quad \text{provided } x > 0.$$

Rule 4

$$\int k\cdot F(x)\,dx = k\cdot \int F(x)\,dx, \quad \text{where } k \text{ is a constant.}$$

Rule 5

$$\int (F(x) + G(x))\,dx = \int F(x)\,dx + \int G(x)\,dx$$

and

$$\int (F(x) - G(x))\,dx = \int F(x)\,dx - \int G(x)\,dx.$$

Rule 6

$$\int e^{kx}\,dx = \frac{1}{k}e^{kx} + c, \quad \text{where } k \text{ is a constant real number.}$$

Examples

1.

$$\int (x^4 + 6x^2 + 9)\,dx = \int x^4\,dx + 6\int x^2\,dx + \int 9\,dx$$
$$= \frac{1}{5}x^5 + 6\cdot\frac{1}{3}x^3 + 9x + c$$
$$= \frac{1}{5}x^5 + 2x^3 + 9x + c.$$

2.

$$\int \left(9\sqrt{x} + \frac{6}{x^2}\right)dx = \int (9x^{1/2} + 6x^{-2})\,dx$$

$$= 9\int x^{1/2}\,dx + 6\int x^{-2}\,dx$$

$$= 9 \cdot \frac{1}{\frac{3}{2}}x^{3/2} + 6 \cdot \frac{1}{-1}x^{-1} + c$$

$$= 6x^{3/2} - \frac{6}{x} + c.$$

3.

$$\int \left(e^{2x} - \frac{1}{x}\right)dx = \int e^{2x}\,dx - \int \frac{1}{x}\,dx$$

$$= \frac{1}{.2}e^{2x} - \ln x + c = 5e^{2x} - \ln x + c.$$

Antidifferentiation by the method of substitution

If one has a great need to antidifferentiate frequently, then he can purchase a book of antiderivative (integral) formulas. In order to use such a table, one would have to know the method of substitution. We would like to illustrate this method. Therefore, we present four more antidifferentiation formulas. In the following rules, assume that u is a differentiable function of x.

Rule 7

$$\int u^n\,du = \frac{1}{n+1}u^{n+1} + c, \quad \text{provided } n \neq {}^-1.$$

Rule 8

$$\int u^{-1}\,du = \int \frac{1}{u}\,du = \ln u + c, \quad \text{provided } u > 0.$$

Rule 9

$$\int e^u\,du = e^u + c.$$

Rule 10

$$\int \ln u\,du = u \cdot \ln u - u + c.$$

Examples

1.

$$\int (3x+1)^5 dx.$$

Let $u=3x+1$. Then, $du/dx=3$, so that $du=3\cdot dx$, or $dx=\frac{1}{3}du$. Making these substitutions yields the integral

$$\frac{1}{3}\int u^5 du = \frac{1}{18}u^6+c = \frac{1}{18}(3x+1)^6+c.$$

2.

$$\int x\sqrt{x^2+1}\ dx.$$

Let $u=x^2+1$. Then, $du/dx=2x$, so that $x\,dx=\frac{1}{2}du$. Making these substitutions yields the integral

$$\frac{1}{2}\int\sqrt{u}\ du = \frac{1}{2}\int u^{1/2} du = \frac{1}{2}\frac{1}{\frac{3}{2}}u^{3/2}+c = \frac{1}{3}u^{3/2}+c = \frac{1}{3}(x^2+1)^{3/2}+c.$$

3.

$$\int\frac{x}{2x^2+1}dx.$$

Let $u=2x^2+1$. Then, $du/dx=4x$, so that $x\,dx=\frac{1}{4}du$. Making these substitutions yields the integral

$$\frac{1}{4}\int\frac{du}{u} = \frac{1}{4}\int\frac{1}{u}du = \frac{1}{4}\ln u+c = \frac{1}{4}\ln(2x^2+1)+c.$$

4.

$$\int x\cdot e^{-x^2/2}dx.$$

Let $u=-x^2/2$. Then, $du/dx=-x$, or $x\,dx=-du$. Making these substitutions yields the integral

$$-\int e^u du = -e^u+c = -e^{-x^2/2}+c.$$

5.

$$\int \ln(5x)\,dx.$$

Let $u=5x$. Then, $du/dx=5$, or $dx=\frac{1}{5}du$. Making these substitutions yields

$$\frac{1}{5}\int \ln u\,du = \frac{1}{5}(u\cdot\ln u - u)+c$$

$$= \frac{1}{5}(5x\cdot\ln(5x)-5x)+c.$$

A program for antidifferentiating polynomials (optional)

In general, it is difficult to write programs for finding antiderivatives of functions. However, it is possible to write programs to find antiderivatives of specific cases of functions. As an illustration of this, we present the following program for finding antiderivatives of polynomials.

Program 9.1 ANTIDIFF

∇ *POLY*$\leftarrow$*C ANTIDIFF COEFS*; *N*; *EXPS*

[1] *N*$\leftarrow\rho$ *COEFS*   *N* is the number of coefficients.

[2] *EXPS*$\leftarrow$*1*$+$*N*$-\iota$*N*   The exponents for the antiderivative.

[3] *POLY*$\leftarrow$*(COEFS*$\div$*EXPS), C*   *POLY* is the vector of coefficients
      ∇ of the antiderivative. *C* is the con-
                                stant of antidifferentiation.

Example 1

Find $\int (4x^3+6x^2-2x+5)dx$, where the constant of antidifferentiation is 3.

      3 *ANTIDIFF* 4 6 ¯2 5
   1 2 ¯1 5 3

Thus, the antiderivative polynomial is $x^4+2x^3-x^2+5x+3$.

Example 2

Find $\int (x^4+6x^2+9)dx$, where *c* is 5.

      5 *ANTIDIFF* 1 0 6 0 9   Note we must account for 0
   2 0 2 0 9 5   coefficients in *COEFS* also.

Thus, we get the antiderivative

$$0.2x^5+0x^4+2x^3+0x^2+9x+5=0.2x^5+2x^3+9x+5.$$

More applications of antidifferentiation

1. Suppose that the marginal revenue from the sale of x items is given by M.R.$=1/(x+1)$. Find the revenue function.

   Since marginal revenue is the derivative of revenue, then revenue is the antiderivative of marginal revenue. Thus,

$$R=\int \frac{1}{x+1}dx.$$

Let $u = x + 1$. Then, $du/dx = 1$, or $du = dx$. Thus,

$$R = \int \frac{1}{u} du = \ln u + c = \ln(x+1) + c.$$

Since, if you sell no items, you receive no revenue, then $0 = \ln 1 + c = 0 + c$, or $c = 0$. Thus,

$$R = \ln(x+1).$$

2. Suppose that the slope of the tangent line to some unknown curve at any point $P(x,y)$ is given by $m = 3x^2(x^3+3)^3$. Suppose also that the curve passes through the point $(1,6)$. Find the equation of the curve.

$$y = \int 3x^2(x^3+3)^3 dx.$$

Let $u = x^3 + 3$. Then, $du/dx = 3x^2$, or $du = 3x^2 dx$. Therefore,

$$y = \int u^3 du = \frac{1}{4}u^4 + c, \quad \text{or} \quad y = \frac{1}{4}(x^3+3)^4 + c.$$

However, since $(1,6)$ lies on the curve, then $6 = \frac{1}{4}(1+3)^4 + c = 64 + c$. Thus, $c = {}^-58$ and the equation of the curve is

$$y = \frac{1}{4}(x^3+3)^4 - 58.$$

3. A projectile is fired vertically upward from a height of 5 feet with a muzzle velocity of 960 feet/second.
   (a) Find the function for the velocity at the end of the tth second. Due to gravity, the acceleration is $A = {}^-32$ feet/second. Since acceleration is the derivative of velocity, then velocity is the antiderivative of acceleration. Therefore,

$$v = \int {}^-32 \, dt = {}^-32t + c.$$

The muzzle velocity is the velocity at $t = 0$. So, $c = 960$. Thus, $v = {}^-32t + 960$.
   (b) Find the function for the height after t seconds. Since velocity is the derivative of height, then height is the antiderivative of velocity. Thus,

$$s = \int ({}^-32t + 960) \, dt = {}^-16t^2 + 960t + k.$$

Since the initial height is 5 feet, then $k = 5$. Thus,

$$s = {}^-16t^2 + 960t + 5.$$

251

EXERCISES

1. Use the rules of antidifferentiation to find the following:

(a) $\int (8x^3 - 6x + 2)\,dx$ (f) $\int 14x(2x^2+1)^6\,dx$

(b) $\int (x^5 + x^3 + 1)\,dx$ (g) $\int e^{\ln x} \cdot \dfrac{1}{x}\,dx$

(c) $\int \left(4e^{2x} - \dfrac{4}{x}\right)\,dx$ (h) $\int \ln 2x\,dx$

(d) $\int \left(\dfrac{2}{\sqrt{x}} - \dfrac{3}{x^2}\right)\,dx$ (i) $\int \dfrac{1}{2x+1}\,dx$

(e) $\int 3\sqrt{4x+1}\,dx$ (j) $\int \dfrac{\ln x}{x}\,dx$

2. An object is moving in a straight line in such a way that its acceleration after t seconds is given by $A = 2t - 4$.
   (a) Find the velocity after t seconds if the initial velocity is 5 feet/second.
   (b) Find the position function if the initial position is $s = 0$.

3. The slope of the tangent line to an unknown curve is given by $m = 1/(2x+1)^2$ and the curve passes through the origin. Find the equation of the curve.

4. The marginal cost of the xth item is given by M.C. $= 10e^{0.5x}$, and the initial cost is $100. Find the total cost function.

9.3 Area under a curve

We now consider a geometry problem, the solution of which will lead us to the definition and geometric interpretation of the definite integral. First, however, we need to discuss the use of the summation symbol Σ.

The summation symbol, Σ

The symbol Σ is frequently used in mathematics when working with a sum of a great many numbers. Its use is described as follows:

$$\sum_{i=a}^{b} F(i) = F(a) + F(a+1) + F(a+2) + F(a+3) + \ldots + F(b),$$

where a and b are integers.

Examples

1.
$$\sum_{i=1}^{10} i = 1+2+3+4+5+6+7+8+9+10 = 55.$$

2.
$$\sum_{i=0}^{4} (i^2+2i) = (0^2+2.0) + (1^2+2\cdot1) + (2^2+2\cdot2) + (3^2+2\cdot3) + (4^2+2\cdot4)$$
$$= 0+3+8+15+24 = 50.$$

252

3.
$$\sum_{i=1}^{100} i^2 = 1^2 + 2^2 + 3^2 + \ldots + 100^2.$$

4.
$$\sum_{i=1}^{n} F(x_i) = F(x_1) + F(x_2) + F(x_3) + \ldots + F(X_n).$$

Area under a curve

Let $y = F(x)$ be a function which is always $\geqslant 0$ for all x in the interval $a \leqslant x \leqslant b$, so that its graph lies entirely above the x axis in that interval. We want to find the area under the curve of $y = F(x)$, above the x axis, and between the vertical lines $x = a$ and $x = b$ (see Figure 9.1).

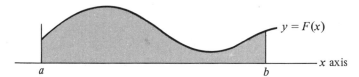

Figure 9.1   Area under a curve.

Let us consider the following scheme for approximating an area such as this:

First divide the interval $a \leqslant x \leqslant b$ into n subintervals each of width $\Delta x = (b - a)/n$. Call the points of subdivision $x_0 = a, x_1 = a + \Delta x, x_2 = a + 2 \cdot \Delta x, \ldots, x_i = a + i \cdot \Delta x, \ldots, x_n = a + n \cdot \Delta x = b$. At each of these points of subdivision, erect a perpendicular to the x axis and extend it upward until it meets the curve $y = F(x)$. In this way, we have sliced the area under the curve into n slices of area (see Figure 9.2). We now need a way of estimating the area of each slice. The area of the typical ith slice can be approximated by the area of the rectangle of width Δx and height $F(x_i)$. The sum $S = \sum_{i=1}^{n} F(x_i) \cdot \Delta x$, of areas of these rectangles is an approximation to the area under the curve.

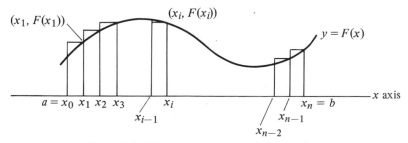

Figure 9.2   The area as a sum of rectangles.

253

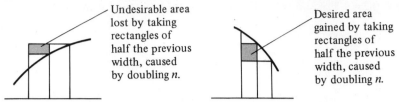

Figure 9.3   Increasing n makes the approximation better.

In order to make this approximation more accurate, take n larger, thereby making more rectangles of smaller width. As n gets larger and larger, $S = \sum_{i=1}^{n} F(x_i) \cdot \Delta x$ gets closer and closer to the actual area under the curve (see Figure 9.3). The actual area is symbolized by

$$\lim_{n \to \infty} \sum_{i=1}^{n} F(x_i) \cdot \Delta x.$$

Example

Approximate the area under the curve $y = F(x) = x^2$ over the interval $1 \leqslant x \leqslant 2$ (see Figure 9.4)

Let $n = 5$. Then,

$$\Delta x = \frac{b-a}{n} = \frac{2-1}{5} = 0.2.$$

Also, $x_1 = a + \Delta x = 1 + 0.2 = 1.2, x_2 = 1 + 2 \cdot 0.2 = 1.4, x_3 = 1 + 3 \cdot 0.2 = 1.6, x_4 =$

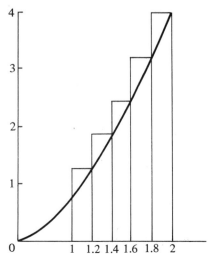

Figure 9.4   Approximation of the area under the curve $y = F(x) = x^2$.

$1+4\cdot0.2=1.8, x_5=b=2$. The area is approximated by the sum

$$S=\sum_{i=1}^{5} F(x_i)\cdot\Delta x$$

$$= F(x_1)\cdot\Delta x + F(x_2)\cdot\Delta x + F(x_3)\cdot\Delta x + F(x_4)\cdot\Delta x + F(x_5)\cdot\Delta x$$

$$= F(1.2)\cdot0.2 + F(1.4)\cdot0.2 + F(1.6)\cdot0.2 + F(1.8)\cdot0.2 + F(2)\cdot0.2$$

$$= (1.44+1.96+2.56+3.24+4.00)\cdot0.2 = (13.20)\cdot0.2 = 2.64.$$

The actual area (found by a more sophisticated technique to be discussed in Section 9.5) is $7/3 = 2.33333$. If one were to repeat the above process with $n = 100$, one would get approximately 2.358. If one did it with $n = 1000$, one would get approximately 2.336. Of course, if $n = 1000$, the process would be far too tedious to do by hand. Therefore, we shall use the following program AREA to accomplish the process described above with $n = 1000$.

Program 9.2 AREA

    ∇ SUM←A AREA B; N; WIDTH; HEIGHTS

                Find the area from A to B.

[1]   N←1000   Use 1000 rectangles. The reader may want to experiment with other values of N.

[2]   WIDTH←(B−A)÷N   The width of each rectangle.

[3]   HEIGHTS←FN(A+ WIDTH× ιN)

                A vector of heights of rectangles using right-hand endpoints of each subinterval.

[4]   SUM← +/WIDTH× HEIGHTS The sum of the areas of the rectangles.

    ∇

Examples

1. Find the area under the curve $y = F(x) = x^2$ over the interval $1 \leqslant x \leqslant 2$.

    ∇ FN [1]

[1]   Y←X∗2 ∇   We need to alter FN to fit our function.

     1 AREA 2
 2.3358335

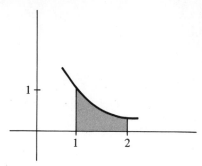

Figure 9.5   The shaded area is approximately 0.694 square units.

2. Find the approximate area under the curve $y = F(x) = 1/x$ in the interval $1 \leqslant x \leqslant 2$ (see Figure 9.5).

    ∇ *FN* [1]

[1]   *Y* ← 1 ÷ *X* ∇ Change *FN* to the new function.

      1 *AREA* 2
.6938972431

3. Approximate the area under the curve $y = F(x) = e^{-x^2}$ from $x = 0$ to $x = 1$ (see Figure 9.6).

    ∇ *FN* [1]

[1]   *Y* ← * (− *X* * 2)∇ APL for $y = e^{-x^2}$.

      0 *AREA* 1
.7475080112

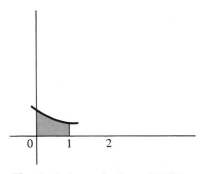

Figure 9.6   The shaded area is about 0.7475 square units.

4. Approximate the area under $y = F(x) = \sqrt{1 + x^2}$ in the interval from 1 to 5 (see Figure 9.7).

    ∇ *FN* [1]

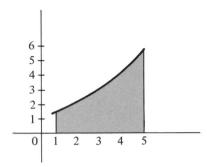

Figure 9.7   The shaded area is about 12.769 square units.

[1] $Y \leftarrow (1 + X * 2) * .5$ ∇

   1 *AREA* 5

12.76900121

EXERCISES

1. Use pencil and paper to do the following:
   (a) Approximate the area under the curve $y = F(x) = x^3 + 2x$ in the interval $0 \leqslant x \leqslant 2$. (Use $n = 4$ rectangles.)
   (b) Estimate the area under the curve $y = F(x) = 3x^2 + 2x + 1$ in the interval $1 \leqslant x \leqslant 2$. (Use $n = 5$ rectangles.)
   (c) Estimate the area under $y = F(x) = -x^2 + 2x + 3$ over the interval $^-1 \leqslant x \leqslant 3$. (Use $n = 4$ rectangles.)

2. Use the program *AREA*, which uses 1000 rectangles, to get better approximations to the areas in Exercise 1.

3. (a) Sketch the graph of the function $y = F(x) = x^2 - 2x - 3$ over the interval $^-1 \leqslant x \leqslant 3$.
   (b) Evaluate $^-1$ *AREA* 3 for this function.
   (c) Compare your answer in Part (b) with the answer to Exercise 2, Part (c).
   (d) Can you explain the sign of this answer?

4. (a) Sketch the curve of $y = F(x) = 4x^3$ in the interval $^-2 \leqslant x \leqslant 2$.
   (b) Evaluate $^-2$ *AREA* 2 for this function.
   (c) Can you explain the reason for this answer?

5. Write a program to evaluate $\sum_{i=1}^{100} i^2$.

9.4 The definite integral

Definition of the definite integral

The *definite integral* of a function $F(x)$ from a to b, symbolized by $\int_a^b F(x)\,dx$, is defined as follows:

$$\int_a^b F(x)\,dx = \lim_{n \to \infty} \sum_{i=1}^n F(x_i) \cdot \Delta x,$$

257

where the interval $a \leqslant x \leqslant b$ is divided into n subintervals each of width $\Delta x = (b-a)/n$ and $x_i = a + i \cdot \Delta x$ for $i = 1, 2, \ldots, n$.

This is precisely the same quantity that we used in the previous section to compute the area under a curve $y = F(x)$ over the interval $a \leqslant x \leqslant b$. Thus, if $y = F(x)$ is positive for all x in the interval $a \leqslant x \leqslant b$, then $\int_a^b F(x)\,dx$ yields the area under the curve of $y = F(x)$, above the x axis, between the vertical lines $x = a$ and $x = b$. Also, since there was nothing in the program AREA that depended on $F(x)$ being positive in the interval $a \leqslant x \leqslant b$, then we can use the program AREA to compute definite integrals. The following program does just this. We will then use this program INTEGRAL to approximate definite integrals.

Program 9.3 INTEGRAL

     ∇ I←A INTEGRAL B   Note that we merely change the
                               name of the program AREA to IN-
[1]   I←A AREA B ∇   TEGRAL.

Examples

1. Approximate $\int_1^2 x^2\,dx$.

     ∇ FN [1]

[1]   Y←X*2 ∇

     1 INTEGRAL 2
2.3358335

2. Estimate $\int_1^2 (1/x)\,dx$.

     ∇ FN [1]

[1]   Y←1÷X ∇

     1 INTEGRAL 2
.6938972431

3. Approximate $\int_0^1 e^{-x^2}\,dx$.

     ∇ FN [1]

[1]   Y← *(−X*2) ∇

     0 INTEGRAL 1
.74750800112

In each of the above examples, we have computed the areas under the curves. This is because each of the curves lies completely on or above the x axis over the prescribed intervals. One might wonder what the geometric significance is of the definite integral if the function does not have its graph lying entirely on or above the x axis over the designated interval.

More on the geometric interpretation of the definite integral

To illustrate the complete picture of the geometric interpretation of the definite integral, let us consider the function of Example 2 in Section 6.5: $y = F(x) = x^4 - x^3 - 7x^2 + x + 6$ (see Figure 9.8).

    ∇ *FN* [1]

[1] $Y \leftarrow (X*4) + (-X*3) + (-7 \times X*2) + X + 6$ ∇

Altering FN to fit this function.

    $^-1$ *INTEGRAL* 1
7.7333326667

Since the function is positive for all x in the interval $^-1 \leqslant x \leqslant 1$, then as previously stated, $\int_{-1}^{1} F(x)\,dx$ yields the area under the curve of $y = F(x)$, above the x axis, between $x = ^-1$ and $x = 1$. This area is approximately 7.73.

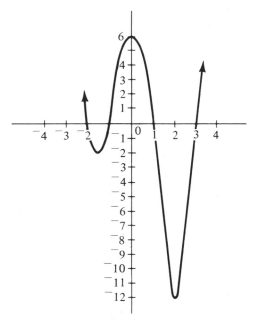

Figure 9.8   Note that the graph of this function is sometimes above and sometimes below the x axis.

Let us now use *INTEGRAL* to approximate $\int_{-2}^{-1} F(x)\,dx$ and $\int_{1}^{3} F(x)\,dx$. Notice that in the intervals $^-2 \leqslant x \leqslant ^-1$ and $1 \leqslant x \leqslant 3$, the function is negative for all x. Its graph lies entirely below the x axis in these intervals.

　　　$^-2$ *INTEGRAL* $^-1$
$^-1.883331417$

　　　1 *INTEGRAL* 3
$^-16.266664933$

It thus appears that if the function is always negative in an interval $a \leqslant x \leqslant b$, so that its graph lies below the x axis in this interval, then $\int_{a}^{b} F(x)\,dx$ yields a negative answer. This answer represents the negative of the area between the curve of $y = F(x)$, the x axis, and the lines $x = a$ and $x = b$.

Let us now consider the integral $\int_{-2}^{3} F(x)\,dx$. Notice that in the interval $^-2 \leqslant x \leqslant 3$, the function is sometimes positive and sometimes negative, so that its graph is sometimes above the x axis and sometimes below it.

　　　$^-2$ *INTEGRAL* 3
$^-10.41655208$

If a function is sometimes negative and sometimes positive in an interval $a \leqslant x \leqslant b$, then $\int_{a}^{b} F(x)\,dx$ yields the *net area* above the x axis between the lines $x = a$ and $x = b$. If the result is a negative number, as in this example, this indicates that there is more area under the x axis than above it. As a demonstration of the validity of these statements, let us add the integrals above.

　　　($^-2$ *INTEGRAL* $^-1$) + ($^-1$ *INTEGRAL* 1) + (1 *INTEGRAL* 3)
$^-10.41655208$

To get the *total area* between a curve and the x axis between $x = a$ and $x = b$, we need to find the x intercepts and add the absolute values of the integrals between these x intercepts. Thus, the total area between the graph of the above function $y = F(x)$, the x axis, and the lines $x = ^-2$ and $x = 3$ is approximated by the following:

　　　$(|^-2$ *INTEGRAL* $^-1$) + $(|^-1$ *INTEGRAL* 1) + $(|1$ *INTEGRAL* 3)
25.88331341

Areas between two curves

If a curve $y = G(x)$ is above a curve $y = F(x)$ for all x in the interval $a \leqslant x \leqslant b$ (see Figure 9.9), then the total area between these curves and the

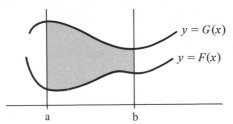

Figure 9.9   Area between two curves.

lines $x = a$ and $x = b$ is found by evaluating

$$\int_a^b \underset{\substack{\text{Upper}\\\text{curve}}}{G(x)\ dx} - \int_a^b \underset{\substack{\text{Lower}\\\text{curve}}}{F(x)\ dx} = \int_a^b (G(x) - F(x))\,dx.$$

This formula will work whether both ·curves are above the x axis, both below it, or one above and the other below. The reader can easily convince himself of this by drawing some sketches and using the above geometric interpretations of the definite integral in each case. This will be left as a good exercise for the reader.

Example 1

Find the area between the curves $y = G(x) = \sqrt{x}$ and $y = F(x) = x^2$. We can use a graph (Figure 9.10) to see which curve is the upper curve and which is the lower curve, and to help us in finding the values of a and b.

Thus, this area is found by computing

$$\int_0^1 \sqrt{x}\ dx - \int_0^1 x^2 dx = \int_0^1 (\sqrt{x} - x^2)\,dx.$$

We can use the program **INTEGRAL** to estimate this area.

∇ **FN** [1]

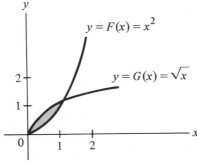

Figure 9.10   Note that these curves meet when $x = 0$ and $x = 1$, since $\sqrt{x} = x^2$ only for these values of x. Also, $\sqrt{x} \geqslant x^2$ for all x in the interval $0 \leqslant x \leqslant 1$.

[1] $Y \leftarrow (X*.5)-(X*2)$ ∇

　　0 *INTEGRAL* 1
.33333266344

Thus, the area between these curves is approximately 0.333.

Example 2

Find the area between the curves $y = F(x) = 2x^2$ and $y = G(x) = x^4 - 2x^2$.
Once again, a sketch (Figure 9.11) helps determine which curve is the
upper curve and what are the values of a and b.

These curves meet when $x = {}^-2$, $x = 0$, and $x = 2$, since $2x^2 = x^4 - 2x^2$
when $4x^2 = x^4$ or when x has these values. Also, the graph of $y = 2x^2$ is
always above the graph of $y = x^4 - 2x^2$ for all x in the interval ${}^-2 \leqslant x \leqslant 2$.
Thus, the area is given by

$$\int_{-2}^{2} 2x^2 \, dx - \int_{-2}^{2} (x^4 - 2x^2) \, dx = \int_{-2}^{2} (4x^2 - x^4) \, dx.$$

This can be approximated using the program *INTEGRAL* as follows:

　　∇ *FN* [1]

[1] $Y \leftarrow (4 \times X*2)-(X*4)$ ∇

　　${}^-2$ *INTEGRAL* 2
8.533290667

Thus, the area between these curves is about 8.53 square units.

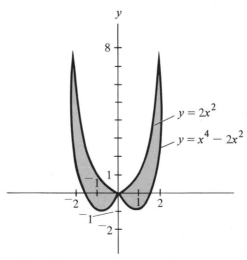

Figure 9.11   Area between two curves.

EXERCISES

1. Use the program *INTEGRAL* to approximate the following:

(a) $\int_1^2 (3x^2 + 2x + 1)\,dx$ (c) $\int_4^5 \sqrt{1 + x^2}\,dx$

(b) $\int_{-1}^1 (x^3 - 3x^2)\,dx$ (d) $\int_1^4 (\ln x)\,dx$

2. (a) Sketch the graph of the following function showing the x intercepts: $y = F(x) = x^3 - 4x$.
   (b) Find the net area between this graph, the x axis, and the first and third x intercepts.
   (c) Find the total area between this graph, the x axis, and the first and third x intercepts.

3. (a) Sketch the graph of the following function showing the x intercepts: $y = F(x) = x^3 - 9x^2 + 23x - 15$.
   (b) Find the net area between this graph, the x axis, and the first and third intercepts.
   (c) Find the total area between this graph, the x axis, and the first and third intercepts.

4. Find the total area between the following pairs of curves:
   (a) $y = x^3$ and $y = x$ (in the first quadrant)
   (b) $y = 2 - x^2$ and $y = x^2$
   (c) $y = x^2$ and $y = 2x + 8$

5. Draw some sketches and use the geometric interpretations of the definite integral to explain why

$$\int_a^b G(x)\,dx - \int_a^b F(x)\,dx,$$

where $G(x)$ is always above $F(x)$, in the interval $a \leqslant x \leqslant b$, will always yield the area between the curves $y = G(x)$ and $y = F(x)$, and the lines $x = a$ and $x = b$, regardless of whether both or none or one of the curves of $G(x)$ an $F(x)$ are above the x axis.

9.5 The fundamental theorem of calculus

We now consider a theorem of such importance that it is referred to as the *fundamental theorem of calculus*.

Let $y = F(x)$ be defined and continuous in the interval $a \leqslant x \leqslant b$, and let $G(x)$ be any antiderivative of $F(x)$. Then,

$$\int_a^b F(x)\,dx = G(b) - G(a).$$

This theorem establishes a relationship between the concept of antidifferentiation (the inverse process of differentiation) and definite integration (which has to do with infinitely large sums of areas of rectangles).

Although it has been our policy throughout this text to avoid rigorous mathematical proofs, we do feel that some justification of this theorem would be desirable, since it is not intuitively obvious. Therefore, we shall prove this theorem in the special case where $F(x) \geqslant 0$ for all x in the interval $a \leqslant x \leqslant b$.

Let x be in the interval $a \leqslant x \leqslant b$. Define the function $A(x)$ to be the area under the curve $y = F(x)$ above the x axis between $x = a$ and $x = x$ (see Figure 9.12). Let $h \geqslant 0$ be such that $a \leqslant x + h \leqslant b$. Then, $A(x+h) - A(x)$ gives the area under the curve between $x = x$ and $x = x + h$.

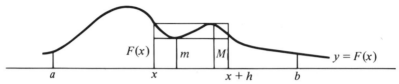

Figure 9.12   A function $y = F(x)$ in the interval $a \leqslant x \leqslant x + h \leqslant b$.

Let m be the minimum height of the curve between $x = x$ and $x = x + h$, and let M be the maximum height between $x = x$ and $x = x + h$. Then,

$$m \cdot h \leqslant A(x+h) - A(x) \leqslant M \cdot h, \quad \text{or} \quad m \leqslant \frac{A(x+h) - A(x)}{h} \leqslant M.$$

Now, as $h \to 0$, $m \to F(x)$ and $M \to F(x)$. Or,

$$\lim_{h \to 0} m = \lim_{h \to 0} M = F(x).$$

However, since $(A(x+h) - A(x))/h$ is between m and M, then

$$\lim_{h \to 0} \frac{A(x+h) - A(x)}{h} = F(x).$$

Thus, $A'(x) = F(x)$ and $A(x)$ is an antiderivative of $F(x)$.

By the definition of $A(x)$, $A(b)$ is the area under the curve between $x = a$ and $x = b$. Also, $A(a) = 0$, since this is the area under the curve from a to a. Recall that $\int_a^b F(x) \, dx$ also gives the area under the curve from $x = a$ to $x = b$. Therefore,

$$\int_a^b F(x) \, dx = A(b) - A(a).$$

Now let $G(x)$ be any other antiderivative of $F(x)$. Then, since any two antiderivatives of a function differ only by a constant, $A(x) = G(x) + c$, for some constant c. Therefore,

$$\int_a^b F(x) \, dx = A(b) - A(a) = (G(b) + c) - (G(a) + c) = G(b) - G(a).$$

This completes the proof of this special case of the Fundamental theorem of calculus.

This theorem enables us to easily get exact answers to definite integrals, provided we can easily find an antiderivative of the function being integrated. For many functions, however, it is not easy to find an antiderivative. There even exist functions for which there is no antiderivative! For functions such as these, we can always use the program *INTEGRAL* to get a good approximation to the definite integral.

Example 1

Evaluate $\int_{1}^{2} x^2 dx$.

Since $\frac{1}{3} x^3$ is an antiderivative of x^2, then by the Fundamental theorem,

$$\int_{1}^{2} x^2 dx = \frac{1}{3} x^3 \Big|_{1}^{2} = \left(\frac{1}{3} \cdot 2^3\right) - \left(\frac{1}{3} \cdot 1^3\right) = \frac{8}{3} - \frac{1}{3} = \frac{7}{3}.$$

Using the program *INTEGRAL*, we got 2.3358, which is a good approximation to this integral.

Example 2

Evaluate $\int_{1}^{2} \frac{1}{x} dx$.

Since $\ln x$ is an antiderivative of $1/x$, then

$$\int_{1}^{2} \frac{1}{x} dx = \ln x \Big|_{1}^{2} = (\ln 2) - (\ln 1) = (\ln 2) - 0 = \ln 2.$$

The program *INTEGRAL* yielded 0.693897.

    ⊛ 2
0.6931471806   ln 2 in APL.

Thus, the program *INTEGRAL* yields a good approximation to this integral.

Example 3

Evaluate $\int_{1}^{5} \sqrt{1+x^2} \, dx$.

We would have a very difficult task in trying to find an antiderivative for the function $\sqrt{1+x^2}$, although it does have an antiderivative. Therefore, we will have to settle for the approximation we can get using the program *INTEGRAL*.

    ∇ *FN* [1]
[1]   Y←(1 + X * 2) * .5 ∇

    1 *INTEGRAL* 5
12.76900121

Example 4

Find $\int_0^1 e^{-x^2}\,dx$.

The function $F(x)=e^{-x^2}$ has no elementary antiderivative. Thus, we have no choice but to settle for an approximation for this integral.

$\nabla\,FN\,[1]$

[1] $Y \leftarrow *(-X*2)\ \nabla$

0 *INTEGRAL* 1
.7475880112

Example 5

Find the area under the curve $y = F(x)=3x^2+4x+1$, above the x axis, between the vertical lines $x=2$ and $x=4$.

Since this curve lies entirely above the x axis between $x=2$ and $x=4$, then $\int_2^4 (3x^2+4x+1)\,dx$ yields the desired area. By the Fundamental theorem,

$$\int_2^4 (3x^2+4x+1)\,dx = (x^3+2x^2+x)\Big|_2^4$$

$$= (4^3+2.4^2+4)-(2^3+2.2^2+2)=82.$$

Example 6

Find the net area between the curve $y = F(x)=x^4-x^3-7x^2+x+6$, the x axis, and $x=-2$ and $x=3$.

$$\int_{-2}^3 (x^4-x^3-7x^2+x+6)\,dx = \left(\tfrac{1}{5}x^5-\tfrac{1}{4}x^4-\tfrac{7}{3}x^3+\tfrac{1}{2}x^2+6x\right)\Big|_{-2}^3$$

$$= \left(\tfrac{243}{5}-\tfrac{81}{4}-\tfrac{189}{3}+\tfrac{9}{2}+18\right)-\left(\tfrac{32}{5}-\tfrac{16}{4}+\tfrac{56}{3}+2-12\right)$$

$$= -10\tfrac{5}{12} = -10.4166666667.$$

Since this answer is negative, then more of the curve must be below the x axis than above it.

Example 7

Find the area between the curves $y = \sqrt{x}$ and $y = x^2$. This was Example 1 of the area between curves in the last section. This area can be found by evaluating the definite integral

$$\int_0^1 (\sqrt{x}-x^2)\,dx = \int_0^1 (x^{0.5}-x^2)\,dx = \left(\frac{1}{1.5}x^{1.5}-\frac{1}{3}x^3\right)\Big|_0^1$$

$$= \left(\frac{2}{3}-\frac{1}{3}\right)-(0-0)=\frac{1}{3}.$$

266

EXERCISES

1. Use the Fundamental theorem of calculus to compute the following:

(a) $\int_2^3 (4x^3 - 6x^2)\,dx$

(d) $\int_1^4 \left(\frac{2}{\sqrt{x}} - \frac{3}{x^2}\right)\,dx$

(b) $\int_1^3 (x^5 + x^3 + 1)\,dx$

(e) $\int_0^4 \sqrt{2x+1}\,dx$

(c) $\int_0^1 e^x\,dx$

(f) $\int_0^1 \frac{1}{2x+1}\,dx$

2. Find the area under the curve $y = 9 - x^2$, above the x axis, between $x=0$ and $x=3$.

3. Find the area between the curves $y = 8 - x^2$ and $y = x^2$.

4. (a) Find the net area between the x axis and the curve $y = x^3 - x$, between $x = ^-1$ and $x=1$.
   (b) Find the total area.

5. Use the definition of definite integral, and the Fundamental theorem to show that

$$\lim_{n\to\infty} \frac{e^{1/n} + e^{2/n} + e^{3/n} + \cdots + e^{(n-1)/n} + e^{n/n}}{n} = e - 1.$$

9.6 More applications of integration

Example 1

Suppose that the velocity at the end of t seconds of an object moving in a straight line is given by

$$V = \frac{dS}{dt} = 5 + 10t - t^2 \text{ feet/second.}$$

Find the total distance, D, traveled by the object during the first 5 seconds.

Since velocity, V, is the derivative of position, S, then position, S, is the antiderivative of velocity, V. Thus, the position function is given by

$$S = F(t) = \int (5 + 10t - t^2)\,dt = 5t + 5t^2 - \frac{1}{3}t^3 + c,$$

for some constant c. The distance traveled is therefore,

$$D = F(5) - F(0) = \left(25 + 125 - \frac{125}{3} + c\right) - c = 108\tfrac{1}{3} \text{ feet.}$$

This can also be found using the definite integral

$$D = \int_0^5 (5 + 10t - t^2)\,dt = \left(5t + 5t^2 - \frac{1}{3}t^3\right)\Big|_0^5 = 108\tfrac{1}{3} \text{ feet.}$$

267

The distance traveled during the 5th second is given by

$$\int_4^5 (5+10t-t^2)\,dt = \left(5t+5t^2-\frac{1}{3}t^3\right)\Bigg|_4^5 = 29\tfrac{2}{3}\ \text{feet}.$$

Example 2

The marginal cost of the xth item is given by $M.C.=dc/dx=10/\sqrt{x}$. Find the total cost as x goes from 25 to 100 items.

$$\Delta C = \int_{25}^{100}\frac{10}{\sqrt{x}}\,dx = \int_{25}^{100}10x^{-1/2}\,dx = 20x^{1/2}\Bigg|_{25}^{100}$$

$$= 20\sqrt{100} - 20\sqrt{25} = 200-100 = \$100.$$

Example 3

A company is trying to decide whether or not to purchase a new computer. The salesman tells them that if they purchase the new computer, they will save on the cost of their operation at the rate of

$$\frac{dS}{dt} = 20000t+20000 \quad \text{(in dollars per year)},$$

where t is the number of years they will have the computer and S is the total savings after t years. How much will they save during the first 5 years as a result of purchasing the computer?

This can be computed by the following definite integral:

$$\int_0^5 (20,000t+20,000)\,dt = 10,000t^2+20,000t\big|_0^5 = \$350,000.$$

If the computer costs \$480,000, how long does it take before the computer saves the company enough money to pay for itself?

Let x denote the length of time needed for the computer to save the company the \$480,000 it cost. Then,

$$\int_0^x (20,000t+20,000)\,dt = 10,000t^2+20,000t\big|_0^x$$

$$= 10,000x^2+20,000x = 480,000.$$

Or,

$$x^2+2x-48 = (x+8)\cdot(x-6) = 0.$$

Thus, $x=6$ years.

The previous three examples all illustrate that if we are given a function for the rate of change of a quantity y per unit change in a quantity x, then we can use the definite integral to find the total change in y between two values of x.

268

Most calculus texts emphasize the use of the definite integral in solving geometrical problems. We have already considered the use of the definite integral in solving problems dealing with area. We would now like to consider the use of the definite integral in solving problems dealing with volume.

Volumes of solids of revolution

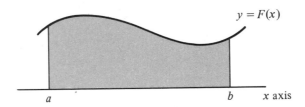

A *solid of revolution* is a solid formed by revolving a plane region about a line. Suppose the region bounded by the curve $y = F(x)$, the x axis, and the lines $x = a$ and $x = b$ is revolved about the x axis, forming a solid of revolution (Figure 9.13a). We would like to find the volume of this solid.

As we did when we found areas, we begin by dividing the interval $a \leqslant x \leqslant b$ into n subintervals of length $\Delta x = (b-a)/n$, and label the points of subdivision $x_1, x_2, \ldots, x_i, \ldots, x_n$. At each of these points of subdivision, we

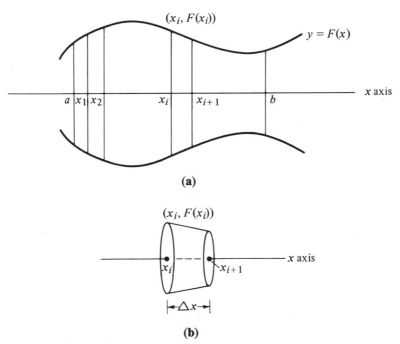

Figure 9.13   (a) A solid of revolution. (b) One crosswise slice of the solid.

pass a plane perpendicular to the x axis, thereby slicing our solid into slices of width Δx.

Consider the typical ith such slice of the solid (Figure 9.13b). This slice resembles a right circular cylinder, provided Δx is small. The volume of a right circular cylinder is found by the formula $V = \pi \cdot r^2 \cdot h$, where r is the radius of the base and h is the height. If we let $r = F(x_i)$ and $h = \Delta x$, then a pretty good approximation to the volume of this ith slice of the solid of revolution can be obtained by using the formula $V_i = \pi \cdot (F(x_i))^2 \cdot \Delta x$; especially if Δx is small.

Adding these approximations of the volumes of the slices, one gets the following approximation to the total volume of the solid of revolution:

$$\sum_{i=1}^{n} V_i = \sum_{i=1}^{n} \pi \cdot (F(x_i))^2 \cdot \Delta x.$$

This approximation can be made better by making n larger, thereby making Δx smaller. The actual volume can be obtained by computing $\lim_{n \to \infty} \sum_{i=1}^{n} \pi \cdot (F(x_i))^2 \cdot \Delta x$. However, from our definition of definite integral, this limit can be obtained by computing $\int_{a}^{b} \pi \cdot (F(x))^2 \, dx$.

Example

Find the volume of the solid of revolution obtained by revolving the region between the curve of $y = F(x) = 3\sqrt{x}$, the x axis, $x = 1$ and $x = 4$ about the x axis (see Figure 9.14).

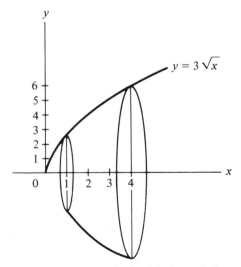

Figure 9.14   Another solid of revolution.

Using the formula developed above, we get

$$V = \int_1^4 \pi \cdot (3\sqrt{x}\,)^2\,dx = \pi \cdot \int_1^4 9x\,dx = \pi \cdot \frac{9}{2}x^2\Big|_1^4 = \pi \cdot \frac{9}{2}(16-1)$$

$$= \frac{135\pi}{2} \text{ cubic units.}$$

In a manner similar to this, we could develop definite integral formulas for computing lengths of curves, surface areas of solids of revolution, centers of mass, and many other geometrical quantities. However, we will leave this task to the standard calculus course. We wished only to illustrate the technique for developing such formulas.

EXERCISES

1. The velocity of a projectile (in feet per second) is given by $V = dS/dt = 640 - 32t$. Find the total distance covered by the projectile during the first 10 seconds.

2. The marginal cost of the xth item is given (in dollars per item) by $M.C. = dC/dx = \sqrt{x+10}$. Find the total change in cost as x goes from 6 to 15.

3. The rate of change of sales of a new product is given by $dS/dx = 100 \cdot e^{0.01x}$, where x is the number of days the product has been on the market. Find the total sales during the first 100 days.

4. The rate of increase of bacteria in a certain culture is given by $F(x) = 200 \cdot e^{0.02x}$, where x is in minutes. Find the total number of bacteria in the culture when x is one hour.

5. A company is trying to decide whether or not to buy a new machine which costs $5000. If they purchase the machine, they will save on production costs at a rate given (in dollars per month) by $dS/dt = 100t + 150$. In how many months will the machine have paid for itself?

6. Write a program for approximating the volumes of solids of revolution obtained by revolving a region bounded by a curve $y = F(x)$, the x axis, and $x = a$ and $x = b$ about the x axis. (You may use the program *INTEGRAL* as a subprogram.)

7. Find the volume of the solid of revolution obtained by revolving the region bounded by the curve $y = \sqrt{4x}$, the x axis, and $x = 1$ and $x = 4$ about the x axis.

8. Find the volume of the solid of revolution obtained by revolving the region bounded by the curve $y = \sqrt{1-x^2}$, the x axis, and $x = -1$ and $x = 1$ about the x axis.

9. Find the volume of the solid of revolution obtained by revolving the region bounded by the curve $y = F(x) = 4 - x^2$ and the x axis about the x axis.

10. The length of a curve $y = F(x)$ from a point $(a, F(a))$ to a point $(b, F(b))$ is given by the formula $\int_a^b \sqrt{1 + (F'(x))^2}\,dx$. Use this formula to find the length of the curve $y = F(x) = x^{3/2}$ from $(0,0)$ to $(4,8)$.

10 Probability

Originally, probability theory was developed to describe games of chance. However, in recent years, the theory of probability has become extremely useful in solving practical problems involving the chances of succeeding at various endeavors. Often businessmen or engineers must decide which of several different alternatives has the greatest probability of success. Probability may be used in any situation in which one is not certain of the outcome. In this chapter, we shall study some of the more important aspects of elementary probability theory.

10.1 Axioms of probability

In this text, an *experiment* is any situation in which we are not certain of the outcome.

Sample spaces and events

A *sample space* for an experiment is the set of all possible single outcomes of one performance of the experiment. We shall use the letter S to denote the sample space for our experiment.

An *event*, E, for the experiment, is any subset of the sample space S. We say that an event has occured if any outcome in the event occurs.

Examples

1. Consider the experiment of rolling a die. Let 1 be used to denote the occurrence of one dot facing upward on the die, 2 to denote the occurrence of two dots facing upward, and so on. Then, the sample space for this experiment can be represented by $S = \{1, 2, 3, 4, 5, 6\}$ (or $S \leftarrow 1\ 2\ 3\ 4\ 5\ 6$, in APL).

The event that an even number occurs is given by $E = \{2,4,6\}$ (or $E \leftarrow 2\ 4\ 6$, in APL). The event that an odd number occurs is $O = \{1,3,5\}$. The event that a number greater than 6 occurs is given by $\emptyset$, since this event is impossible.

2. Consider the event of rolling a die and flipping a coin. The sample space for this experiment is the set

$$S = \{(1,H),(1,T),(2,H),(2,T),(3,H),(3,T),$$
$$(4,H),(4,T),(5,H),(5,T),(6,H),(6,T)\},$$

where H represents heads and T tails.

The event that a 5 occurs is the subset of S

$$F = \{(5,H),(5,T)\}.$$

The event that a head occurs is the subset

$$H = \{(1,H),(2,H),(3,H),(4,H),(5,H),(6,H)\}.$$

3. A dart is thrown at a circular target of radius 12 inches in such a way that the target is never missed. The sample space for this experiment is the set $S = \{x | 0 \leqslant x \leqslant 12\}$, where x is the distance of the dart from the center of the target.

If the bullseye has a radius of 3 inches, then the event that the bullseye is hit is the set $B = \{x | 0 \leqslant x \leqslant 3\}$.

Some important special events

Since S is a subset of S, then S is an event. It is usually referred to as the *certain event* because, by the definition of sample space, we are certain to get an outcome which is in S.

Also, since the empty set $\emptyset$ is a subset of S, then $\emptyset$ is an event. Since it is impossible to get any elements in $\emptyset$, $\emptyset$ is called the *impossible event*.

Let A and B be events in S. The event $A \cap B$ is the event A "and" B, and it occurs when an outcome that is in both A and B occurs.

The event $A \cup B$ is the event A "or" B, and it occurs when an element in A or in B or in both occurs.

The event A' is the event "not" A (the complement of A) and it occurs when an outcome not in A, but in the sample space, occurs.

Example

Consider Example 2 above in which the sample space was

$$S = \{(1,H),(1,T),(2,H),(2,T),(3,H),(3,T),$$
$$(4,H),(4,T),(5,H),(5,T),(6,H),(6,T)\}.$$

Let

$$F = \{(5,H),(5,T)\} \quad \text{and}$$
$$\underline{H} = \{(1,H),(2,H),(3,H),(4,H),(5,H),(6,H)\}$$

as above. Then, the event that a 5 *and* a head occur is given by $F \cap \underline{H} = \{(5, H)\}$. The event that a 5 *or* a head occurs is $F \cup \underline{H} = \{(5, T), (1, H), (2, H), (3, H), (4, H), (5, H), (6, H)\}$. The event that a head does not occur is $\underline{H}' = \{(1, T), (2, T), (3, T), (4, T), (5, T), (6, T)\}$. This event, $\underline{H}'$ is equal to the event $\underline{T}$ that a tail occurs. The event that a negative number occurs is $\varnothing$, since this is an impossible event.

Definition of a probability function on a sample space

Let S be the sample space for an experiment. A probability function on S is a function P defined on the set of all events in S having the following properties:

1. If E is an event, then $0 \leqslant P(E) \leqslant 1$. ($P(E)$ is read as "the probability of E.")
2. $P(S) = 1$. (Since S is the certain event, we assign to S the largest probability.)
3. $P(\varnothing) = 0$. (Since $\varnothing$ is the impossible event, we assign to $\varnothing$ the smallest probability.)
4. If $E_1, E_2, \ldots, E_n$ are pairwise disjoint events, (i.e., $E_i \cap E_j = \varnothing$, for $i \neq j$), then

$$P(E_1 \cup E_2 \cup \ldots \cup E_n) = P(E_1) + P(E_2) + \ldots + P(E_n).$$

(Although it is not explicitly stated in our definition, we would also hope that $P(E)$ would be a measure of the likelihood of occurrence of E.)

Example 1

In the experiment of rolling a *fair* die, the sample space is $S = \{1, 2, 3, 4, 5, 6\}$. If the die is fair, then each of the elements in the sample space would be *equally likely*. Thus, $P(1) = P(2) = P(3) = P(4) = P(5) = P(6)$. Let us call this common value p. Since $P(S) = 1$, and $S = \{1\} \cup \{2\} \cup \{3\} \cup \{4\} \cup \{5\} \cup \{6\}$, then by Axiom 4 in the preceding definition,

$$P(S) = P(1) + P(2) + P(3) + P(4) + P(5) + P(6) = 6p = 1.$$

Therefore, $p = 1/6$.

Let E be the event that an even number is rolled. Then, $E = \{2, 4, 6\}$, and by Axiom 4,

$$P(E) = P(2) + P(4) + P(6) = \frac{1}{6} + \frac{1}{6} + \frac{1}{6} = \frac{3}{6} = \frac{1}{2}.$$

Let $F = \{5, 6\}$ be the event that a number greater than 4 is rolled. Then, $P(F) = P(5) + P(6) = 2/6 = 1/3$.

This example suggests the following probability rule for sample spaces containing equally likely outcomes.

Probability rule 1

If a sample space contains N *equally likely* outcomes, and if & is one of these outcomes, then $P(\&) = 1/N$. Also, if an event E in S contains M of these outcomes, then

$$P(E) = \frac{M}{N} = \frac{\text{The number of outcomes "favorable" to } E}{\text{The total number of outcomes in } S}.$$

[*Note*: In APL, this would be $PE \leftarrow (\rho E) \div (\rho S)$].

Example 2

In the experiment of rolling a die and flipping a coin, the sample space consisted of 12 equally likely outcomes, provided the experiment was done fairly. Therefore, $P((1, H)) = 1/12$ and $P((1, T)) = 1/12$, etc. In the event of rolling a five, $P(F) = 2/12$, since F has two elements. In the event $\underline{H}$ of flipping a head, $P(\underline{H}) = 6/12$, since $\underline{H}$ has 6 elements.

Example 3

Consider the experiment of rolling a pair of dice. Assume the experiment is done fairly. Suppose that we are interested in the total rolled on the two dice. Then, the sample space would be the set $S = \{2, 3, 4, 5, 6, 7, 8, 9, 10, 11, 12\}$. There are 11 elements in this sample space. However, as anyone who has ever played dice knows, they are not all equally likely. It is much harder to roll a total of 2 than it is to roll a total of 7. Therefore, $P(2) \neq P(7) \neq (1/11)$. The above Rule 1 only applies to sample spaces with equally likely outcomes. In order to compute the probabilities of the elements of this sample space, we shall consider an alternative sample space in which the outcomes are equally likely. Suppose we could distinguish between the two dice. Let S be the following set of ordered pairs, where the first coordinate is the number rolled on the first die and the second coordinate is the number rolled on the second die.

$$\underline{S} = \left\{ \begin{array}{l} (1,1),\ (1,2),\ (1,3),\ (1,4),\ (1,5),\ (1,6),\ (2,1),\ (2,2), \\ (2,3),\ (2,4),\ (2,5),\ (2,6),\ (3,1),\ (3,2),\ (3,3),\ (3,4), \\ (3,5),\ (3,6),\ (4,1),\ (4,2),\ (4,3),\ (4,4),\ (4,5),\ (4,6), \\ (5,1),\ (5,2),\ (5,3),\ (5,4),\ (5,5),\ (5,6),\ (6,1),\ (6,2), \\ (6,3),\ (6,4),\ (6,5),\ (6,6) \end{array} \right\}.$$

The outcomes in this alternative sample space are equally likely. Therefore, we could use Rule 1 to find that the probability of each of these outcomes is $1/36$. Let T_2 be the event that a total of 2 is rolled. Then, $T_2 = \{(1,1)\}$. Thus, $P(T_2) = P(2) = 1/36$. Let $T_7 = \{(1,6), (2,5), (3,4), (4,3), (5,2), (6,1)\}$ be the event that a total of 7 is rolled. Then, by Rule 1, $P(T_7) = P(7) = 6/36 = 1/6$. In a similar manner, the probability of each of the elements of the sample space S can be computed. These are listed below for future reference.

T_i	2	3	4	5	6	7	8	9	10	11	12
$P(T_i)$	$\frac{1}{36}$	$\frac{2}{36}$	$\frac{3}{36}$	$\frac{4}{36}$	$\frac{5}{36}$	$\frac{6}{36}$	$\frac{5}{36}$	$\frac{4}{36}$	$\frac{3}{36}$	$\frac{2}{36}$	$\frac{1}{36}$

Example 4

Suppose $S=\{a,b,c,d,e\}$ is a sample space. Suppose also that $P(a)=1/12$, $P(b)=2/12$, $P(c)=1/12$, and $P(d)=3/12$.

(a) Find $P(e)$.

Since, by Axiom 2,

$$P(S)=1=P(a)+P(b)\ +P(c)\ +P(d)+P(e)$$
$$\text{(Axiom 4)}$$

$$=\frac{7}{12}+P(e),\quad \text{then } P(e)=\frac{5}{12}.$$

(b) Let G be the event $G=\{a,c,e\}$. Find $P(G)$. Since these outcomes in G are not equally likely, we cannot use Rule 1. However, since $G=\{a\}\cup\{c\}\cup\{e\}$, then by Axiom 4,

$$P(G)=P(a)+P(c)+P(e)+\frac{1}{12}+\frac{1}{12}+\frac{5}{12}=\frac{7}{12}.$$

This example suggests the following rule for finding probabilities of events in sample spaces in which all of the outcomes are *not* equally likely.

Probability rule 2

Let E be an event from sample space S. Then,

$$P(E)=\sum_{o_i\in E} P(o_i).$$

In other words, add the probabilities of all elements in E.

Example 5

Three teams, team A, team B, and team C are entered in a tournament. Team A is twice as likely to win as team B. Team B is twice as likely to win as team C. Find the probability of each team winning the tournament.

Let A represent the event that team A wins, B represent that team B wins, and C represent that team C wins. We want to find $P(A)$, $P(B)$, $P(C)$.

Call $P(C)=x$. Then, $P(B)=2\cdot P(C)=2x$ and $P(A)=2\cdot P(B)=4x$. Then,

$$P(S)=P(A)+P(B)+P(C)=4x+2x+x=7x=1.$$

Thus, $P(A)=4/7$, $P(B)=2/7$, and $P(C)=1/7$.

EXERCISES

1. A whole number is chosen at random between 1 and 12, inclusive.
   (a) Find the probability of choosing an odd number.
   (b) Find the probability of choosing a number greater than 9.
   (c) Find the probability of choosing a number between 5 and 8, exclusive.
   (d) Find the probability of choosing a number between 5 and 8, inclusive.

2. An experiment has sample space $S = \{x_1, x_2, x_3, x_4, x_5, x_6\}$. Suppose $P(x_1) = P(x_5)$, $P(x_2) = 1/16$, $P(x_3) = 3/16$, $P(x_4) = 2/16$, and $P(x_6) = 4/16$.
   (a) Find $P(x_1)$ and $P(x_5)$.
   (b) Let $E = \{x_1, x_3, x_5\}$; find $P(E)$.

3. Consider the experiment of drawing a card at random from a standard deck of 52 playing cards. Find the probability
   (a) Of drawing an ace.
   (b) Of drawing a spade.
   (c) Of drawing an ace of spades.
   (d) Of drawing an ace or a spade.
   (e) Of drawing a spade or a club.

4. Three candidates are running for an office. Candidates A and B are considered equally likely to win. Candidate C is only half as likely to win as A or B. Find the probability of each candidate winning the electron.

5. Let $S = \{1, 2, 3, 4, 5\}$. Let $P(x) = k/x$, where k is a constant and $x = 1, 2, 3, 4, 5$ respectively. Evaluate k.

6. Suppose that 3 coins (a nickel, dime, and quarter) are tossed.
   (a) Write a solution space for this experiment containing equally likely outcomes.
   (b) Find the probability that all three coins are heads.
   (c) Find the probability that at least one coin is a tail.
   (d) Find the probability that exactly one coin is a tail.

10.2 More rules of probability

Probability rule 3

For any event A, $P(A) = 1 - P(A')$ or $P(A') = 1 - P(A)$.

The reason for this rule is as follows: $A \cap A' = \emptyset$ and $A \cup A' = S$, so by Axiom 4, $P(S) = P(A) + P(A') = 1$ (see Figure 10.1).

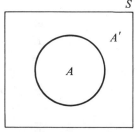

Figure 10.1 $P(S) = P(A) + P(A')$.

277

Examples

1. If the probability of rain, $P(R)$, is .8, then the probability of no rain, $P(R')$, is $1-0.8=0.2$.
2. Three men are trying to decide which of them should buy the coffees. In order to decide this, they decide to each toss a coin, and the odd man, if there is an odd man, will buy. Find the probability that there will be an odd man.

   Since there are two ways in which the first man can toss his coin, H or T, and for each of these ways, there are two ways in which the second man can toss his coin, and for each of these, there are two ways in which the third man can toss his coin, then there are 8 ways in which the three coins can be tossed. There are only two ways in which there would not be an odd man, (H,H,H) or (T,T,T). If we let O denote the event of an odd man, then

$$P(O)=1-P(O')=1-\frac{2}{8}=\frac{6}{8}-\frac{3}{4}.$$

   This example illustrates that in some instances it is easier to compute $P(A')$ and use the formula $P(A)=1-P(A')$ then to compute $P(A)$ directly.

Probability rule 4

Let A and B be two events. Then, $P(A\cup B)=P(A)+P(B)-P(A\cap B)$. [*Note*: If $A\cap B=\varnothing$, then $P(A\cap B)=0$, and this rule becomes $P(A\cup B)=P(A)+P(B)$, in accordance with Axiom 4.]

   The Venn diagram in Figure 10.2 will help in explaining this rule. If $A\cap B\neq\varnothing$, then the elements in $A\cap B$ would be counted twice, as part of A and as part of B, if one were to use Axiom 4. Therefore, we subtract $P(A\cap B)$ in order to subtract one of these counts.

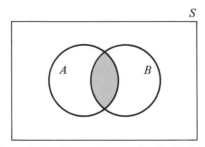

Figure 10.2   Venn diagram for Rule 4.

Examples

1. In the experiment of drawing a card at random from a standard deck of 52 playing cards, find the probability of drawing an ace or a heart.

   Let A denote the drawing of an ace, and H denote the drawing of a heart. We are seeking $P(A \cup H)$. By Rule 4,

$$P(A \cup H) = P(A) + P(H) - P(A \cap H) = \frac{4}{52} + \frac{13}{52} - \frac{1}{52} = \frac{16}{52}.$$

If we did not subtract $P(A \cap H)$, then we would be counting the ace of hearts twice, as an ace and as a heart.

2. If the probability of rain is $P(R) = 0.3$, the probability of cooler is $P(C) = 0.6$, and the probability of rain and cooler is $P(R \cap C) = 0.2$, then the probability of rain or cooler is $P(R \cup C) = P(R) + P(C) - P(R \cap C) = 0.3 + 0.6 - 0.2 = 0.7$.

3. A man is about to undergo an operation to cure two ailments. The doctors have told him that the probability that he will be completely recovered from the first ailment as a result of the operation is 0.5 and the probability that he will be completely cured of the second ailment is 0.6. Upon pressure from the poor man, the doctors also told him that the probability that he will be completely cured of at least one of the ailments is 0.8. He is now concerned with his probability of recovering from both ailments. Find this probability.

   Let R_1 be the event that he recovers from the first ailment, and R_2 be the event that he recovers from the second ailment. Then, we are given $P(R_1) = 0.5$, $P(R_2) = 0.6$, and $P(R_1 \cup R_2) = 0.8$. We are seeking $P(R_1 \cap R_2)$. From Probability rule 4,

$$P(R_1 \cup R_2) = P(R_1) + P(R_2) - P(R_1 \cap R_2),$$

or

$$0.8 = 0.5 + 0.6 - P(R_1 \cap R_2),$$

so that $P(R_1 \cap R_2) = 0.3$.

Conditional probabilities

Let A and B be two events that depend on each other in some way. The event $(A|B)$, read "A given B" or "A if B," is the event that A occurs given that B occurs. The probability of the event $(A|B)$ is given by the following rule:

Probability rule 5

For two events A and B,

$$P(A|B) = \frac{P(A \cap B)}{P(B)}.$$

This probability of $(A|B)$ is called the *conditional probability* of A given B.

The rationale behind this rule is as follows: Since we are given that B has occurred, we are no longer interested in the whole sample space S, but rather, we are only interested in those outcomes appearing in B (see Figure 10.3). Of these, the outcomes that appear in A would then appear in $A \cap B$.

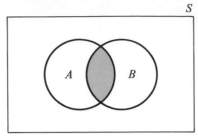

Figure 10.3   Venn diagram for Rule 5.

Thus, using the fact that a probability is a ratio of outcomes favorable to total number of outcomes possible, we arrive at the above rule.

An immediate consequence of this rule is the following rule for computing the joint probability that A and B both occur.

Probability rule 6

$$P(A \cap B) = P(B) \cdot P(A|B).$$

Examples

1. Find the probability of drawing an ace from a well shuffled deck of cards, given that the card was a spade.

   Let A denote ace and S denote spade. We are seeking $P(A|S)$. According to Rule 5, $P(A|S) = \dfrac{P(A \cap S)}{P(S)} = \dfrac{(1/52)}{(13/52)} = \dfrac{1}{13}.$

   This probability can also be computed as follows: Since we know that the card drawn was a spade, we are no longer interested in any card that is not a spade. Thus, we are only interested in the 13 spades of which 1 is an ace. Therefore, $P(A|S) = 1/13$.

2. A basket contains 10 light bulbs of which 7 are good and 3 are defective. Two bulbs are chosen at random one at a time without replacing the first bulb before choosing the second bulb. Find the probability that the second bulb is good given that the first bulb is good.

   Let G_1 denote the event that the first bulb is good and G_2 denote the event that the second bulb is good. We want to find $P(G_2|G_1)$. This can be found as follows: Since we have drawn 1 bulb and it was good, there are now 9 bulbs left of which 6 are good. Therefore, $P(G_2|G_1) = 6/9 = 2/3$.

3. Suppose that the weatherman reports that the probability of rain, $P(R)$, is 0.5, the probability of cooler, $P(C)$, is 0.6, and the probability of rain if it is cooler, $P(R|C)$, is 0.7. Find the probability that it will rain *and* be cooler, $P(R \cap C)$.

   By Probability rule 6, $P(R \cap C) = P(C) \cdot P(R|C) = (0.6) \cdot (0.7) = 0.42.$

Independent events

Two events A and B are said to be *independent* if the occurrence or nonoccurrence of one of them has no effect on whether or not the other will occur. A more mathematical definition for this phenomenon is as follows:

Definition of independent events

A and B are *independent* if and only if $P(A|B) = P(A)$ and $P(B|A) = P(B)$.

A consequence of this definition and Probability rule 6 is the following alternative definition of independence:

Probability rule 7

Two events A and B are independent if and only if $P(A \cap B) = P(A) \cdot P(B)$.

In fact, this rule can be extended to more than two events. Therefore, n events $E_1, E_2, \ldots, E_n$ are independent if and only if $P(E_1 \cap E_2 \cap \ldots \cap E_n) = P(E_1) \cdot P(E_2) \cdot \ldots \cdot P(E_n)$.

Examples

1. A coin is flipped and a die is rolled. Find the probability of a head on the coin and a 5 on the die. Let H denote head on the coin and F denote 5 on the die. Obviously, these events are independent, since the occurrence or nonoccurrence of one of them has no effect on the occurrence or nonoccurrence of the other. Therefore,

$$P(H \cap F) = P(H) \cdot P(F) = \frac{1}{2} \cdot \frac{1}{6} = \frac{1}{12}.$$

Find the probability of head on the coin *or* 5 on the die. By Probability rule 4,

$$P(H \cup F) = P(H) + P(F) - P(H \cap F) = \frac{1}{2} + \frac{1}{6} - \frac{1}{12} = \frac{7}{12}.$$

2. Four missiles are fired independently at a ship. Find the probability that the ship gets hit. (Assume that the probability that each missile hits the ship is $1/2$.)

   A typical outcome of this experiment is (H, H, M, M), where H denotes a hit and M a miss. Thus, there are 16 possible outcomes. The ship will be hit if *at least one* missile hits the ship. In this problem, it might be easier to compute the probability of the complementary event that the ship does not get hit. There is only one outcome for this event, (M, M, M, M). Thus, the probability that the ship does not get hit is $1/16$. Or, using the fact that the missiles are fired independently,

$$P(M, M, M, M) = \frac{1}{2} \cdot \frac{1}{2} \cdot \frac{1}{2} \cdot \frac{1}{2} = \frac{1}{16}.$$

Now, applying Probability rule 3, the probability that the ship does get hit is 1 minus the probability that it does not get hit. Therefore, the probability that the ship gets hit is $1-(1/16)=15/16$.

EXERCISES

1. A card is drawn at random from a standard deck of 52 playing cards. Find the probability that
   (a) It is a king or a queen.
   (c) It is a king, given that it is red.
   (b) It is a red card or a king.
   (d) It is neither red nor a king.

2. Consider the experiment of rolling a pair of dice, one white and one yellow. Find the probability of rolling
   (a) A 1 on the white die or a total of 7.
   (b) A total of 7 if it is known that a 1 was rolled on the white die.
   (c) A total of 7 or 11.
   (d) A total which is not 7.

3. The probability of a new baby being a boy is taken to be .5. If a couple plans on having three babies, find the probability that they will have
   (a) Two girls and then a boy.
   (c) At least one boy.
   (b) All girls.
   (d) All of the same sex.

4. A box contains 8 batteries of which 5 are good and 3 bad. Two batteries are chosen at random without replacing the first before choosing the second. Find the probability that
   (a) Both are good.
   (b) At least one is good.
   (c) The first is bad and the second good.
   (d) The second is good if the first was bad.

5. Repeat Exercise 4 if the first battery is replaced before the second is chosen.

6. Television station WXXX took a survey to determine the percentage of people, according to sex, who watch their evening news telecast. The results of this survey are given in the following table:

	Watch WXXX news	Do not watch WXXX news
Men	120	180
Women	90	110

   If one of these 500 people surveyed were chosen at random, then find
   (a) The probability that the person chosen watches the WXXX news telecast.
   (b) The probability that the person chosen is a woman or watches WXXX news.
   (c) The probability that if the person chosen is a man, then he watches WXXX news.
   (d) The probability that if the person chosen watches WXXX news, then this person is a woman.

7. John estimates the probability that he will get an A in math to be 0.5. He estimates the probability that he will make the Dean's list to be 0.4. He

estimates the probability that he will do both to be 0.3. Find the probability that

(a) He will make the Dean's list if he gets an A in math.

(b) He will get an A in math or make the Dean's list.

(c) He will neither make the Dean's list nor get an A in math.

(d) Are the events "get an A in math" and "make the Dean's list" dependent or independent?

8. The probability that a man will live to be age 65 is 0.6. The probability that his wife will live to be age 65 is 0.7. Assume that their life spans are independent. Find the probability that

(a) Both will live to age 65.

(b) At least one of them will live to age 65.

(c) The wife will live to age 65, but the husband will not.

(d) The husband will live to age 65 if his wife does.

9. The probability that a baseball player will get a hit in any given time at bat is 0.300. If in a particular game, he will bat 4 times, find the probability that

(a) He will go hitless. (c) He will get 4 hits.

(b) He will get at least one hit. (d) He will get exactly one hit.

10. A New York baseball fan follows the fortunes of the Yankees and the Mets. Before the season, it is estimated that the probability that the Yankees will win the pennant is 0.3, and the probability that the Mets will win the pennant is 0.2. Of course, the events that the Yankees win the pennant and the Mets win the pennant are independent. Find the probabilities that

(a) Both the Yankees and the Mets will win their respective pennants.

(b) At least one of them wins its pennant.

(c) Neither of them wins its pennant.

(d) The Yankees will win their pennant if the Mets win their pennant.

10.3 Permutations and combinations

In this section, we shall consider some counting techniques that are quite useful in counting the number of elements in certain sample spaces and events. The basis for these counting techniques is the following rule which we shall call the Counting principle.

The Counting principle

If one procedure can be performed in m ways, and if for each of these ways a second procedure can be performed in n ways, then the two procedures can be performed consecutively in $m \cdot n$ ways.

This principle can also be extended to more than two procedures.

Examples

1. In how many ways can one roll a die and flip a coin?

   One can roll a die in 6 ways, and for each of these, he can flip a coin in two ways. Therefore, he can do both in $6 \cdot 2 = 12$ ways.

283

2. A luncheon menu consists of 5 kinds of sandwiches, 3 kinds of salad, 5 kinds of drinks, and 4 kinds of desserts. In how many ways can one select a lunch consisting of a sandwich, a salad, a drink and dessert?

There are 5 ways to select a sandwich, and for each of these there are 3 ways to select a salad, and for each of these 15 ways of selecting a sandwich and salad, there are 5 ways to select the drink, and for each of the 75 ways of selecting sandwich, salad, and drink, there are 4 ways of selecting dessert. Therefore, there are $5 \cdot 3 \cdot 5 \cdot 4 = 300$ ways of selecting a lunch.

3. In how many ways can 6 people arrange themselves in a line?

There are 6 ways of deciding who will be first in line, and for each of these, there are 5 ways of deciding who will be second, and for each of these, there are 4 ways of deciding who will be third, etc. Thus, there are $6 \cdot 5 \cdot 4 \cdot 3 \cdot 2 \cdot 1 = 720$ ways of arranging 6 people in a line.

This last example illustrates a mathematical principle called "factorial."

Definition of n factorial

For any positive integer n, n *factorial* is defined to be the product $n \cdot (n-1) \cdot (n-2) \cdot \ldots \cdot 2 \cdot 1$. 0 factorial is defined to be 1.

We will denote n factorial by $!n$ in this text. [*Note*: In most texts, n factorial is denoted by $n!$ However, in APL, a monadic operator always appears on the left of the argument. It is for this reason that we prefer $!n$ to $n!$]

Factorial is a keyboard operation in APL. The form is $!N$. (! is an overstrike symbol obtained by typing the quote symbol ' then backspacing and typing a period . resulting in !)

```
      !6
720
```

```
      ×/⍳6
720
```
This is the definition of 6 factorial expressed in APL.

```
      !0
1
```
0 factorial is defined to be 1.

```
      !6 8 0
720 40320 1
```
Factorial with a vector argument.

As illustrated in the above Example 3, the number of distinct arrangements of n objects is given in $!n$.

Example 4

In how many ways can a coach position 5 players on a basketball team?

        !5
120

Permutations

A *permutation* of a set of objects is an arrangement of part or all of the set of objects.

Example

In a club consisting of 7 members, in how many ways can an executive board be formed consisting of a president, a vice-president, a treasurer, and a secretary?

There are 7 ways of selecting the president, and for each of these ways, there are 6 ways of selecting the vice-president, and for each of these ways, there are 5 ways of selecting the treasurer, and for each of these ways, there are 4 ways of selecting the secretary. Thus, by the Counting principle, there are $7 \cdot 6 \cdot 5 \cdot 4 = 840$ ways of forming the executive board. In this problem, we are interested in the number of ways of arranging (the number of permuations of) 7 people, 4 at a time.

The number of permutations of N objects K at a time is given by the formula

$$\frac{!N}{!(N-K)}.$$

Thus, in the above example, the number of permutations of the 7 members into 4 member executive boards can be given by

$$\frac{!7}{!(7-4)} = \frac{7 \cdot 6 \cdot 5 \cdot 4 \cdot 3 \cdot 2 \cdot 1}{3 \cdot 2 \cdot 1} = 840.$$

We now present a program for the number of permutations of N objects K at a time.

Program 10.1 **PERMUTATIONS**

    ∇*Y*←*K PERMUTATIONS N*

[1]   *Y*←(!*N*)÷(!*N*−*K*) ∇   This is just the above formula expressed in APL.

Example

In how many ways can 5 out of 8 people be seated in 5 available chairs?

    5 *PERMUTATIONS* 8
6720

10   Probability

Combinations

Here, we shall be more interested in the number of ways (regardless of order) of choosing K objects from N objects than in the number of permutations (where order is important) of N objects K at a time. In other words, we will be interested in the number of *combinations* of N objects K at a time. The difference between a permutation and a combination is that in a permutation the arrangement of the objects is very important, while in a combination one can rearrange the objects and still have the same combination. Thus, abc and bac are different permutations of the letters a, b, and c; but, they are the same combination of the letters. There are actually 6 permutations of these three letters, namely abc, acb, bac, bca, cab, cba, all of which are the same combination. If we were interested in the number of combinations of letters 3 at a time, we would have to divide the number of permutations of letters 3 at a time by $(!3) = 6$. Similarly, the number of combinations of N objects K at a time equals the number of permutations of N objects K at a time divided by $!K$. Thus, a program for the number of combinations of N objects K at a time would be as follows:

Program 10.2 **COMBINATIONS**

$\nabla C \leftarrow K$ COMBINATIONS N This program is just A.P.L. notation for the formula
$$C = \frac{!N}{!(N-K) \cdot !K}.$$

[1] $C \leftarrow (!N) \div (!N - K) \times (!K)$ ∇

Examples

    2 COMBINATIONS 5
10

    3 COMBINATIONS 8
56

    1 COMBINATIONS 5
5

    0 COMBINATIONS 5
1

Actually, the number of combinations can be found directly in APL by the dyadic use of the operation !. The form is $K!N$. From now on, we will use this notation for the number of combinations of N objects K at a time. (The more common conventional notations for this are $_n^kC$, $C(n,k)$, and $\binom{n}{k}$.)

286

Examples

     2!5

10

     3!8

56

     1!5

5

     0!5

1

     2 3 1 0!5 8 5 5   Vector arguments.

10 56 5 1

Some applications of combinations

1. Find the number of 5-person committees that can be formed in a club consisting of 9 members.

    Unlike an executive board, the particular arrangement of people on a committee is not important. Therefore, we are interested in the number of combinations of 9 members 5 at a time, rather than the number of permutations of 9 members 5 at a time.

     5!9

126   There are 126 such committees.

2. If a club consists of 12 members, 7 of which are women and 5 of which are men, then find the number of possible committees consisting of 3 women and 2 men.

     (3!7)×(2!5)   By the Counting principle, the num-

350   ber of ways of forming the com-
                              mittee is the product of the number
                              of ways of choosing the women and
                            the number of ways of choosing the
                            men.

3. How many samples of 5 can be chosen from a lot of 10 items?
    In a sample, the particular arrangement of objects is not important.

     5!10

252   There are 252 such samples.

4. How many possible 5-card hands can be dealt from a deck of 52 cards?

     5!52

2598960   WOW!

5. In how many ways can 5 rolls of a die result in three 6's?

This is the number of combinations of 5 things (rolls) taken 3 (6's) at a time.

$$\frac{3!5}{10}$$

The possibilities are $666NN$, $66NN6$, $6NN66$, $66N6N$, $6N6N6$, $6N66N$, $NN666$, $N6N66$, $N66N6$, $N666N$, where N denotes a non-6.

This last example illustrates that if any trial of an experiment can result in two possible outcomes (6 or non 6), then in N trials of the experiment, there are $K!N$ ways in which to arrange K of one outcome and $N-K$ of the other.

EXERCISES

1. Find the total number of possible 5-digit telephone numbers beginning with 54 or with 45.

2. Find the number of permutations that can be made from the letters of the word "computer."

3. In how many ways can one choose a card from a deck of 52 playing cards, roll a die, and flip a coin?

4. In how many ways can one answer a 10-question true–false exam?

5. In how many ways can one answer a 5-question multiple-choice exam if there are 3 choices for each question?

6. A little league baseball coach has 15 players on his team.
   (a) How many possible 9-person lineups can he present?
   (b) How many possible 9-person combinations can he field (regardless of position or place in the batting order)?

7. A small company has 15 employees, of which 10 are laborers and 5 are white collar workers. How many possible 6-person committees can be formed consisting of 4 laborers and 2 white collar workers?

8. A little league team plays 10 games. In how many ways can the team win 6 and lose 4?

9. In how many ways can 6 people sit in a row if a certain two people refuse to sit next to each other?

10. A box contains 10 light bulbs of which 7 are good and 3 are bad.
    (a) How many possible samples of 5 light bulbs are there?
    (b) How many of these samples of 5 contain exactly 4 good bulbs?
    (c) How many of these samples of 5 contain at least 4 good bulbs?
    (d) How many of these samples of 5 contain at most 1 bad bulb?

11. In how many ways can one draw from a deck of cards a 5-card hand containing
    (a) All spades?   (c) Three spades and 2 hearts?
    (b) Four aces?   (d) Two pairs?

12. The following program uses a very powerful process in APL, called *recursion*. A program is said to be *recursive* if it refers to itself within itself. That is, the program is repeatedly used as a subprogram within itself.

    ∇ Y←FAC N

    [1]   Y←1   This program computes N factorial.
    [2]   →(N=0)/0
    [3]   Y←N×FAC N−1
           ∇

    To understand how this program works, *TRACE* it finding *FAC* 5.

10.4 The hypergeometric distribution

When there are only two alternative outcomes in any endeavor, we shall call them *success* and *failure*. The success will be the alternative in which we are primarily interested. In this and the next two sections, we will consider three basic probability distributions which deal with the number of successes in a sample, or on repeated trials of an experiment, or in an area or time interval. If we let X denote the number of these successes, then a *probability distribution for X* consists of the values for X together with their associated probabilities.

The first probability distribution we shall consider is known as the *hypergeometric distribution*. The hypergeometric distribution is the probability distribution used in the following situation:

A random sample of M objects is taken from a population of N objects of whch K are classified as successes and the remaining $N-K$ as failures. Let X denote the number of successes chosen in the sample of M objects. X is often referred to as a hypergeometric random variable. The probability that X equals some value x is given by

$$P(X=x) = \frac{(x!K)\cdot((M-x)!(N-K))}{(M!N)}, \quad \text{for } x=0,1,\ldots,M.$$

The reason for this formula is as follows: There is a total of $(M!N)$ ways of getting a random sample of M objects from a population of N objects. There are $(x!K)$ ways of getting x successes from the K successes in the population. There are $((M-x)!(N-K))$ ways of getting the remaining $M-x$ failures from the remaining $N-K$ failures in the population. By the Counting principle, there are, therefore, $(x!K)\cdot((M-x)!(N-K))$ ways of getting x successes and $M-x$ failures in the sample. Now, using Probability rule 1 of Section 10.1., since all of the $(M!N)$ random samples are equally likely, we divide the number of ways of getting x successes and $M - x$ failures by the total number of random samples which comprise our sample space. Thus, we arrive at the above formula.

Example

In a club consisting of 3 men and 5 women, a committee of 4 is to be selected. Let X denote the number of men on the committee.

(a) Find the probability distribution for X. The possible values for X are 0, 1, 2, and 3. Using the above probability formula for X, the associated probabilities for X are as follows:

$$P(X=0) = \frac{(0!3) \cdot (4!5)}{(4!8)},$$

since if one chooses 4 people from 8 people, and if 0 men are chosen, then 4 women must have been chosen. Let us use APL to compute this probability.

   (0!3)×(4!5)÷(4!8)
.07142857143

Similarly,

   (1!3)×(3!5)÷(4!8) $P(X=1) = \dfrac{(1!3) \cdot (3!5)}{(4!8)}$

.4285714286

   (2!3)×(2!5)÷(4!8) $P(X=2) = \dfrac{(2!3) \cdot (2!5)}{(4!8)}$

.4285714286

   (3!3)×(1!5)÷(4!8) $P(X=3) = \dfrac{(3!3) \cdot (1!5)}{(4!8)}$

.07142857143

Figure 10.4 shows this probability distribution.

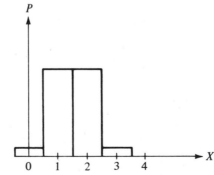

Figure 10.4   Graphical probability distribution.

(b) Find the probability of at least one man being on the committee.

$$P(X \geqslant 1) = P(X=1) + P(X=2) + P(X=3) = 0.9285714286,$$

since at least one man means one or more men, and the maximum possible number of men is 3. Another method for solving this problem is:

$$P(X \geqslant 1) = 1 - P(X=0) = 1 - 0.0714285714 = 0.9285714286,$$

since the complement of at least one man is no men.

Let us now consider a program for computing hypergeometric probabilities.

Program 10.3 *HYPERGEOMETRIC*

$\nabla\ H \leftarrow XK\ HYPERGEOMETRIC\ MN;\ X;\ K;\ M;\ N$

[1]	$X \leftarrow XK$ [1]	X is the number of successes in the sample.
[2]	$K \leftarrow XK$ [2]	K is the number of successes in the population.
[3]	$M \leftarrow MN$ [1]	M is the size of the sample.
[4]	$N \leftarrow MN$ [2]	N is the size of the population.
[5]	$H \leftarrow (X!K) \times ((M-X)!(N-K)) \div (M!N)$ ∇	
		The formula for hypergeometric probabilities in APL.

In this program, *XK* is a vector consisting of the value of X, the number of successes in the sample, followed by K, the number of successes in the population. *MN* is the vector consisting of the value of M, the size of the sample, followed by N, the size of the population.

Examples

1. Let us use this program to compute the probabilities for the number of men on the committee in the previous example.

      0 3 *HYPERGEOMETRIC* 4 8 $P(X=0)$.
.07142857143

      1 3 *HYPERGEOMETRIC* 4 8 $P(X=1)$.
.4285714286

      2 3 *HYPERGEOMETRIC* 4 8 $P(X=2)$.
.4285714286

      3 3 *HYPERGEOMETRIC* 4 8 $P(X=3)$.
.07142857143

2. A basket contains 10 lightbulbs of which 8 are good and 2 are defective. Three lightbulbs are chosen at random from the sample.
   (a) Find the probability of getting exactly one good bulb in the sample. In this problem, a success is a good bulb. Therefore, let X denote the number of good bulbs in the sample. We want $P(X=1)$.

   1 8 *HYPERGEOMETRIC* 3 10
.0666666667

   (b) Find the probability of getting at least two good bulbs in the sample.
   At least two good bulbs is the same as getting two or more good bulbs. Therefore, we are seeking $P(X \geqslant 2)$. However, $P(X \geqslant 2) = P(X=2) + P(X=3)$ in this problem.

   (2 8 *HYPERGEOMETRIC* 3 10) + (3 8 *HYPERGEOMETRIC* 3 10)
.9333333333

3. A company produces batteries which it ships in boxes of 12. Before shipping any box of batteries, an inspector checks each box to make sure that the batteries work. Instead of checking every battery in the box, he uses the following scheme to decide whether or not to accept the box for shipment. He selects 3 batteries at random from each box and tests them. If all 3 tested are good, the box is accepted for shipment. Otherwise, it is rejected.
   (a) If a box of batteries really contains 4 defective batteries, what is the probability that the inspector erroneously accepts the box for shipment?
   In order for him to accept the box for shipment, the three batteries which the inspector selected and tested must have all been good. Therefore, we want $P(X=3)$, where X is the number of good batteries in the sample of 3 selected.

   3 8 *HYPERGEOMETRIC* 3 12
.2545454545

   (b) If the box has 1 defective battery in it, what is the probability that it is rejected?
   In order for the box to be rejected, it is necessary that this one defective battery be included in the random sample of 3 batteries from the box. Thus, we want $P(Y=1)$, where Y denotes the number of defective batteries in the sample.

   1 1 *HYPERGEOMETRIC* 3 12
.25

   (c) If the box contains two defective batteries, what is the probability that it is rejected?

This box will be rejected if $Y=1$ or if $Y=2$. Thus, we want $P(Y=1)+P(Y=2)$.

(1 2 *HYPERGEOMETRIC* 3 12)+(2 2 *HYPERGEOMETRIC* 3 12)
.4545454545

4. A 5-card poker hand is drawn from a deck of 52 cards.
   (a) Find the probability of getting all four aces.
      A success here is getting an ace, of which there are 4, all of which are drawn. A failure is getting a non-ace, of which there are 48, of which 1 is drawn. Let X denote the number of aces drawn. We want $P(X=4)$.

   4 4 *HYPERGEOMETRIC* 5 52
   .0000184689   Not very likely.

   (b) Find the probability of drawing all spades. Let Y denote the number of spades drawn. We want $P(Y=5)$.

   5 13 *HYPERGEOMETRIC* 5 52
   .0004951980792

   (c) Find the probability of getting all in one suit.
      Since there are 4 suits in all, all of which are equally likely, we need only to multiply the probability of getting all spades by 4.

   4×(5 13 *HYPERGEOMETRIC* 5 52)
   .001980792317

EXERCISES

1. A box of one dozen donuts contains 6 plain and 6 jelly donuts. If one chooses 4 donuts at random, what is the probability that he chooses
   (a) Two of each?   (c) At least one jelly donut?
   (b) All plain donuts?   (d) At most one jelly donut?

2. A Congressional committee is composed of 6 Republicans and 4 Democrats. A 3-person subcommittee is to be formed at random. What is the probability that this subcommittee is composed of
   (a) All Democrats?   (c) More Republicans than Democrats?
   (b) All Republicans?   (d) At least one member of each political party?

3. A 7-card rummy hand is drawn at random from a deck of 52 cards. Find the probability of drawing
   (a) Four aces.   (c) All hearts.
   (b) Four of one denomination.   (d) All of one suit.

4. A bag contains 20 jelly beans of which 9 are red, 6 are black, and 5 are white. Six jelly beans are chosen at random. Find the probability of choosing the following: [*Note*: This is an extension of the hypergeometric distribution to

three categories for which the reader will have to develop his own probability formula.]

(a) All red.
(b) Three red, 2 black, and 1 white.

(c) Two of each color.
(d) All white.

10.5 The binomial distribution

The *binomial distribution* is used when one is interested in the number of successes, X, that occur in N independent trials of an experiment, where on each trial, the probability of success, P, and the probability of failure, $1-P$, remain the same. X is called a *binomial random variable*. The probability function for X is as follows:

$$P(X=x)=(x!N)\cdot P^x\cdot (1-P)^{N-x}, \quad \text{for } x=0,1,2,\ldots,N.$$

To understand the reasoning behind this formula, consider the following example:

Example

Let X denote the number of 3's rolled in 5 rolls of a fair die. Let "3" denote a 3 rolled on any particular roll of the die, and let N denote a non-3 on any particular roll. As in Example 5 of applications of combinations in Section 10.3., there are $(2!5)=10$ ways in which to arrange 2 3's and 3 N's:

$$(3,3,N,N,N),(3,N,3,N,N),(3,N,N,3,N),(3,N,N,N,3),(N,N,N,3,3),$$

$$(N,N,3,N,3),(N,N,3,3,N),(N,3,N,3,N),(N,3,3,N,N),(N,3,N,N,3).$$

Since each roll of the die is independent of the other rolls, then each of the above arrangements has the same probability, namely,

$$P(3,3,N,N,N)=P(3)\cdot P(3)\cdot P(N)\cdot P(N)\cdot P(N)$$

$$=\frac{1}{6}\cdot\frac{1}{6}\cdot\frac{5}{6}\cdot\frac{5}{6}\cdot\frac{5}{6}=\left(\frac{1}{6}\right)^2\cdot\left(\frac{5}{6}\right)^3.$$

Thus, $P(X=2)=(2!5)\cdot(\frac{1}{6})^2\cdot(\frac{5}{6})^3$, as in the formula above.

Since the computations in the binomial distribution can be quite tedious, we would like to have a program for computing binomial probabilities.

Program 10.4 BINOMIAL

```
    ∇ DISTRIBUTION←P BINOMIAL N;X
[1]   X←0, ιN
[2]   DISTRIBUTION←(X!N)×(P*X)×(1−P)*(N−X)
    ∇
```

This program generates the entire binomial distribution, where P is the probability of success on any trial and N is the number of trials.

Example

.25 *BINOMIAL* 4 $P=0.25, N=4.$
.3164 .421875 .2109375 .046875 .00390625

The result is the entire distribution of binomial probabilities for $P=$ 0.25, $N=4$. To compute any particular probabilities from this distribution, just select the appropriate term from this distribution. For example:

(.25 *BINOMIAL* 4)[1]   This is $P(X=0)$.
.3125

(.25 *BINOMIAL* 4)[2]   This is $P(X=1)$.
.421875

(.25 *BINOMIAL* 4)[3]   This is $P(X=2)$.
.2109375

(.25 *BINOMIAL* 4)[4]   This is $P(X=3)$.
.046875

(.25 *BINOMIAL* 4)[5]   This is $P(X=4)$.
.00390625

Notice that $P(X=k)$ is always the $k+1$ term in the distribution, since the first term corresponds to $P(X=0)$.

Example 1

(a) Find the probability of two 3's in 5 rolls of a die.

(.1666666667 *BINOMIAL* 5)[3]
.1647510293

(b) Find the probability of at least one 3 in 5 rolls of a die.
We need $P(X \geqslant 1)$. One way of computing this would be to compute $P(X=1)+P(X=2)+P(X=3)+P(X=4)+P(X=5)$. However, a simpler way to find $P(X \geqslant 1)$ is to find the probability of the complementary event, $P(X<1)=P(X=0)$, and use Probability rule 3. Thus,

$$P(X \geqslant 1)=1-P(X=0).$$

1−(.1666666667 *BINOMIAL* 5)[1]
.5981224288

Example 2

Based on past records, it is found that the probability that a patient will recover from a certain difficult operation performed by a leading specialist

295

in the field is 0.75. If this specialist will perform this operation on 6 patients this week, what is the probability that

(a) All 6 will recover?

Since this doctor is a specialist, he will not let the success or failure of previous operations affect his chances of succeeding in the next operation. Therefore, we can assume that the operations are independent, each with probability of success 0.75. Let X denote the number of the 6 operations which are successful. We want $P(X=6)$.

　　(.75 *BINOMIAL* 6)[7]
.1779785156

We should not expect that all 6 operations will be successful!

(b) At most one fails to recover?

Let Y denote the number who fail to recover. On any given operation, the probability of a failure to recover is $1-0.75=0.25$. We want $P(Y \leqslant 1)$. This is obtained by adding $P(Y=0)$ to $P(Y=1)$.

　　(.25 *BINOMIAL* 6)[1]+(.25 *BINOMIAL* 6)[2]
.5339355469

Thus, there is a slightly better than even chance that at most one patient fails to recover.

The next example illustrates the use of the binomial distribution in sampling *with* replacement. If one uses sampling without replacement, of course, then one uses the hypergeometric distribution.

Example 3

A card is drawn at random from a deck of 52 well shuffled cards, then replaced. The deck is then reshuffled, another card drawn and then replaced. This process is continued until 5 cards have been drawn.

(a) Find the probability of getting all hearts.

Since each card is replaced and the deck reshuffled before the next card is drawn, the trials of this experiment are independent. The probability of a heart on each draw is the same, namely $13/52=0.25$.

　　(.25 *BINOMIAL* 5)[6]
.0009765625

(b) Find the probability of getting two hearts.

　　(.25 *BINOMIAL* 5)[3]
.2636718570

(c) Find the probability of getting at least one heart. This is the complement of getting 0 hearts.

　　1-(.25 *BINOMIAL* 5)[1]
.7626953125

Example 4

This last example illustrates the use of the binomial distribution as an estimate of the hypergeometric distribution when the population is large and the sample small.

Of 1000 delegates at a convention, 500 are Republicans and 500 Democrats. Five delegates are chosen at random to form a committee. Find the probability that there is a majority of Republicans on the committee.

Since this is sampling *without* replacement, we should use the hypergeometric distribution. However, the size of the resulting combinations would be extremely large: (5!1000) is too large to be easily computed even by the computer. Also, each time a delegate is chosen, the probability that he will be a Republican is pretty close to 0.5. For example, suppose the first two delegates chosen are Republicans. The probability that the third will be a Republican is 498/998, which is approximately 0.5. If the first two are Democrats, the probability that the third will be Republican is 502/998, which is still approximately 0.5. Therefore, to simplify the problem, we shall assume that the probability of a Republican being chosen for each successive seat on the committee is 0.5. Let X denote the number of Republicans on the committee. There is a majority of Republicans on the committee if $X \geqslant 3$. Thus, we want $P(X \geqslant 3)$. Using **BINOMIAL**, we can get a good estimate of this probability.

(.5 *BINOMIAL* 5)[4]+(.5 *BINOMIAL* 5)[5]+(.5 *BINOMIAL* 5)[6]

.5

Thus, there is an even chance that there will be a majority of Republicans on the committee. The same is true for Democrats.

EXERCISES

1. A baseball player is a consistent .300 hitter. (That is, each official time up, the probability he gets a hit is 0.3.) If he bats 4 times in a game, find the probability that he
   (a) Gets 2 hits.   (c) Gets at least 1 hit.
   (b) Gets no hits.   (d) Goes "4 for 4."

2. Suppose that the probability that a tulip bulb will germinate is 0.8. If one plants 10 tulip bulbs, find the probability that
   (a) All will germinate.
   (b) At least 8 will germinate.
   (c) At most 1 will not germinate.

3. When a baby is born, the probability that it will be a boy is 0.5. If a family has 6 children, find the probability
   (a) They are all boys.
   (b) They are 3 boys and 3 girls.

(c) They have at least 1 boy.

(d) If the first 5 are girls, then the next will be a boy.

4. Ten men have a weekly raffle. They each put their name on a ticket and deposit the ticket in a hat. Each week, a ticket is drawn to determine the winner. The winning name is then put back in the hat for the next week's drawing.
(a) Find the probability that in 5 weeks, Mr. X will win twice.
(b) Find the probability that in 5 weeks, Mr. X will not win at all.
(c) Find the probability that Mr. X will win twice in a row.

5. It is estimated that 3000 of the 10000 residents of a town favor fluoridation of their drinking water. If 12 residents are selected at random, find the probability that
(a) Fewer than 4 of them favor fluoridation.
(b) A majority of them favor fluoridation.

10.6 The Poisson distribution

The final probability distribution which we shall consider in this chapter is known as the Poisson distribution. *The Poisson distribution* is used when one is interested in the number of successes, X, occurring in a time interval or a region. X is called a *Poisson random variable*. Using techniques too advanced for this text, it can be shown that the probability distribution for X is

$$P(X=x)=\frac{(e^{-u})\cdot(u^x)}{(!x)}, \quad \text{where } x=0,1,2,\ldots$$

and where u is the average number of successes during this given time interval or region.

Example

Suppose the average number of telephone calls coming into a switch board per minute is 3.

(a) Find the probability that during a given minute, no telephone calls come into the switchboard.
The value of u is 3 calls per minute. Let X denote the number of calls coming into the switchboard during this particular minute. We want $P(X=0)$.

$$P(X=0)=\frac{(e^{-3})\cdot(3^0)}{(!0)}=e^{-3}.$$

```
* ‾3
.0497870837
```

This is e^{-3} in APL.

(b) Find the probability of at least one call during a given minute.
At least 1 call is the complement of no calls. Therefore, $P(X \geqslant 1) = 1 - P(X=0)$. Thus, we get $P(X \geqslant 1) = 1 - e^{-3} = 0.95021$.

(c) Find the probability of no calls during a 5-minute interval.
Since the average number of calls for a 1-minute interval is 3, then the average number of calls for a 5-minute interval is $5 \cdot 3 = 15$. Therefore, in this problem $u = 15$. Now, let Y be the actual number of calls during this 5-minute interval. We want $P(Y=0)$.

$$P(Y=0) = \frac{(e^{-15}) \cdot (15^0)}{(!0)} = e^{-15}.$$

In APL, we get

```
   *-15
3.059023205 E-7
```
                                              Or, 0.000000305902.

Before considering more examples, let us consider a program for computing Poisson probabilities.

Program 10.5 **POISSON**

```
   ∇ P←X POISSON U
[1]   P←(*-U)×(U*X)÷(!X)
   ∇
```

Example 1

Let us use the program **POISSON** to do the previous example.

(a) Find $P(X=0)$ where $u=3$.

```
   0 POISSON 3
.04978706837
```

(b) Find $P(X \geqslant 1)$ where $u=3$.

```
   1-0 POISSON 3
.9502129316
```

(c) Find $P(Y=0)$ where $u=15$.

```
   0 POISSON 15
3.0590232057 E-7
```

Example 2

The average number of bacteria per square inch of a culture is 5.

(a) Find the probability that 4 square inches of the culture contains 10 of these bacteria.
$u = 4 \cdot 5$ is the average number of these bacteria in 4 square inches. Let

X be the actual number of bacteria in the 4-square-inch culture. We are seeking $P(X=10)$.

        10 *POISSON* 20
.005816306518

(b) Find the probability that there are at least 10 bacteria in this culture. Here, we want $P(X \geqslant 10)$. Since there is no upper bound on X in the Poisson distribution, we can do this problem more easily by finding $1 - P(X \leqslant 9)$. In order to do this, we need to compute

$$1 - (P(X=0) + P(X=1) + P(X=2) + P(X=3) + P(X=4)$$
$$+ P(X=5) + P(X=6) + P(X=7) + P(X=8) + P(X=9)).$$

This would be quite tedious to compute.

Problems of this type are actually more common than problems involving the probability of a single value of X when dealing with the Poisson distribution. Therefore, it would be quite useful to have a program for finding $P(X \leqslant C)$, for some value of C. Such a probability is called a *cumulative probability*, since it accumulates the previous probabilities:

$$P(X \leqslant C) = P(X=0) + P(X=1) + P(X=2) + \cdots + P(X=C).$$

The following program computes this cumulative probability for the Poisson distribution.

Program 10.6 CUMPOISSON

```
     ∇ SUM←C CUMPOISSON U;X
[1]    X←0,ιC                    X is the vector 0 1 2 ...C.
[2]    SUM←+/X POISSON U         This sums up 0 POISSON U, 1
     ∇                           POISSON U,...,C POISSON U.
```

We can now complete Example 2 above.

        1-9 *CUMPOISSON* 20
.9950045877

Example 3

On the average there are 2 typographical errors per page in a manuscript. In 8 pages, find the probability of

(a) Not more than 10 errors.
    $u = 16$ is the average number of errors in 8 pages. Let X denote the actual number of errors in 8 pages. We are seeking $P(X \leqslant 10)$.

        10 *CUMPOISSON* 16
.07739601577

300

(b) More than 20 errors.
 This is $P(X>20)=1-P(X\leqslant20)$.

 1−20 *CUMPOISSON* 16
.1318319657

(c) Find the probability of from 11 to 20 errors, inclusive.
 This is $P(11\leqslant X\leqslant20)=P(X\leqslant20)-P(X\leqslant10)$.

 (20 *CUMPOISSON* 16)−(10 *CUMPOISSON* 16)
.7907720185

Finally, we should point out that the Poisson distribution may be used to approximate the binomial distribution, if N is large.[1] Recall that N is the number of trials in the binomial distribution, and P is the probability of success on any one trial. The value of u used in the Poisson approximation of the binomial is $u=N{\cdot}P$, the expected or most likely number of successes in the N trials.

Example 4

Let X denote the number of 6's in 300 rolls of a fair die.

(a) Find $P(X\leqslant25)$.
 Since $N=300$ and $P=1/6$, then $u=N{\cdot}P=50$ is the expected number of 6's in 300 rolls of the die. Now, (25!300) and !300 are beyond the capacity of the computer. Witness:

 25!300
DOMAIN ERROR

 !300
DOMAIN ERROR

 Thus, although the binomial distribution should really be used, we will have to use the Poisson approximation.

 25 *CUMPOISSON* 50
7.160717367 *E*⁻5

(b) Find the probability of more than 75 sixes.

 1−75 *CUMPOISSON* 50   Since $P(X>75)=1-P(X\leqslant75)$.
0.0003719691515

---

[1]It can be shown (although it is beyond the scope of this text to formally do so) that as $N\to\infty$, then

$$(x!N){\cdot}P^x{\cdot}(1-P)^{N-x}\to\frac{(e^{-N{\cdot}P}){\cdot}(N{\cdot}P)^x}{(!x)}.$$

EXERCISES

1. The average number of days per winter in which school is cancelled due to snow in a certain north eastern city of the United States is 4. For any given winter, find
   (a) The probability of no cancellations.
   (b) The probability of 2 cancellations.
   (c) The probability of at least 2 cancellations.
   (d) The probability of at most 5 cancellations.

2. The old Baxters live on a certain obscure country road. One of their favorite pastimes is to sit on their front porch and wave at cars. Cars pass by the old Baxters' home at the rate of 5 per hour. If they sit on their porch for two hours, find the probability that
   (a) They see no cars.   (c) They see no more than 4 cars.
   (b) They see at least 3 cars.   (d) They see exactly 1 car.

3. Repeat Problem 2 if the old Baxters sit outside for only half an hour.

4. The average number of defects in 100 feet of cord put out by a certain new company is 6. Find the probability
   (a) Of no defects in 50 feet of cord.
   (b) Of no more than 10 defects in 200 feet of cord.
   (c) Of at least 8 defects in 150 feet of cord.

5. The probability that a person who lives to be 75 will die during his 75th year is estimated to be 0.01. Of 1000 people chosen at random entering their 75th year, find the probability that
   (a) Fewer than 5 will die during their 75th year.
   (b) More than 20 will die during their 75th year.

6. A card is drawn at random from a well shuffled deck of cards, then replaced, and the cards reshuffled, another drawn, and so on until 100 cards have been drawn. Find the approximate probability that
   (a) At least 20 are hearts.
   (b) At most 10 are hearts.

<h1>Statistics 11</h1>

Statistics deals with arriving at conclusions about some unknown characteristics of a large population based on a random sample of observations from the population. For example, we might want to know the average weight of a man of age 50. Or we might want to know the probability that a man of age 50 is overweight. In this chapter, we will deal with the statistical tools for answering questions such as these.

11.1 Random samples and frequency distributions

The *population* for a statistical investigation is the sample space for all observations under investigation. A *random sample of size N* from the population is a collection of N observations taken from the population in random manner so that any collection of N observations from the population is just as likely to be chosen as any other.

One way of choosing a random sample is to assign a number to each element of the population and then to choose a random sample of these numbers using a table of random numbers. In APL, we have a built-in random number generator. To randomly choose N numbers from ιM, without repetition, type $N?M$.

Examples

```
   4?9
7 3 2 8
```

Choose 4 numbers from the vector $\iota 9$.

Suppose that a club contains 50 members of which we wish to choose 5 at random to form a committee. We could then assign a number from 1 to 50 to each member of the club. To randomly pick 5 of these numbers, we could use

        5?50
42 30 13 45 8

The people holding these numbers would now form the committee.

We could even use this ? function, hypothetically, to randomly choose N cards from a deck of cards. The following program, presented just for fun, accomplishes this.

Program 11.1 DEAL (optional)

```
     ∇ DEAL N
[1]    CARDS←N?52                Choose N numbers from ι52 without
                                 replacement.
[2]    SUITS←'SHDC'
[3]    DENOMINATIONS←'A23456789TJQK'
[4]    ⍉ (2,N)ρDENOMINATIONS[1+13|CARDS], SUITS[1+4|CARDS]
     ∇
```

This program uses the residue operation | to assign a *SUIT* and *DE-NOMINATION* to each number chosen at random from ι52.

Example

```
     DEAL 5
TH                               Ten of hearts.
JH                               Jack of hearts.
2D                               Two of diamonds.
9S                               Nine of spades.
9C                               Nine of clubs.
```

In many samples, it may be permissible to have repeated values. For example, suppose that a random sample of 10 fish is taken from a lake and their weights recorded to the nearest ounce. It is reasonable to expect some repetition in these weights. A typical such sample might be

        10, 15, 16, 9, 10, 6, 9, 7, 12, 15.

The monadic use of the random generator ? allows repetition.

Examples

```
     ? 10                        A random number from ι10.
7

     ? 10 10 10 10 10            Five random numbers, each from 1
5 7 1 3 10                       to 10.

     ?(8ρ10)                     Or, ? 10 10 10 10 10 10 10 10.
6 5 8 2 8 6 3 1
```

Organizing a sample

Now that we have considered some ways of generating a random sample, we turn to organizing the elements in the sample in such a way as to make the data in the sample more meaningful and easier to handle.

Suppose for example, that the following data represent the weights of a random sample of 25 fish caught in a certain pond:

> *SAMPLE*←13 9 12 8 9 6 15 12 7 8 5 14 8 6 9 12 9 13 7
>
> *SAMPLE*←*SAMPLE*,14 10 7 18 12 6

> This illustrates the use of catenation on large samples.

These weights are not organized, and, therefore, it is not very easy to draw any conclusions from this sample. The first thing we might like to do is to sort the elements in the sample in order of magnitude from smallest to largest. To do this, we can make use of the APL "grade-up function," ⍋. (This is an overstrike of △ and |.)

Examples

Let

> *V*←6 8 4 9 3

> ⍋*V*
> 5 3 1 2 4

> ⍋*V* yields a permutation vector of indices which would rearrange the elements of *V* in order of magnitude from smallest to largest.

> *V*[⍋*V*]
> 3 4 6 8 9

> This actually accomplishes this rearranging of elements.

> *V*[⍒*V*]
> 9 8 6 4 3

> ⍒ is the "grade-down function." It causes the rearranging of elements from largest to smallest.

It might be useful to have a program to accomplish this sorting of the elements of a sample into ascending order.

Program 11.2 **SORT**

> ∇ *ASCENDING*←*SORT SAMPLE*

[1]   *ASCENDING*←*SAMPLE*[⍋ *SAMPLE*] ∇

> *SORT SAMPLE*
> 5 6 6 6 7 7 7 8 8 8 9 9 9 9 10 12 12 12 12 13 13 14 14 15 18

305

Sorting the sample of 25 fish weights above.

From this more organized form of the sample, we can tell at a glance such things as that 5 ounces is the smallest weight in the sample, 18 is the largest, and 9 and 12 occur the most often. We could even set up a frequency distribution. A *frequency distribution* is merely a list of the values in the sample together with their frequency of occurrence. In our example, a frequency distribution would be as follows:

Weights	Frequencies
5	1
6	3
7	3
8	3
9	4
10	1
11	0
12	4
13	2
14	2
15	1
16	0
17	0
18	1

A pictorial representation of a frequency distribution can be obtained by drawing a *bar graph*. To draw a bar graph, such as in Figure 11.1, merely draw a bar whose height represents the frequency of each weight (or score) above that weight (or score).

Let us now consider a program which sets up a frequency distribution for a sample and also plots a bar graph. First, we need a subprogram, **GRAPH**, which will be used in plotting the bar graph.

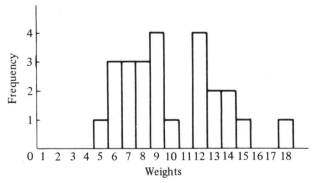

Figure 11.1   Bar graph of a frequency distribution.

Program 11.3 GRAPH

```
    ∇ G ← GRAPH INTEGER
[1]   G ← INTEGER ρ ' * ' ∇          Print INTEGER *'s.

      GRAPH 5
* * * * *

      GRAPH 10
* * * * * * * * * *
```

Program 11.4 DISTRIBUTE

```
    ∇ DISTRIBUTE SAMPLE; SCORE; P; FREQUENCY
```
[1]	SAMPLE ← SORT SAMPLE	Put the sample in order of magnitude.
[2]	SCORE ← SAMPLE [1]	Initially, SCORE is the smallest value in SAMPLE.
[3]	PICK: P ← SCORE = SAMPLE	P is a vector of 0's and 1's.
[4]	FREQUENCY ← + / P	FREQUENCY is the total number of 1's in P. That is, the number of times SCORE is in SAMPLE.
[5]	SCORE; ' WITH A FREQUENCY '; FREQUENCY	
[6]	GRAPH FREQUENCY	
[7]	SCORE ← SCORE + 1	Increase SCORE by 1.
[8]	→ (SCORE ≤ SAMPLE[ρSAMPLE]) / PICK	
	∇	If this new SCORE is ≤ the largest element in SAMPLE, branch to PICK. Otherwise, end the program.

Let us apply this program to our sample of fish weights.

```
    SAMPLE                      Recalling this SAMPLE.
13 9 12 8 9 6 15 12 7 8 5 14 8 6 9 12 9 13 7 14 10 7 18 12 6

    DISTRIBUTE SAMPLE
5 WITH A FREQUENCY 1
*
6 WITH A FREQUENCY 3
* * *
7 WITH A FREQUENCY 3
* * *
8 WITH A FREQUENCY 3
* * *
9 WITH A FREQUENCY 4
* * * *
```

10 *WITH A FREQUENCY* 1
*
11 *WITH A FREQUENCY* 0
12 *WITH A FREQUENCY* 4
* * * *
13 *WITH A FREQUENCY* 2
* *
14 *WITH A FREQUENCY* 2
* *
15 *WITH A FREQUENCY* 1
*
16 *WITH A FREQUENCY* 0
17 *WITH A FREQUENCY* 0
18 *WITH A FREQUENCY* 1
*

In order to get the full benefit from the bar graph, one would have to turn the paper 90 degrees. If there is a large range of scores in a large sample, then perhaps one would be more interested in the frequencies of scores occurring in various intervals than he would in the frequencies of each individual score.

Another example

Suppose the following scores were obtained on a test:

> *SAMPLE*←81 90 84 98 85 67 48 60 70 68 85 72 72 92 73 74
> *SAMPLE*←*SAMPLE*, 92 73 82 90 78 85 54 87 58 64 50 70 94
> *SAMPLE*←*SAMPLE*, 76

In order to get a better idea of the distribution of scores in this sample, we could sort them using the program *SORT*.

> *SORT SAMPLE*
> 48 50 54 58 60 64 67 68 70 70 72 72 73 73 74 76 78 81 82 84
> 85 85 85 87 90 90 92 92 94 98

The teacher would probably be more interested in the frequencies of scores in the 40's, 50's, 60's, 70's, 80's, and 90's than in the frequencies of each individual score. Therefore, we need a program for setting up a frequency distribution with intervals. The intervals in such a distribution should include every score and should not overlap. Consider the following program for constructing such a distribution.

Program 11.5 DISTRIBUTION

> ∇ *FREQUENCIES*← *INTERVALS DISTRIBUTION SAMPLE*

[1] $FREQUENCIES \leftarrow +/((^{-}1\downarrow INTERVALS)\circ.\leqslant SAMPLE)\wedge$
$((1\downarrow INTERVALS)\circ.> SAMPLE)$

∇

In order to explain this extremely powerful 1-line program, we shall consider a simple example.

Example 1

Suppose our sample consists of the following six scores:

$SAMPLE \leftarrow 55\ 56\ 63\ 65\ 68\ 72$

and suppose we wish to know the frequencies of scores in the 50's, 60's, and 70's. Then, *INTERVALS* will be

$INTERVALS \leftarrow 50\ 60\ 70\ 80$	These intervals are 50–59, 60–69, 70–80.
$^{-}1\downarrow INTERVALS$ 50 60 70	Drop the last component of *INTERVALS*.
$(^{-}1\downarrow INTERVALS)\circ.\leqslant SAMPLE$ 1 1 1 1 1 1 0 0 1 1 1 1 0 0 0 0 0 1	Each component in turn of $(^{-}1\downarrow IN\text{-}TERVALS)$ is compared to each component of *SAMPLE*, using $\leqslant$. Refer to the appendix for a more detailed explanation of outer product.
$1\downarrow INTERVALS$ 60 70 80	Drop the first component of *INTER-VALS*.
$(1\downarrow INTERVALS)\circ.> SAMPLE$ 1 1 0 0 0 0 1 1 1 1 1 0 1 1 1 1 1 1	The outer product with $>$.
$((^{-}1\downarrow INTERVALS)\circ.\leqslant SAMPLE)\wedge((1\downarrow INTERVALS)\circ.> SAMPLE)$ 1 1 0 0 0 0 0 0 1 1 1 0 0 0 0 0 0 1	Recall that $1\wedge 1$ yields 1, while $1\wedge 0$, $0\wedge 1$, $0\wedge 0$ all yield 0.
$+/((^{-}1\downarrow INTERVALS)\circ.\leqslant SAMPLE)\wedge$ $((1\downarrow INTERVALS)\circ.> SAMPLE)$[1] 2 3 1	Recall that for matrices, $+/A$ adds the columns of A horizontally.

Thus, for this example,

50 60 70 80 *DISTRIBUTION* 55 56 63 65 68 72
2 3 1

---

[1]Space limitations have forced us to show this instruction on two lines. In practice, *it must be typed on one line.*

There are 2 in the 50's, 3 in the 60's, and 1 in the 70's.

Let us also consider a program for plotting a bar graph of the distribution.

Program 11.6 BAR

```
        ∇ INTERVALS BAR SAMPLE; FREQUENCIES; I
[1]     FREQUENCIES←INTERVALS DISTRIBUTION SAMPLE
[2]     I←0
[3]     INCREASE: I←I+1
[4]     GRAPH FREQUENCIES[I]
[5]     →(I<ρFREQUENCIES)/INCREASE
        ∇
```

This program merely uses the program *GRAPH* on each component of FREQUENCIES.

Example 2

Let us return to the teacher with the test scores:

```
    SAMPLE←81 90 84 98 85 67 48 60 70 68 85 72 72 92 73 74
    SAMPLE←SAMPLE, 92 73 82 90 78 85 54 87 58 64 50 70
    SAMPLE←SAMPLE, 94 76

    INTERVALS←40 50 60 70 80 90 100

    INTERVALS DISTRIBUTION SAMPLE
1 3 4 9 7 6
```

Thus, there is 1 score in the 40's, 3 in the 50's, 4 in the 60's, 9 in the 70's, 7 in the 80's, and 6 in the 90's.

```
    INTERVALS BAR SAMPLE
*
* * *
* * * *
* * * * * * * * *
* * * * * * *
* * * * * *
```

A bar graph for the distribution of test scores.

EXERCISES

1. Use ? to generate the following random samples:
    (a) Five distinct numbers from 1 to 12.
    (b) Five numbers from 1 to 12 with repetition allowed.
    (c) Five numbers between 6 and 15.
    (d) Five numbers between ⁻5 and 5.

2. Devise a scheme for choosing 100 people at random in a city of 10000 people.

3. Professor X is on a committee which is making up questions for a national mathematics exam. To test the questions, he decides to give the exam to one of his classes at his own college. Is this a random sample? Discuss.

4. The following are test scores for 28 students in a class:

    SCORES←92 94 80 83 56 72 85 90 86 96 65 87 63 65 82 74 88

    SCORES←SCORES, 83 83 60 93 86 60 57 51 94 76 66

    (a) Use the program SORT to put these scores in order.
    (b) If 90's are A's, 80's are B's, 70's are C's, 60's are D's, and below 60 F's, use the program DISTRIBUTION to find how many of each grade are assigned.

5. Write a program REARRANGE for rearranging a vector of numbers in order of magnitude from the largest to smallest.

6. The ages of students in a certain college classroom as of their previous birthdays are as follows:

    AGES←19 19 24 20 19 18 18 23 21 21 30 23 18 18 20 18 20

    AGES←AGES, 18 20 20 20 19 19 17 20 23 19 20 19 19 19 20

Use the program DISTRIBUTE to determine the number of students of each age in the classroom.

7. The following represent college board scores for a certain class of high school seniors:

520	375	350	475
610	430	530	510
455	540	500	495
380	555	490	480
720	610	590	520
580	650	600	540
645	720	500	650
425	790	510	750

Use the program DISTRIBUTION to put these college board scores into a frequency distribution with intervals 350–399, 400–449, 450–499, 500–549,...,750–800.

8. Use the program BAR to plot a bar graph of the distribution of college board scores in Exercise 7.

11.2 Measures of central tendency

A *statistic* is a number computed from a sample. When presented with a set of data (a sample), it is often desirable to have a single number, a statistic, which describes the "typical" or "average" value in the sample. This kind of statistic is often referred to as "a measure of central tendency." The most common measures of central tendency are the mean, the median, and the mode. In this section, we will consider these three measures of central tendency and the relative merits of each.

The mean

The most commonly used measure of central tendency is the *mean* or average of the elements in the sample. If the elements in the sample are $x_1, x_2, \ldots, x_n$, then the sample mean, usually denoted by $\overline{X}$, is given by

$$\overline{X} = \frac{x_1 + x_2 + \ldots + x_n}{n} = \frac{\displaystyle\sum_{i=1}^{n} x_i}{n}.$$

Example

A test is given to 10 students with the resulting scores:

$$70, 64, 81, 83, 52, 91, 70, 66, 75, 80.$$

The mean score is

$$\overline{X} = \frac{70 + 64 + 81 + 83 + 52 + 91 + 70 + 66 + 75 + 80}{10} = \frac{732}{10} = 73.2.$$

If the sample is large, then it becomes quite tedious computing the mean by hand. Therefore, we present the following simple program for computing the mean.

Program 11.7 *MEAN*

```
    ∇ X←MEAN SAMPLE
[1]   X←(+/SAMPLE)÷(ρ SAMPLE) ∇
```

This is merely the sum of the elements in the sample divided by the number of elements in the sample, which is precisely the definition of the mean.

Examples

1. Compute the mean test score above using *MEAN*.

```
    MEAN 70 64 81 83 52 91 70 66 75 80
73.2
```

2. Compute the average weight of the fish caught in the pond in the example from the last section.

```
    SAMPLE←13 6 5 12 10 9 15 14 9 7 12 12 8 13 18 8 7 6 7
    SAMPLE←SAMPLE, 12 9 8 9 14 6
    MEAN SAMPLE
9.96
```

Thus, the average fish weights just under 10 pounds.

The median

Another measure of central tendency is the *median*, which we shall denote by Md. It is often referred to as the "most central" or "middlemost" value in the sample. Suppose the values in the sample are arranged in order of magnitude, from smallest to largest. If there is an odd number of values in the sample, then Md is the value in the center of this rearranged sample, below which and above which lie half of the values. If there is an even number of values in the sample, then Md is the average of the two central values in this rearranged sample.

Examples

1. Find the median weight for the 25 fish weights above.

   First, we need the program *SORT* to sort the elements of *SAMPLE* in order of magnitude.

   SORT SAMPLE
   5 6 6 6 7 7 7 8 8 8 9 9 9 9 10 12 12 12 12 13 13 14 14 15 18

   Since there is an odd number of weights in this sample, namely 25 of them, then Md is the central, or 13th value. Thus, the median is

   (SORT SAMPLE) [13]
   9   The median is 9.

2. Find the median test score in the other Example 2 of this section.

   SCORES←70 64 81 83 52 91 70 66 75 80
   SORT SCORES
   52 64 66 70 70 75 80 81 83 91

   Since there is an even number of scores in this sample, then the median is the average of the two most central scores. Thus,

   $$Md = \frac{70 + 75}{2} = 72.5.$$

Although the mean is the more commonly used measure of central tendency, the median sometimes gives a better indication of the "typical" value in the sample than the mean. This is especially true when there are extremely large or small values in the sample when compared to the other values in the sample. These extreme values exert an inordinate influence on the mean, while they do not effect the median.

Example

Suppose we have the following 7 test scores:

$$22, 82, 87, 90, 91, 97, 98.$$

The median of these scores is 90 while the mean is only 81. In this example, the median, 90, is a better indication of the typical score than the mean of 81. The mean has been "dragged down" by the extremely poor score of 22, while the median has not been affected by it.

The mode

The other measure of central tendency which we would like to consider is the *mode*. The mode is the value in the sample which occurs with the greatest frequency. The mode is easy to find and sometimes gives a very good indication of the typical value in the sample.

Examples

Suppose the scores of 15 students on a 10-point quiz were:

$$0, 0, 2, 5, 7, 7, 7, 7, 7, 7, 8, 8, 9, 10, 10.$$

The mode is 7. In this example, this is a pretty good indication of the typical score on the quiz.

Suppose the scores were

$$0, 0, 0, 2, 5, 6, 7, 7, 7, 8, 8, 9, 10, 10, 10.$$

Here, there are three modes: 0, 7, and 10. This illustrates an important disadvantage of the mode as a measure of central tendency; namely that in some cases there may be more than one mode.

Suppose the scores were

$$0, 0, 0, 2, 5, 6, 7, 7, 7, 8, 8, 9, 10, 10, 10, 10.$$

Here, the mode is 10, since it occurs with the greatest frequency. However, 10 is not a very good indication of the typical score on this quiz.

As stated before, the mean is the most commonly used measure of central tendency. It is easily computed and takes into consideration the entire distribution. If the sample is large, then the mean will not be overly influenced by a few extreme values. Also, the mean is a reliable measure of central tendency. That is, if repeated large samples are taken, the values of the means computed from these samples will not vary very much.

11.3 Measures of dispersion

It is possible for two samples to have the same mean and yet be quite different. For example, consider the following sets of scores for a test given to 10 students:

> Sample I: 52, 64, 66, 70, 70, 73, 80, 81, 83, 91
>
> Sample II: 25, 45, 50, 73, 80, 85, 87, 90, 95, 100
>
> Sample III: 73, 73, 73, 73, 73, 73, 73, 73, 73, 73.

These three samples all have the same mean, 73. However, they differ greatly in the amount of dispersion of scores.

One measure of dispersion is called the *range*. The range is just the largest value minus the smallest value in the sample. In sample I above, the range is $91 - 52 = 39$. In sample II, it is $100 - 25 = 75$. In sample III, it is $73 - 73 = 0$. The range is very easy to compute and does give an indication of the spread of values. However, it only depends on the two most extreme values and indicates nothing about the dispersion of values between these two extremes.

The two most commonly used measures of dispersion are the variance and the standard deviation. The *variance*, usually denoted by S^2, of a sample $x_1, x_2, \ldots, x_n$ is defined to be

$$S^2 = \frac{\sum_{i=1}^{n} \left(x_i - \overline{X} \right)^2}{n},$$

where $\overline{X}$ is the mean of the sample. The variance measures the spread of values in the sample away from the mean $\overline{X}$. It is the mean of the squares of the differences between the sample values and the mean. The more spread out the sample values are from the mean $\overline{X}$, the larger the value of S^2. The following program can be used to compute the variance of a sample.

Program 11.8 VARIANCE

```
    ∇ SPREAD← VARIANCE SAMPLE
[1]   SPREAD←MEAN (SAMPLE− MEAN SAMPLE)∗2 ∇
```

(Another formula for variance can be used to simplify the computation when they are done by hand or on calculators. It is

$$S^2 = \frac{\sum_{i=1}^{n} x_i^2}{n} - \overline{X}^2.$$

This formula is easily derived from the definition of variance.)

Before considering some examples, we would also like to consider the standard deviation. The *standard deviation* of a sample is just the square root of the variance. Thus, it is now quite easy to write a program for standard deviation.

Program 11.9 STDEV

```
    ∇ S← STDEV SAMPLE
[1]   S←(VARIANCE SAMPLE)∗.5 ∇
```

As with the variance, the larger the value of the standard deviation, the more spread out are the values of the sample away from the sample mean.

315

Examples

Let us find the variances and standard deviations in the samples of test scores at the beginning of this section.

>   SAMPLE1←52 64 66 70 70 (3 80 81 83 91

>   VARIANCE SAMPLE1

112.6

>   STDEV SAMPLE1

10.6113

>   SAMPLE2←25 45 50 73 80 85 87 90 95 100
>   VARIANCE SAMPLE2

550.8

>   STDEV SAMPLE2

23.4691

>   SAMPLE3←73 73 73 73 73 73 73 73 73 73
>   VARIANCE SAMPLE3

0

>   STDEV SAMPLE3

0

Thus, sample 2 has the greatest spread of values away from its mean of 73.

Let us also find the variance and standard deviation for the 25 fish weights of the fish in the example of Section 11.1.

>   SAMPLE←13 6 5 12 10 9 15 14 9 7 12 12 8 13 18 8 7 6 7

>   SAMPLE←SAMPLE, 12 9 8 9 14 6

>   VARIANCE SAMPLE

10.8384

>   STDEV SAMPLE

3.29218

The values of the variance and standard deviation of a sample do not really mean too much unless there is another sample with which the given sample is being compared. They are used in a relative sense, rather than in an exclusive sense.

EXERCISES

1. Write a program for finding the median of a sample.

2. Write a program for finding the mode of a sample.

3. Write a program for finding the range of a sample.

4. Find the mean, median, mode, range, variance, and standard deviation for the 28 test scores in Exercise 4 of Section 11.1.

5. Repeat Exercise 4 on the 30 ages of Exercise 6 of Section 11.1.

6. Repeat Exercise 4 on the 32 college board scores of Exercise 7 of Section 11.1.

7. The grade-point averages of 20 students selected at random are

2.34	3.25	4.00	1.96	2.68
2.80	3.46	2.26	2.88	1.60
3.42	3.44	2.60	2.85	1.84
2.74	2.56	3.04	2.30	2.00

Find the mean grade-point average and the standard deviation of the grade-point averages.

8. Find the mean, median, and mode for the following samples. Which one would give the best measure of central tendency in each case? Why?
(a) *SAMPLE1* ← 12 15 9 14 10 20 15 14 75
(b) *SAMPLE2* ← 2 3 3 5 5 5 6 7 7 8
(c) *SAMPLE3* ← 7 4 6 7 4 4 8 7

9. Suppose that it is desired to know the average weight of a college freshman at a large university. Describe a way of getting a good estimate of this average weight.

10. A machine produces parts of mean length 5 inches. In terms of the standard deviation of the parts produced by the machine, discuss why it is important to keep the machine well oiled and maintained.

11. Write a program for computing variance by the alternative formula presented in Section 11.3.

12. A professor uses the following scheme for assigning letter grades to tests: Scores that are more than two standard deviations above the mean are called A's, between 1 and 2 standard deviations above the mean are called B's, within 1 standard deviation of the mean on either side are called C's, between 1 and 2 standard deviations below the mean are called D's, and more than 2 standard deviations below the mean are called F's. Find the number of each letter grades on a test with the scores:

SCORES ← 85 55 82 94 66 78 92 57 67 60 63 72 60 69 37 90

SCORES ← *SCORES*, 88 56 60 70 68 51 85 98 66 92 75 61 39 70

Would you consider his grading fair?

13. The floor function, ⌊, when used monadically, yields the largest *integer* ⩽ the right argument N. Thus,

⌊3.14

3

⌊5.98

5

⌊7

7

Consider the following program *ROUND*, which rounds a number off to 4 decimal places:

   ∇ *Q←ROUND N*

[1]   *Q←*(10∗⁻4)×⌊0.5+(*N*×10∗4) ∇

(a) Explain this program by rewriting it in three steps.
(b) Evaluate the following:

   ROUND 2.354725804

   ROUND 3.141592654

   ROUND 2.718281828

   ROUND ⁻7.389056099

[*Note*: This program will be used in the next section.]

14. Generalize the program in Exercise 13 to one that will round off a number *N* to *P* places.

11.4 The normal distribution

Perhaps most of the readers are, at least vaguely, familiar with the famous "bell-shaped" curve which is characteristic of the normal distribution. In this section, we shall examine some of the characteristics of a normal distribution. In the next section, we will consider an important application of the normal distribution in statistics.

The hypergeometric, binomial, and Poisson distributions studied in the previous chapter are all examples of *discrete* distributions. In these distributions, the values of X, the number of successes, were always integers. In a discrete distribution, X can only assume a finite, or at most countably infinite number of values. "Countably infinite" means that although there are an infinite number of values, they can be counted, as in the case of the set of nonnegative integers. A distribution which is not discrete is said to be *continuous*. In a continuous distribution, X can assume an uncountably infinite number of values. For example, the distribution of weights of fish in a river is a continuous distribution, since a fish could weigh 5 pounds, 5.1 pounds, 5.11 pounds, or 5.111 pounds, and so on. Unless one rounds off, it is possible that the weight of a fish could be any positive real number less than some maximum weight.

Since in a continuous distribution, there are an uncountably infinite number of values of X, the probability of X equaling any particular value is usually taken to be 0. This doesn't mean that it is impossible for X to attain this value. It merely means that one would not be wise to wager that X would attain this particular value out of the uncountably infinite number of values it can attain. With a continuous distribution, one is more interested in the probability that X lies in some interval of values, as

$P(a \leqslant X \leqslant b)$ or $P(X \leqslant a)$ or $P(X > b)$, than in the probability that it attains a particular value.

Associated with any continuous distribution is a function $F(x)$, called the *probability density function* for X. The graph of such a probability density function is a continuous curve which always lies above the x axis. The total area between this curve and the x axis is defined to be 1. To compute probabilities such as $P(a \leqslant X \leqslant b)$, one finds the area between the curve of $F(x)$, the x axis, and $x = a$ and $x = b$. Since $P(x = a) = P(x = b) = 0$, then $P(a \leqslant X \leqslant b) = P(a < X < b)$.

Perhaps the most important example of a continuous distribution is the *normal distribution*. The probability density function for the normal distribution looks like

$$F(x) = \frac{1}{\sqrt{2\pi} \cdot \sigma} \cdot e^{-\frac{1}{2}\left(\frac{x-u}{\sigma}\right)^2} \quad \text{where } -\infty < x < \infty.$$

u is called the mean for X, and is the mean or average value in the distribution of values for X. σ is called the standard deviation of X, and is a measure of the spread of values of X away from the mean, u. The larger σ is, the more spread out are the values of X away from u. A graph of the density function with mean u and standard deviation σ appears in Figure 11.2.

As one can see, the graph of the density function for a normal distribution is symmetrical about the vertical line $x = u$, and the maximum value of $F(x)$ occurs at $x = u$. The points on the graph corresponding to $u - \sigma$ and $u + \sigma$ are the points of inflection for the curve. There are as many normal distributions as there are values of u and σ.

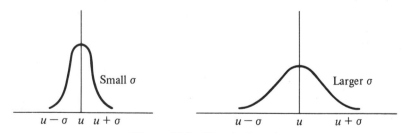

Figure 11.2   Density function.

To simplify the work when using normal distributions, the normal distribution with mean 0 and standard deviation 1, called the *standard normal distribution*, is usually used. The letter Z is usually used for the standard normal random variable. By making the substitution $Z = (X - u)/\sigma$, it is possible to change any probability problem in a normal distribution with mean u and standard deviation σ to a standard normal distribution probability problem problem (Figure 11.3). For example, to find $P(a \leqslant X \leqslant b)$ for a normal distribution with mean u and standard

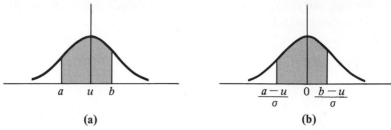

Figure 11.3 (a) Normal distribution with mean u, standard deviation σ. (b) The standard normal distribution.

deviation σ, one needs to find the area under the curve of

$$F(x) = \frac{1}{\sqrt{2\pi} \cdot \sigma} \cdot e^{-\frac{1}{2}\left(\frac{x-u}{\sigma}\right)^2}$$

between $x = a$ and $x = b$. In other words, one needs to evaluate the definite integral

$$\int_a^b \frac{1}{\sqrt{2\pi} \cdot \sigma} \cdot e^{-\frac{1}{2}\left(\frac{x-u}{\sigma}\right)^2} \cdot dx.$$

If we make the substitution $z = (x - u)/\sigma$, then $dz/dx = 1/\sigma$, or $dx = \sigma \cdot dz$. Also, when $x = a$, then $z = (a - u)/\sigma$, and when $x = b$, then $z = (b - u)/\sigma$. Thus, the above integral becomes

$$\int_{(a-u)/\sigma}^{(b-u)/\sigma} \frac{1}{\sqrt{2\pi}} \cdot e^{-\frac{1}{2} \cdot z^2} \cdot dz,$$

which is $P((a-u)/\sigma \leqslant Z \leqslant (b-u)/\sigma)$.
   This function

$$F(z) = \frac{1}{\sqrt{2\pi}} \cdot e^{-\frac{1}{2} z^2}$$

has no antiderivative, so that the Fundamental theorem of calculus cannot be used to evaluate this integral. However, we can use the program **INTEGRAL** of Chapter 9 to approximate this integral. First, however, we need a subprogram for computing the values of the standard normal density function, since the program **INTEGRAL** calls for a subprogram **FN**.

∇ *FN* [1]
[1] $Y \leftarrow (* (-X * 2) \div 2) \div ((O2) * .5) \nabla$

> Altering **FN** to fit the standard normal density function.

Examples

1. Find $P(0 \leqslant Z \leqslant 1)$.
    In order to do this, we need to find the area under the standard normal curve $F(z)$ from $z = 0$ to $z = 1$. Or, we need to evaluate $\int_0^1 F(z) \cdot$

dz. We can use the program *INTEGRAL* to evaluate an approximation to this integral as follows:

0 *INTEGRAL* 1
.3416651824

Thus, about 34 percent of the area under the normal curve lies within one standard deviation to the right of the mean.

2. Find $P(^-1 \leqslant Z \leqslant 0)$.

$^-1$ *INTEGRAL* 0
.3416651824

Notice that $P(0 \leqslant Z \leqslant 1) = P(^-1 \leqslant Z \leqslant 0)$. This is reasonable to expect due to the symmetry of the standard normal curve about its mean of 0 (see Figure 11.4).

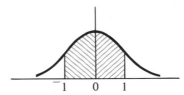

Figure 11.4   The area under the standard normal curve. Note that the area from 0 to 1 is the same as the area from $^-1$ to 0.

In fact, using this symmetry, we need only deal with probabilities of the form $P(0 \leqslant Z \leqslant c)$ for $c > 0$ to evaluate any probability for the standard normal distribution. Thus, we present the following simple program for computing $P(0 \leqslant Z \leqslant c)$.

Program 11.10 **NORMAL**

```
    ∇ P←NORMAL C
[1]   P←0 INTEGRAL C
[2]   P←ROUND P
    ∇
```

This rounds the probability off to 4 decimal places using the subprogram ROUND from Exercise 13 of the previous set of exercises.

Example 1

Suppose that the weights of fish in a pond are normally distributed with a mean of 5 pounds and a standard deviation of 1 pound. If a fish is caught at random from this pond and its weight is X find the following probabilities:

(a) Find $P(4 \leqslant X \leqslant 6)$.

321

Since $u=5$ and $\sigma=1$, then if we standardize this distribution using $Z=(X-u)/\sigma$, this problem becomes one of finding

$$P\left(\frac{4-5}{1}\leqslant\frac{X-5}{1}\leqslant\frac{6-5}{1}\right)=P(^{-}1\leqslant Z\leqslant 1).$$

Due to the symmetry of the standard normal distribution, since $P(^{-}1\leqslant Z\leqslant 0)=P(0\leqslant Z\leqslant 1)$, then $P(^{-}1\leqslant Z\leqslant 1)=2\times P(0\leqslant Z\leqslant 1)$. Thus, we get

2 × NORMAL 1

.6834

So, about 68 percent of the fish caught weigh between 4 and 6 pounds.

(b) Find $P(X\leqslant 5)$.

Standardizing this, we get

$$P\left(\frac{X-5}{1}\leqslant\frac{5-5}{1}\right)=P(Z\leqslant 0).$$

Since the total area under the standard normal curve is 1, and the distribution is symmetrical about $z=0$, then the area to the left of the mean 0 is 0.5000 (see Figure 11.5). Thus, $P(X\leqslant 5)=0.5$ or 50 percent of the fish caught weigh $\leqslant 5$ pounds.

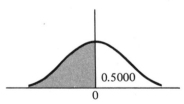

0.5000

0

Figure 11.5 Probability of $5\geqslant X$ is represented by the shaded area below the curve; $P=0.5000$.

(c) Find $P(X\leqslant 7)$.

Standardizing this yields

$$P\left(\frac{X-5}{1}\leqslant\frac{7-5}{1}\right)=P(Z\leqslant 2)=P(Z\leqslant 0)+P(0\leqslant Z\leqslant 2)$$

$$=0.5000+P(0\leqslant Z\leqslant 2).$$

Thus, we get (see Figure 11.6)

.5000 + NORMAL 2

.9777

Almost 98 percent of the fish weigh $\leqslant 7$ pounds.

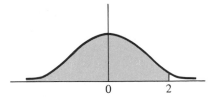

Figure 11.6   Probability of $7>X$ is shaded area below curve; $P=0.9777$.

(d) Find $P(X>7)$.

Standardizing this, we get

$$P\left(\frac{X-5}{1}>\frac{7-5}{1}\right)=P(Z>2)=1-P(Z\leqslant 2),$$

using Probability rule 3 of Chapter 10. Thus, we get

$1-.9777$

$.0223$

Another way of doing this problem is to use the fact that the total probability to the right of $z=0$ is 0.5000. Thus, $P(Z>2)=0.5000-P(0\leqslant Z\leqslant 2)$.

$.5000-$ *NORMAL 2*

$.0223$ So, about 2 percent of the fish weigh more than 7 pounds.

(e) Find $P(X<3)$.

Thus, we want

$$P(X<3)=P\left(\frac{X-5}{1}<\frac{3-5}{1}\right)=P(Z<^-2).$$

However, since the standard normal curve is symmetrical about $z=0$, then $P(Z<^-2)=P(Z>2)=0.0223$, by Part (d). Thus about 2 percent of the fish weigh less than 3 pounds (see Figure 11.7).

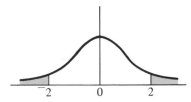

Figure 11.7   Probability of $3>X$ is shaded area below curve; $P=0.0223$.

(f) Find $P(4\leqslant X\leqslant 7)$.

$$P\left(\frac{4-5}{1}\leqslant\frac{X-5}{1}\leqslant\frac{7-5}{1}\right)=P(^-1\leqslant Z\leqslant 2)=P(^-1\leqslant Z\leqslant 0)+P(0\leqslant Z\leqslant 2)$$

$$=P(0\leqslant Z\leqslant 1)+P(0\leqslant Z\leqslant 2).$$

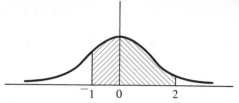

Figure 11.8   Probability of $7 \geqslant X \geqslant 4$ is shaded area below curve; $P=0.8194$.

(*NORMAL* 1)+(*NORMAL* 2)
.8194

Almost 82 percent weigh between 4 and 7 pounds (see Figure 11.8).

Example 2

The scores on a standardized exam are normally distributed with the average score being 500 with standard deviation 100. If a student is selected at random and given this exam, find the probability that his score will be

(a) Greater than 700.

Let X be this student's score. We want $P(X > 700)$. Since $u = 500$ and $\sigma = 100$, then standardizing this problem yields

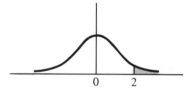

Figure 11.9 $P(X > 700) = 0.0223$.

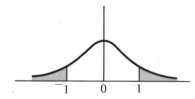

Figure 11.10 $P(X < 400) = 0.1583$.

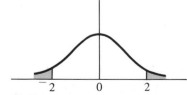

Figure 11.11 $P(X < 300) + P(X > 700) = 0.0446$.

$$P\left(\frac{X-500}{100} > \frac{700-500}{100}\right) = P(Z>2) = 0.5000 - P(0 \leqslant Z \leqslant 2).$$

.5000 – *NORMAL 2*

.0223

Thus, $P=0.0223$ (see Figure 11.9).

(b) Less than 400.

We want

$$P(X<400) = P\left(\frac{X-500}{100} < \frac{400-500}{100}\right) = P(Z<^{-}1) = P(Z>1)$$

$$= 0.5000 - P(0 \leqslant Z \leqslant 1).$$

.5000 – *NORMAL 1*

.1583

Thus, $P=0.1583$ (see Figure 11.10).

(c) Under 300 or over 700.

We want

$$P(X<300) + P(X>700)$$

$$= P\left(\frac{X-500}{100} < \frac{300-500}{100}\right) + P\left(\frac{X-500}{100} > \frac{700-500}{100}\right)$$

$$= P(Z<^{-}2) + P(Z>2) = 2 \cdot P(Z>2)$$

$$= 2 \cdot 0.0223 = 0.0446 \quad \text{(by Part (a)).}$$

Thus, $P=0.0446$ (see Figure 11.11).

EXERCISES

1. A soda machine gives an average drink of 12 ounces with a standard deviation of 0.5 ounces. If a cup holds 13 ounces, find the probability that a cup overflows, assuming the distribution of drinks dispersed by the machine is normal.

2. A machine produces pipes which are normally distributed with an average length of 8 inches and a standard deviation of 0.25 inches. Find the probability that a pipe chosen at random will be between 7.75 and 8.25 inches long.

3. The I.Q.'s of applicants to Picky U. are normally distributed with a mean I.Q. of 120 and a standard deviation of 15. To be accepted at Picky U., an applicant must have an I.Q. of at least 100. Find the probability that an applicant will be rejected on the basis of his I.Q.

4. The amount of time required to play 18 holes of golf on a weekday at Divot C.C. is normally distributed with a mean of 4 hours and a standard deviation of 20 minutes. Find the probability that a golfer will require between $4\frac{1}{2}$ and 5 hours to play a round of golf.

5. The set of grades on a mathematics exam given to 500 students are approximately normally distributed with an average grade of 74 and a standard deviation of 10.
   (a) If the lowest passing grade is 60, find the probability that a student passes the exam.
   (b) How many of the 500 passed the exam?

6. If the scores on a test are truly normally distributed, and if to get an A, one's score should be 2 or more standard deviations above the mean, to get a B, between 1 and 2 standard deviations above the mean, to get a C, within 1 standard deviation of the mean, to get a D, between 1 and 2 standard deviations below the mean, and to get an F, more than 2 standard deviations below the mean; find the percentages of A's, B's, C's, D's, and F's on the exam.

7. The tips received by a waitress are normally distributed with an average tip of $1.25 and a standard deviation of $0.30. Find the probability of a tip less than $0.50.

8. If the average height of a student is normally distributed with a mean of 5 feet 8 inches and standard deviation of 2 inches, and if 100 students are in a room, find the approximate number of students one would expect to be more than 6 feet tall.

11.5 The sampling distribution of the mean

A *sampling distribution* for a statistic is the hypothetical distribution of values that one would get for that statistic if he took repeated samples from the same population and computed that statistic for each sample. In this section, we will consider the sampling distribution for the sample mean, $\overline{X}$.

One of the important theorems in theoretical statistics is the so-called Central limit theorem which deals with the sampling distribution of the sample mean. Roughly speaking, the Central limit theorem says the following:

Central limit theorem

If a population is normal with mean u and standard deviation σ, then the sampling distribution for the sample mean $\overline{X}$ computed from a random sample of N objects from this population is also a normal distribution with mean $u_{\overline{X}} = u$ and standard deviation $\sigma_{\overline{X}} = \sigma / \sqrt{N}$.

Even if the population is not normal, the distribution of $\overline{X}$ is still approximately normal with the mean and standard deviation indicated above, as long as the size of the sample is large. The larger the sample, the better the approximation. This indicates the power of the normal distribution.

Since the mean for the sampling distribution of $\overline{X}$, $u_{\overline{X}}$, is the same as the population mean, u, then $\overline{X}$ is a good estimate of the population mean if u is unknown. Also, since the standard deviation of $\overline{X}$, $\sigma_{\overline{X}}$, is $\sigma / \sqrt{N}$, where σ

is the population standard deviation, and since as N gets larger, $\sigma/\sqrt{N}$ gets smaller, then the larger the sample, the better $\overline{X}$ is as an estimate of u.

Example 1

College board scores are normally distributed with a mean of 500 and a standard deviation of 100. If a random sample of 25 students are selected and given the college board examination, and the sample mean $\overline{X}$ for this sample computed,

(a) Find $P(\overline{X} > 550)$.

Since $u = 500$ and $\sigma = 100$, then for $\overline{X}$, $u_{\overline{X}} = u = 500$, and $\sigma_{\overline{X}} = 100/\sqrt{25} = 20$. Also, since the population of college board scores is normally distributed, then we can assume that $\overline{X}$ has the normal distribution with mean 500 and standard deviation 20.

Thus, standardizing the distribution of $\overline{X}$, we get

$$P(\overline{X} > 550) = P\left(\frac{\overline{X} - 500}{20} > \frac{550 - 500}{20}\right) = P(Z > 2.5)$$

$$= 0.500 - P(0 \leqslant Z \leqslant 2.5).$$

Thus, we get

.5000 – *NORMAL* 2.5
.0062

So, it is very unlikely that the sample mean $\overline{X}$ will be greater than 550.

(b) Find $P(\overline{X} < 480)$.

Standardizing this, we get

$$P(\overline{X} < 480) = P\left(\frac{\overline{X} - 500}{20} < \frac{480 - 500}{20}\right) = P(Z < {}^-1) = P(Z > 1)$$

$$= 0.5000 - P(0 \leqslant Z \leqslant 1).$$

So, we get

.5000 – *NORMAL* 1
.1583

(c) Find $P(480 \leqslant \overline{X} \leqslant 520)$.

Thus,

$$P(480 \leqslant \overline{X} \leqslant 520) = P\left(\frac{480 - 500}{20} \leqslant \frac{\overline{X} - 500}{20} \leqslant \frac{520 - 500}{20}\right) = P({}^-1 \leqslant Z \leqslant 1)$$

$$= 2 \times P(0 \leqslant Z \leqslant 1).$$

2 × *NORMAL* 1
.6834

Example 2

It is not known what kind of a distribution the lifetime of a lightbulb possesses. However, extensive testing has indicated that the mean lifetime is 1000 hours with a standard deviation of 120 hours. If a random sample of 100 lightbulbs is taken and $\bar{X}$ computed, find the probability that the sample mean $\bar{X}$ will exceed 980 hours.

Since $N = 100$, which is large (usually any N greater than 30 is considered to be large), then the distribution of $\bar{X}$ will be approximately normal with mean $u_{\bar{X}} = u = 1000$ and standard deviation $\sigma_{\bar{X}} = \sigma / \sqrt{N} = 120 / \sqrt{100} = 12$. Thus,

$$P(\bar{X} > 980) = P\left(\frac{\bar{X} - 1000}{12} > \frac{980 - 1000}{12}\right) = P(Z > {}^-1.667)$$

$$= 0.5000 + P(0 \leqslant Z \leqslant 1.667).$$

Thus, we get (see Figure 11.12).

.5000 + *NORMAL* 1.6667
.9515

About 95 percent of the lightbulbs will live for more than 980 hours.

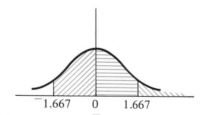

Figure 11.12 $P(\bar{X} > 980 \text{ hours}) = 0.9515$.

EXERCISES

1. The strength of a certain kind of wooden beam is normally distributed with mean 1200 psi (pounds per square inch) and standard deviation 50 psi. If 25 of these beams are selected at random and their strengths tested and the sample mean $\bar{X}$ computed, find the following probabilities:
   (a) $P(\bar{X} > 1210)$ (b) $P(\bar{X} < 1200)$ (c) $P(1990 \leqslant \bar{X} \leqslant 1205)$

2. The type of distribution of weights of male college freshmen is not known. However, it is known that the approximate mean weight is 160 pounds with standard deviation 20 pounds. If a random sample of 225 of these college freshmen is selected and the average weight $\bar{X}$ computed, then find
   (a) $P(\bar{X} > 165)$ (b) $P(\bar{X} < 156)$

3. Suppose that the waiting time for a bus at a certain bus stop is normally distributed with mean 12 minutes and standard deviation 3 minutes. In 25 days, find the probability that the average waiting time will be at least 10 minutes.

4. The average weight of a box of Crunchy cereal is 12 ounces with a standard deviation 0.5 ounces. If 100 boxes are selected at random and $\overline{X}$ computed, where $\overline{X}$ is the average weight of the 100 boxes selected, then find
   (a) $P(\overline{X} > 11.9)$ (b) $P(11.9 \leqslant \overline{X} \leqslant 12.1)$

5. The average distance that Mr. Jones hits a golf ball with a driver is 230 yards with a standard deviation 25 yards. In two rounds of golf, Mr. Jones uses his driver 30 times. Find the probability that during these two rounds, his average drive will be at least 225 yards.

12 The trigonometric functions

Originally, trigonometry was developed to solve problems involving triangles. As such, it is a very important tool for surveyors, machinists, engineers, astronomers, and navigators. As we shall see, the trigonometric functions are readily available in APL. Therefore, we shall study some of the basic properties of the trigonometric functions and explore their use in solving triangles. This chapter is not intended as a complete study of the trigonometric functions. Many of the topics usually studied in a course in trigonometry will be omitted.

12.1 Angles

An *angle* is formed when a ray (half a line) is rotated in a plane with its endpoint (called its *vertex*) fixed. We shall use Greek letters such as θ (theta) or α (alpha) or β (beta) to denote angles. If the rotation is counterclockwise, the angle is positive. If the rotation is clockwise, the angle is negative (see Figure 12.1). There are two common units of measurement of an angle, the degree and the radian.

Degree measurement of an angle

During one complete revolution in a counterclockwise direction, a ray sweeps out all angles from 0 degrees to 360 degrees. Therefore, one quarter of a complete revolution corresponds to 90 degrees. Other angles are indicated in Figure 12.2. (It is customary to denote N degrees by $N°$.)

Radian measurement of an angle

Place the vertex of an angle θ at the center of a circle of radius R. The ray being rotated to form this angle sweeps out an arc $\overset{\frown}{AB}$ on the circumference of the circle. Let the length of this arc be denoted by S. Then, the

330

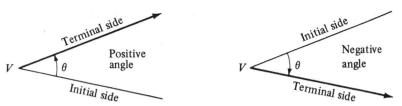

Figure 12.1   Positive and negative angles.

radian measure of this angle θ is defined to be the quotient S/R radians (Figure 12.3). Therefore, an angle of 1 radian is an angle for which $S = R$. Note (1) that if θ is a negative angle, the number of radians in is $-S/R$ and (2) that a radian has no units, since it is a length divided by a length.

Relationship between degree measurement and radian measurement of an angle

In one complete revolution of 360 degrees, the radian measure of the angle will be S/R, where S is the circumference of the circle, $2\pi R$. Therefore, in a complete revolution of 360 degrees, there are $2\pi R/R = 2\pi$ radians. Thus,

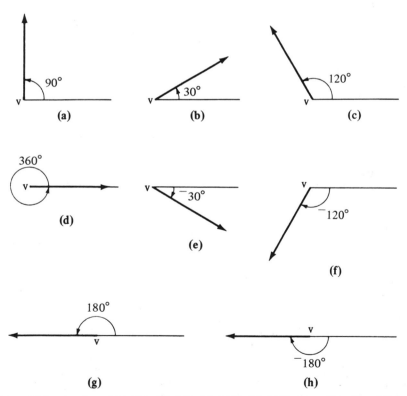

Figure 12.2   Angles of (a) 90°, (b) 30°, (c) 120°, (d) 360°, (e) −30°, (f) −120°, (g) 180° and (h) −180°.

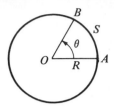

Figure 12.3   Radian measurement of an angle

we get

$$360 \text{ degrees} = 2\pi \text{ radians}$$

$$180 \text{ degrees} = \pi \text{ radians}$$

$$1 \text{ degree} = \frac{\pi}{180} \text{ radians}$$

$$1 \text{ radian} = \frac{180}{\pi} \text{ degrees}.$$

So, to change from degrees to radians, multiply the number of degrees by $\pi/180$ radians. To change from radians to degrees, multiply the number of radians by $180/\pi$ degrees.

Examples

1. Change $5\pi/3$ radians to degrees.

$$\frac{5\pi}{3} \cdot \frac{180}{\pi} = 300 \text{ degrees}.$$

2. Change 75 degrees to radians.

$$75 \cdot \frac{\pi}{180} = \frac{5\pi}{12} \text{ radians}.$$

Angles in APL

The basic unit of angular measurement in APL, as in most computer languages, is the radian. In APL, ○X corresponds to $\pi \cdot X$ in standard mathematical notation.

Examples

```
      ○1
3.141592654
```
π radians.

```
      ○2
6.283185308
```
2π radians.

```
      ○.5
1.570296327
```
$\pi/2$ radians.

It might be useful to have a program for changing from degrees to radians.

Program 12.1 **DEGREES**

  ∇ RADIANS ← DEGREES ANGLE
[1]   **RADIANS ← ANGLE × O(1 ÷ 180) ∇**

Radians = Angle (in degrees) times $(\pi/180)$.

Examples

  DEGREES 180
3.141592654   180 degrees = π radians.

  DEGREES 45
.785148164   45 degrees = $\pi/4$ radians.

  DEGREES 1
.01745329252   1 degree = $\pi/180$ radians.

Also, it might be useful to have a program for changing from radians to degrees.

Program 12.2 **RADIANS**

  ∇ DEGREES ← RADIANS ANGLE
[1]   **DEGREES ← (180 ÷ O1) × ANGLE ∇**

Degrees = $(180/\pi)$ times Angle (in radians)

Examples

  RADIANS O1
180   π radians = 180 degrees.

  RADIANS O(1 ÷ 6)
30                              $\pi/6$ radians = 30 degrees.

  RADIANS 1
57.29577951   1 radian = $180/\pi$ degrees.

The standard position of an angle

An angle is said to be in *standard position* with reference to the coordinate axes if its vertex is at the origin and its initial side is along the positive x axis.

Consider the following examples of angles in standard position:

1. As in Figure 12.4, an angle whose terminal side is in the first quadrant when it is placed in standard position is said to be an angle in the *first quadrant*.

333

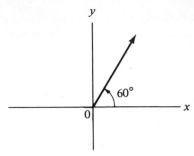

Figure 12.4 $\pi/3$ radians = 60 degrees in standard position.

2. Figure 12.5 shows an angle whose terminal side is in the second quadrant when it is placed in standard position; this is said to be an angle in the *second quadrant*.

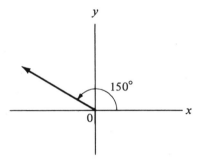

Figure 12.5 $5\pi/6$ radians = 150 degrees in standard position.

3. Figure 12.6 shows an angle whose terminal side is in the third quadrant when it is placed in standard position; this is said to be an angle in the *third quadrant*.

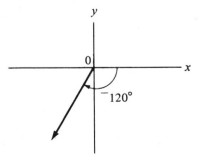

Figure 12.6 $^-2\pi/3$ radians = $^-120$ degrees in standard position.

4. Figure 12.7 shows an angle whose terminal side is in the fourth quadrant when it is placed in standard position; this is said to be an angle in the *fourth quadrant*.

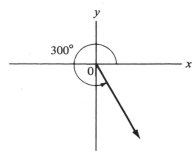

Figure 12.7 $5\pi/3$ radians $=300$ degrees in standard position.

EXERCISES

1. Sketch the following angles in standard position.
   (a) $\pi/6$ radians $=30$ degrees.
   (b) $7\pi/6$ radians $=210$ degrees.
   (c) $^-\pi/3$ radians $=^-60$ degrees.
   (d) $^-11\pi/6$ radians $=^-330$ degrees.
   (e) $5\pi/2$ radians $=450$ degrees.

2. Change the following from degrees to radians:
   (a) 45 degrees   (c) $^-9$ degrees   (e) $^-20$ degrees
   (b) 18 degrees   (d) 240 degrees
   Check your answers at an APL terminal using the program *DEGREES*.

3. Change the following from radians to degrees:
   (a) $3\pi/5$ radians   (c) 2 radians   (e) $^-1.5$ radians
   (b) $^-4\pi/3$ radians   (d) $\pi/12$ radians
   Check your answers at an APL terminal using the program *RADIANS*.

4. Prove that the length S of an arc of a circle of radius R swept out by an angle of θ radians is given by $S=R\cdot\theta$.

5. Write a program for finding the length S of an arc of a circle of radius R subtended by an angle of θ radians.

12.2 The trigonometric functions

Let θ be an angle in standard position and let $P(x,y)$ be a point on its terminal side. Let $r=\sqrt{x^2+y^2}$ be the distance from the origin to P. Then, the 6 trigonometric functions of θ are defined as follows (see Figure 12.8):

$$\sin(\theta)=\frac{y}{r} \quad \text{(sin is an abbreviation for "sine")}$$

$$\cos(\theta)=\frac{x}{r} \quad \text{(cos is an abbreviation for "cosine")}$$

335

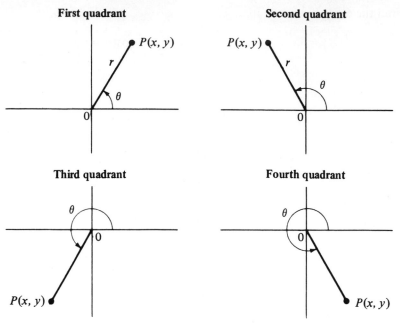

Figure 12.8 $x, y, r,$ and θ in all 4 quadrants.

$$\tan(\theta) = \frac{y}{x} \quad (\text{tan is an abbreviation for "tangent"})$$

$$\cot(\theta) = \frac{x}{y} \quad (\text{cot is an abbreviation for "cotangent"})$$

$$\sec(\theta) = \frac{r}{x} \quad (\text{sec is an abbreviation for "secant"})$$

$$\csc(\theta) = \frac{r}{y} \quad (\text{csc is an abbreviation for "cosecant"}).$$

Thus, the 6 trigonometric functions of θ are just all of the possible ratios of the variables $x, y,$ and r.

Examples

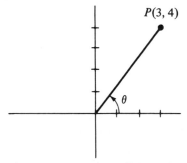

Figure 12.9   One possible angle.

1. Find the 6 trigonometric functions for the angle θ in Figure 12.9.

$$x = 3, \quad y = 4 \qquad \sin(\theta) = \frac{4}{5} \quad \cos(\theta) = \frac{3}{5}$$

$$r = \sqrt{3^2 + 4^2} = 5 \quad \tan(\theta) = \frac{4}{3} \quad \cot(\theta) = \frac{3}{4} \;.$$

$$\sec(\theta) = \frac{5}{3} \quad \csc(\theta) = \frac{5}{4}$$

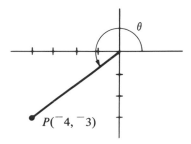

Figure 12.10   Another possible angle.

2. Find the 6 trigonometric functions for the angle θ in Figure 12.10.

$$x = {}^-4, \quad y = {}^-3, \quad r = 5.$$

$$\sin(\theta) = \frac{{}^-3}{5} \qquad \cos(\theta) = \frac{{}^-4}{5} \qquad \tan(\theta) = \frac{{}^-3}{{}^-4} = \frac{3}{4}$$

$$\cot(\theta) = \frac{{}^-4}{{}^-3} = \frac{4}{3} \qquad \sec(\theta) = \frac{{}^-5}{4} \qquad \csc(\theta) = \frac{{}^-5}{3}.$$

Some basic trigonometric identities

An *identity* is a statement which is true for any values of the variables in the domains of the functions involved. The trigonometric functions can be related by a huge number of identities. We shall concern ourselves with only a few of the most basic identities in this text. The following identities are derived immediately from the definitions of the trigonometric functions.

The reciprocal identities

$$\sin(\theta) = \frac{1}{\csc(\theta)} \quad \text{and} \quad \csc(\theta) = \frac{1}{\sin(\theta)}$$

$$\tan(\theta) = \frac{1}{\cot(\theta)} \quad \text{and} \quad \cot(\theta) = \frac{1}{\tan(\theta)}$$

$$\cos(\theta) = \frac{1}{\sec(\theta)} \quad \text{and} \quad \sec(\theta) = \frac{1}{\cos(\theta)}\;.$$

337

These are easily derived as follows:

$$\frac{1}{\csc(\theta)} = \frac{1}{r/y} = \frac{y}{r} = \sin(\theta).$$

The others are derived similarly and are left as exercises.

The quotient identities

$$\tan(\theta) = \frac{\sin(\theta)}{\cos(\theta)} \quad \text{and} \quad \cot(\theta) = \frac{\cos(\theta)}{\sin(\theta)}.$$

These are derived as follows:

$$\frac{\sin(\theta)}{\cos(\theta)} = \frac{y/r}{x/r} = \frac{y}{x} = \tan(\theta).$$

The identity for cot is derived similarly.

The Pythagorean identities

$$\sin^2(\theta) + \cos^2(\theta) = 1$$
$$\tan^2(\theta) + 1 = \sec^2(\theta)$$
$$1 + \cot^2(\theta) = \csc^2(\theta).$$

These are derived as follows:

$$\sin^2(\theta) + \cos^2(\theta) = \left(\frac{y}{r}\right)^2 + \left(\frac{x}{r}\right)^2 = \frac{x^2 + y^2}{r^2} = \frac{r^2}{r^2} = 1.$$

Dividing both sides of this identity by $\cos^2(\theta)$, one obtains

$$\frac{\sin^2(\theta)}{\cos^2(\theta)} + \frac{\cos^2(\theta)}{\cos^2(\theta)} = \frac{1}{\cos^2(\theta)}.$$

Using identities above,

$$\tan^2(\theta) + 1 = \sec^2(\theta).$$

The other Pythagorean identity is derived similarly, and is left as an exercise.

Example

Suppose we know that $\sin(\theta) = 2/5$ and that θ is in the first quadrant. Find the other five trigonometric functions of θ.

$$\sin^2(\theta) + \cos^2(\theta) = \left(\frac{2}{5}\right)^2 + \cos^2(\theta) = 1,$$

so that

$$\cos^2(\theta) = 1 - \left(\frac{4}{25}\right) = \frac{21}{25}, \quad \text{or} \quad \cos(\theta) = \pm\frac{\sqrt{21}}{5}.$$

But, since θ is in the first quadrant, x, y, and r are all positive. So, $\cos(\theta) = \sqrt{21}/5$. Now,

$$\tan(\theta) = \frac{\sin(\theta)}{\cos(\theta)} = \frac{2/5}{\sqrt{21}/5} = \frac{2}{\sqrt{21}} \qquad \cot(\theta) = \frac{\cos(\theta)}{\sin(\theta)} = \frac{\sqrt{21}}{2}$$

$$\sec(\theta) = \frac{1}{\cos(\theta)} = \frac{1}{\sqrt{21}/5} = \frac{5}{\sqrt{21}} \qquad \csc(\theta) = \frac{1}{\sin(\theta)} = \frac{1}{2/5} = \frac{5}{2}.$$

Signs of the trigonometric functions in the various quadrants

If we know the signs of $\sin(\theta)$ and $\cos(\theta)$, then we can use the quotient and reciprocal identities to determine the signs of the other four trigonometric functions. Let us, therefore, consider the signs of $\sin(\theta)$ and $\cos(\theta)$ in the four quadrants. (See Figure 12.8 to help identify the quadrants.)

Quadrant I.

Since r is a distance, the distance from the origin to P, then r is always positive. In quadrant I, x, y, and r are all positive; so, $\sin(\theta) = y/r$ and $\cos(\theta) = x/r$ are positive.

Quadrant II.

In quadrant II, x is negative, while y and r are positive. Thus, $\sin(\theta)$ is positive and $\cos(\theta)$ is negative.

Quadrant III.

In quadrant III, x and y are negative, while r is positive. Thus, $\sin(\theta) = y/r$ and $\cos(\theta) = x/r$ are both negative.

Quadrant IV.

In quadrant IV, x is positive, y is negative, and r is positive. Thus, $\sin(\theta)$ is negative and $\cos(\theta)$ is positive.

Example

Suppose we know that $\sin(\theta) = {}^-3/5$ and that θ is in the third quadrant. Find the other 5 trigonometric functions of θ.

$$\sin^2(\theta) + \cos^2(\theta) = \left(\frac{^-3}{5}\right)^2 + \cos^2(\theta) = 1,$$

so that

$$\cos^2(\theta) = 1 - \frac{9}{25} = \frac{16}{25} \quad \text{or} \quad \cos(\theta) = \pm\frac{4}{5}.$$

However, since θ is in the third quadrant, we know that $\cos(\theta)$ is negative. Thus, $\cos(\theta) = -4/5$. Now, using the quotient and reciprocal identities, we get

$$\tan(\theta)=\frac{\sin(\theta)}{\cos(\theta)}=\frac{3}{4} \qquad \cot(\theta)=\frac{\cos(\theta)}{\sin(\theta)}=\frac{4}{3}$$

$$\sec(\theta)=\frac{1}{\cos(\theta)}=\frac{^-5}{4} \qquad \csc(\theta)=\frac{1}{\sin(\theta)}=\frac{^-5}{3}.$$

Evaluation of some trigonometric functions of some special angles

We now consider some techniques for evaluating some trigonometric functions of some special angles for which it is quite easy to compute the trigonometric functions.

Example 1

Evaluate $\sin(30°)$.

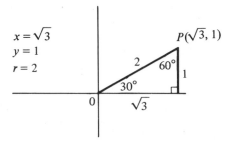

Figure 12.11   Example 1.

As shown in Figure 12.11, first drop a perpendicular from the point P on the terminal side of the 30-degree angle to the x axis, forming a right triangle with angles of 30, 60, and 90 degrees. Now, by an important theorem in geometry, in a 30, 60, 90-degree triangle, the length of the hypoteneuse is twice the length of the side opposite the 30-degree angle. Let the length of the side opposite the 30-degree angle be 1. Then, the length of the hypotheneuse is 2. Using the Pythagorean theorem, the length of the side opposite the 60-degree angle must be $\sqrt{3}$. Thus, in the diagram above, $x=\sqrt{3}$, $y=1, r=2$. Therefore, we get

$$\sin(30°)=\frac{y}{r}=\frac{1}{2}.$$

Example 2

Evaluate $\tan(45°)$.

Form a right triangle as in the previous example. Since this triangle has two 45-degree angles, it is an isosceles triangle. Let the lengths of the two equal sides be 1. By the Pythagorean theorem, the length of the

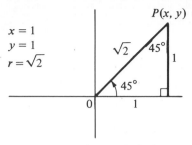

Figure 12.12   Example 2.

hypoteneuse will be $\sqrt{2}$. So, $x=1$, $y=1$, and $r=\sqrt{2}$. Thus (see Figure 12.12),

$$\tan(45°) = \frac{y}{x} = \frac{1}{1} = 1.$$

Example 3

Evaluate sin(90°).

The point $(0, 1)$ is on the terminal side of the 90-degree angle (see Figure 12.13). The distance from the origin to this point is 1. Thus, $x=0$, $y=1$, $r=1$, and

$$\sin(90°) = \frac{y}{r} = \frac{1}{1} = 1.$$

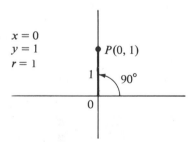

Figure 12.13   Example 3.

Example 4

Evaluate cos(120°).

As before, drop a perpendicular from the point P on the terminal side of the 120-degree angle to the x axis. Since there are 180 degrees in a straight angle, then the angle between the terminal side of the 120-degree angle and the x axis is 60 degrees. Since there are 180 degrees in a triangle, the other angle in the triangle formed must be 30 degrees. Using the theorem about a 30, 60, 90-degree triangle, the length of the hypoteneuse

341

must be twice the length of the side opposite the 30-degree angle. Let the length of the side opposite the 30-degree angle be 1. Then, the length of the hypoteneuse is 2, and by the Pythagorean theorem the length of the side opposite the 60-degree angle is $\sqrt{3}$. However, since our angle is in the second quadrant, x must be negative. Thus, $x = {}^-1$, $y = \sqrt{3}$, $r = 2$, (see Figure 12.14) and

$$\cos(120°) = \frac{x}{r} = \frac{{}^-1}{2}.$$

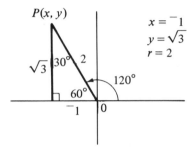

Figure 12.14   Example 4.

Example 5

Evaluate sec(210°).

As before, form a right triangle, resulting in a 30, 60, 90-degree triangle. Since the 210-degree angle is in the third quadrant, then x and y must both be negative. Thus, as in Figure 12.15, $x = {}^-\sqrt{3}$, $y = {}^-1$, $r = 2$, and

$$\sec(210°) = \frac{r}{x} = \frac{2}{{}^-\sqrt{3}} = \frac{{}^-2}{\sqrt{3}}.$$

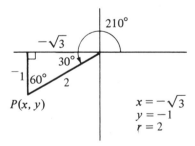

Figure 12.15   Example 5.

Example 6

Evaluate cos(180°).

A point on the terminal side of the 180-degree angle is $P({}^-1,0)$. The distance from this point to the origin is 1. Thus, $x = {}^-1$, $y = 0$, and $r = 1$ (see

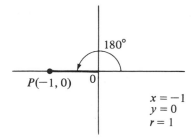

Figure 12.16   Example 6.

Figure 12.16). So,

$$\cos(180°) = \frac{x}{r} = {}^-1.$$

In the next section, we shall consider using APL to evaluate trigonometric functions of angles.

EXERCISES

1. Prove the reciprocal and Pythagorean identities not proved in the text.

2. Find the 6 trigonometric functions for the angle whose terminal side passes through the point $P({}^-12,5)$.

3. Find the 6 trigonometric functions for the angle whose terminal side passes through the point $P(8,{}^-6)$.

4. Suppose that $\cos(\theta) = {}^-4/5$ and that θ is in the second quadrant. Find the other 5 trigonometric functions of θ.

5. Suppose that $\tan(\theta) = 4/3$ and that θ is in the third quadrant. Find the other 5 trigonometric functions of θ.

6. Evaluate the 6 trigonometric functions of 60 degrees.

7. Evaluate the 6 trigonometric functions of 0 degrees.

8. Evaluate the 6 trigonometric functions of 45 degrees.

9. Evaluate the 6 trigonometric functions of 150 degrees.

10. Evaluate the 6 trigonometric functions of 225 degrees.

11. Write a program *SINE* for evaluating the sine of an angle whose terminal side passes through the point (x,y).

12. Write a program *COSINE* for evaluating the cosine of an angle whose terminal side passes through the point (x,y).

13. Using the programs *SINE* and *COSINE*, write a program *TANGENT* for evaluating the tangent of an angle whose terminal side passes through the point (x,y).

14. Repeat Exercise 13 for *COTANGENT, SECANT,* and *COSECANT.*

15. Use the programs you wrote in the previous exercises to do Exercises 2 and 3.

12.3 The trigonometric functions in APL

Recall that when the large circle ○ is used monadically in APL, the result is π times the number on the right. Thus,

    ○1
3.141592654

When used dyadically, the results are as follows:[1]

      1○*ANGLE* yields sin(*ANGLE*),   where *ANGLE* is in radians.

      2○*ANGLE* yields cos (*ANGLE*)

Examples

    1○(○1÷6) $\sin(\pi/6)=\sin(30°)$.
.5

    2○(○1÷6) $\cos(\pi/6)=\cos(30°)$.
.8660254038

Let us use the large circle and the quotient and reciprocal identities to write programs for the six trigonometric functions of an angle θ.

Program 12.3 SIN

      ∇ *T* ← *SIN THETA*
[1]   *T*←1○ *THETA* ∇   Since 1○ *THETA* yields sin(*THETA*).

Program 12.4 COS

      ∇ *T* ← *COS THETA*   Since 2○ *THETA* yields cos(*THETA*).
[1]   *T*←2○ *THETA* ∇

Program 12.5 TAN

      ∇ *T* ← *TAN THETA*
[1]   *T*←(*SIN THETA*)÷(*COS THETA*) ∇

                    Since $\tan(\theta)=\sin(\theta)/\cos(\theta)$.

---

[1]Other results of the dyadic use of the large circle include: 0○*X* yields $(1-X*2)*5$, 3○*X* yields tan *X*, 4○*X* yields $(1+X*2)*5$ and (for those who know of the "hyperbolic functions") 5○*X* yields sinh *X*, 6○*X* yields cosh *X*, and 7○*X* yields tanh *X*. Functions involving ○ are called "circular functions."

Program 12.6 COT

```
    ∇T←COT THETA
[1]   T←(COS THETA)÷(SIN THETA) ∇
```

Since $\cot(\theta)=\cos(\theta)/\sin(\theta)$.

Program 12.7 SEC

```
    ∇T←SEC THETA
[1]   T←1÷(COS THETA) ∇
```

Since $\sec(\theta)=1/\cos(\theta)$.

Program 12.8 CSC

```
    ∇T←CSC THETA
[1]   T←1÷(SIN THETA) ∇
```

Since $\csc(\theta)=1/\sin(\theta)$.

Examples

When using these programs, the angle *THETA* must be in radians.

```
    SIN 1
0.8414709848
```
Sine of 1 radian.

```
    COSO1÷3
.5
```
Cosine of $\pi/3$.

```
    TAN 2
⁻2.185039863
```
Tangent of 2 radians.

If the angle is in degrees, we can use the program *DEGREES* to change it to radians. Then, evaluate the trig function of the angle. This can be done in one step as in the following examples:

```
    SIN DEGREES 30
.5
```
$\sin(30°)$.

```
    COS DEGREES 120
⁻.5
```
$\cos(120°)$.

```
    TAN DEGREES 72
3.077683537
```
$\tan(72°)$.

```
    SEC DEGREES ⁻60
2
```
$\sec(-60°)$.

```
    COT DEGREES 200
2.747477419
```
$\cot(200°)$.

```
    CSC DEGREES ⁻120
⁻1.154700538
```
$\csc(-120°)$.

```
    TAN DEGREES 90
8.002733596E13
```
$\tan(90°)$.

Actually, tan(90°) doesn't exist, since tan(90°) is the quotient of sin(90°) and cos(90°), and cos(90°)=0, so that tan(90°) involves division by 0. The large answer above is a computer estimate of tan(90°). It is essentially infinity.

EXERCISES

1. Use the programs in this section to evaluate:
   (a) sin(1.5 radians)   (d) sec(5π/3 radians)
   (b) tan(π radians)   (e) cot(π/12 radians)
   (c) cos(⁻2 radians)   (f) csc(0 radians)

2. Use the programs in this section and the program *DEGREES* to evaluate the following:
   (a) tan(75°)   (d) sec(70.8°)
   (b) sin(105°)   (e) csc(⁻150°)
   (c) cos(⁻60°)   (f) cot(112.5°)

3. There are 60 minutes (60′) in one degree. An angle is often given in degrees and minutes, as 30°15′. In order to use the program *DEGREES*, it is necessary to change this angle to 30.25° by converting the minutes to a fraction of a degree and expressing it as a decimal. Compute the following using the appropriate programs:
   (a) tan(30°15′)   (c) sec(112°12′)
   (b) sin(75°6′)   (d) cos(⁻25°10′)

4. If $\theta = 25$ degrees, test the validity of the following trigonometric identities:
   (a) $\sin(2\theta) = 2 \cdot \sin(\theta) \cdot \cos(\theta)$
   (b) $\cos(2\theta) = \cos^2(\theta) - \sin^2(\theta)$
   (c) $\tan(2\theta) = 2 \cdot \tan(\theta)/(1 - \tan^2(\theta))$

5. If $\alpha = 25$ degrees and $\beta = 35$ degrees, test the validity of the following identities:
   (a) $\sin(\alpha + \beta) = \sin(\alpha) \cdot \cos(\beta) + \cos(\alpha) \cdot \sin(\beta)$
   (b) $\cos(\alpha + \beta) = \cos(\alpha) \cdot \cos(\beta) - \sin(\alpha) \cdot \sin(\beta)$
   (c) $\tan(\alpha + \beta) = (\tan(\alpha) + \tan(\beta))/(1 - \tan(\alpha) \cdot \tan(\beta))$

12.4 Graphs of the trigonometric functions

A function $y = F(x)$ is *periodic* if there exists a positive constant p such that $F(x+p) = F(x)$, for all x in the domain of F. The smallest such constant p is called the *period* of F.

Now, by the two identities in Exercise 5 of the previous exercise set,

$$\sin(x + 2\pi) = \sin(x) \cdot \cos(2\pi) + \cos(x) \cdot \sin(2\pi)$$

$$= \sin(x) \cdot 1 + \cos(x) \cdot 0 = \sin(x)$$

$$\cos(x + 2\pi) = \cos(x) \cdot \cos(2\pi) - \sin(x) \cdot \sin(2\pi)$$

$$= \cos(x) \cdot 1 - \sin(x) \cdot 0 = \cos(x).$$

Thus, sin(x) and cos(x) are periodic. Their period is 2π.

Since $\sec(x)$ and $\csc(x)$ are reciprocals of $\cos(x)$ and $\sin(x)$, then these two functions are also periodic with period 2π. Thus, the graphs of these four functions repeat themselves every interval of length 2π.

Also,

$$\tan(x+\pi)=\frac{\sin(x+\pi)}{\cos(x+\pi)}$$

$$=\frac{\sin(x)\cdot\cos(\pi)+\cos(x)\cdot\sin(\pi)}{\cos(x)\cdot\cos(\pi)-\sin(x)\cdot\sin(\pi)}$$

$$=\frac{\sin(x)\cdot^-1+\cos(x)\cdot0}{\cos(x)\cdot^-1-\sin(x)\cdot0}=\frac{\sin(x)}{\cos(x)}=\tan(x).$$

Thus, $\tan(x)$ is periodic with period π, and its graph repeats itself every interval of length π. Since $\cot(x)$ is just the reciprocal of $\tan(x)$, then $\cot(x)$ is also periodic with period π.

Because $\sin(x)$ is periodic with period 2π, we have only to graph it over the interval $0\leqslant x\leqslant 2\pi$ and then recopy the graph over all other intervals of length 2π. Consider the following program *TRIGPAIRS* which prints out a table of ordered pairs to be used in graphing $y=\sin(x)$.

Program 12.9 TRIGPAIRS

```
     ∇A TRIGPAIRS B; X; Y; Z
[1]  X←A
[2]  Y←SIN X
[3]  Z←RADIANS X
[4]  Z;' '; X;' '; Y
[5]  X←X+○1÷12          (Add π/12=15 degrees each time.)
[6]  →(X≤B)/2
     ∇
```

Graph of $y=\sin x$

Let us use *TRIGPAIRS* to generate a table of pairs (x,y) to be used in graphing $y=\sin(x)$.

0 *TRIGPAIRS* ○2   Pairs from $x=0$ to $x=2\pi$.

0	0	0
15	.2617993878	.2588190451
30	.5235997756	.5
45	.7853981634	.7071067812
60	1.047197551	.8660254038
75	1.308996939	.9659258263
90	1.50796327	1
105	1.832595715	.9659258263
120	2.094395102	.8660254038

When graphing a trig function, the angle must be in radians. We have included in this program a column for the angles in degrees also (the first column).

347

135	2.35619449	.7071067812
150	2.617993878	.5
165	2.879793266	.2588190451
180	3.141592654	5.341316986 E^-14
195	3.403392041	⁻ .2588190451
210	3.665191429	⁻ .5
225	3.926990817	⁻.7071067812
240	4.188790205	⁻ .8660254038
255	4.450589593	⁻ .965928263
270	4.71238898	⁻1
285	4.974188368	⁻ .9659258263
300	5.235987756	⁻ .8660254038
315	5.497787144	⁻ .7071067812
330	5.759586532	⁻ .5
345	6.021385919	⁻ .2588190451
360	6.283185307	⁻2.489348869 E^-13

x (in degrees)	x (in radians)	y values

Using this table, we can plot the points (x,y) obtaining (in Figure 12.17) the graph of $y = \sin(x)$. From the graph and the table, we can see that the range of $y = \sin(x)$ is $^-1 \leqslant x \leqslant 1$. The domain is all the real angles.

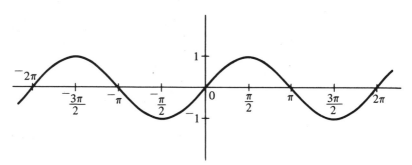

Figure 12.17   Graph of $y = \sin(x)$.

Graph of $y = \tan(x)$

If we alter line 2 of *TRIGPAIRS*, we can use it to generate a table of pairs for $y = \tan(x)$.

```
        ∇ TRIGPAIRS [2]
[2]     Y← TAN X ∇

        0 TRIGPAIRS o2
0   0   0
15      .2617993878      .2679491924
```

30	.5235987756	.5773502692	
45	.7853981634	1	
60	1.047197551	1.732050808	
75	1.308996939	3.732050808	
90	1.570796327	5.101739508 E13	
105	1.832595715	⁻3.732050808	Actually tan(90°) and tan(270°)
120	2.094395102	⁻1.732050808	do not exist, since tan(90°)=
135	2.35619449	⁻1	sin(90°)/cos(90°)=1/0.
150	2.617993878	⁻.5773502692	
165	2.879793266	⁻.2679491924	
180	3.141592654	⁻5.341316986 E⁻14	
195	3.403392041	0.2679491924	
210	3.665191429	0.5773502692	
225	3.926990817	1	
240	4.188790205	1.732050808	
255	4.450589593	3.732050808	
270	4.71238898	7.700734292 E12	
285	4.974188368	⁻3.732050808	
300	5.235987756	⁻1.732050808	
315	5.497787144	⁻1	
330	5.759586532	⁻.5773502692	
345	6.021385919	⁻.2679491924	
360	6.283185307	⁻2.489348869 E⁻13	

Using these pairs, the graph of $y=\tan(x)$ is as shown in Figure 12.18. From this graph and table, we can see that the range of $y=\tan(x)$ is all real numbers.

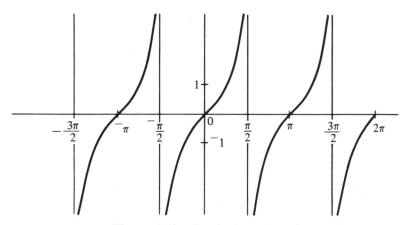

Figure 12.18 Graph of $y=\tan(x)$.

Graph of y = sec(x)

Again, we can alter *TRIGPAIRS* to fit this new function.

∇ *TRIGPAIRS* [2]
[2]   *Y ← SEC X* ∇

0 *TRIGPAIRS* ○2

0	0	1
15	.2617993878	1.03527618
30	.5235987756	1.154700538
45	.7853981634	1.414213562
60	1.047197551	2
75	1.308996939	3.863703305
90	1.570796327	5.101739508 E13
105	1.832595715	‾3.863703305
120	2.094395102	‾2
135	2.35619449	‾1.414213562
150	2.617993878	‾1.154700538
165	2.879793266	‾1.03527618
180	3.141592654	‾1
195	3.403392041	‾1.03527618
210	3.665191429	‾1.154700538
225	3.926990817	‾1.414213562
240	4.18870205	‾2
255	4.450589593	‾3.863703305
270	4.71238898	‾7.700734292 E12
285	4.974188368	3.863703305

Actually, sec(90°) doesn't exist, since   sec(90°) = 1/cos(90°) = 1/0.

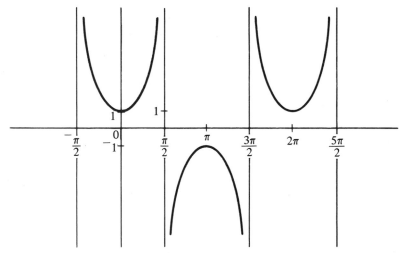

Figure 12.19   Graph of *y* = sec(*x*).

300	5.235987756	2
315	5.497787144	1.414213562
330	5.759586532	1.154700538
345	6.021385919	1.03527618
360	6.283185307	1

Thus, the graph of $y = \sec(x)$ looks as shown in Figure 12.19. From the graph and table, we see that the range of $y = \sec(x)$ includes all $y \leqslant {}^-1$ or $y \geqslant 1$.

Exercises

Use the program *TRIGPAIRS* to generate a table for the other three trigonometric functions. Then, graph them and determine their ranges.

12.5 The inverse trigonometric functions

Two functions which mean the same except that the roles of the independent and dependent variables are reversed are called *inverse functions*. In Chapter 7, we saw that the logarithmic function with base b is the inverse function for the exponential function with base b. The trigonometric functions also have inverses.

The inverse sine function

$y = \arcsin(x)$ (read as "y is the angle whose sine is x") means the same as $x = \sin(y)$. (Some texts denote this function by $y = \sin^{-1}(x)$.) However, there are many y's corresponding to each x. For example, if $x = 1/2$, then we could have $y = \pi/6$ or $y = 5\pi/6$ or $y = {}^-7\pi/6$ or many other values. In order to be a function, there must be a unique value for y corresponding to each value of x in the domain. Thus, as it now stands, $y = \arcsin(x)$ is not a function. However, if the values of $\arcsin(x)$ are restricted to the range $(-\pi/2) \leqslant y \leqslant (\pi/2)$, then there will be a unique value of y corresponding to each x in the domain $^-1 \leqslant x \leqslant 1$. Thus, the complete definition of the inverse sine function is

$$y = \arcsin(x) \text{ means the same as } x = \sin(y),$$

where $\dfrac{-\pi}{2} \leqslant \arcsin x \leqslant \dfrac{\pi}{2}$ or $^-90$ degrees $\leqslant \arcsin x \leqslant 90$ degrees.

[*Note*: The values of y in this restricted range are often referred to as the *principal values* of the arcsin function.]

Example

Evaluate $\arcsin({}^-1/2)$.

$y = \arcsin({}^-1/2)$ means the same as $^-1/2 = \sin(y)$. There are many y's with this sine. However, the only one in the restricted range is $y = {}^-30$ degrees or $y = {}^-\pi/6$.

351

The inverse sine function is also available in APL. Arc sin(x) is given by ‾1○X in APL. For example, to find arc sin(‾1/2), we use

    ‾1○‾.5

‾0.52355987756   The answer is given in radians.

    RADIANS (‾1○‾.5)   Changing the answer to degrees.

‾30

It will be convenient for us in the next two sections to have a program for the inverse sine function which gives the answer in degrees.

Program 12.10 ARCSIN

    ∇ Y←*ARCSIN X*

[1] Y←‾1○X The APL arc sine function.

[2] Y←*RADIANS Y*   Changing from radians to degrees.
    ∇

Examples

    ARCSIN ‾.5

‾30   ‾30 degrees.

    ARCSIN (3∗5)÷2   arc sin($\sqrt{3}$ /2)

60   60 degrees.

The inverse cosine function

$y = \arccos(x)$ means the same as $x = \cos(y)$, where $0 \leqslant y \leqslant \pi$, or 0 degrees $\leqslant y \leqslant$ 180 degrees. Notice that the principal values for the inverse cosine function are different from those for the inverse sine function. The reason for this is that cosine is positive throughout the interval $(-\pi/2) \leqslant y \leqslant (\pi/2)$. Thus, in this interval, there would not be a unique value of y for every x. For example, if $x = 1/2$, then there are two y's in this interval such that $\cos(y) = 1/2$, namely $y = -\pi/3$ and $y = \pi/3$. However, in the interval $0 \leqslant y \leqslant \pi$, there does exist a unique y for each x in the domain.

We will find it valuable to have a program for the inverse cosine function which gives the answer in degrees.

Program 12.11 ARCCOS

    ∇ Y←*ARCCOS X*
[1] Y←‾2○X
[2] Y←*RADIANS Y*
    ∇

Examples

    ARCCOS ‾.5

120   120 degrees.

ARCCOS (3 * .5) ÷ 2	arc cos($\sqrt{3}$ /2).
30	30 degrees.

The inverse tangent function

$y = \arctan(x)$ means the same as $x = \tan(y)$, where $(-\pi/2) < y < (\pi/2)$, or $^-90$ degrees $< y < 90$ degrees. Notice that $^-90$ degrees and 90 degrees are not included in the range above. This is because $\tan(y)$ is not defined for these values, since $\tan(y) = \sin(y)/\cos(y)$ and $\cos(y)$ is 0 for $^-90$ degrees and 90 degrees.

Program 12.12 ARCTAN

```
     ∇ Y ← ARCTAN X
[1]   Y ← ⁻30 X
[2]   Y ← RADIANS Y
     ∇
```

Examples

ARCTAN ⁻1	
⁻45	⁻45 degrees.
ARCTAN 3 * .5	arc tan $\sqrt{3}$.
60	60 degrees.

We will have no need for the inverse trig functions for the other three trig functions in solving triangles. Therefore, we will omit them. However, if one needs to find the value of one of these inverse trig functions, he can easily convert it to a problem involving one of the above inverse trig functions. For example, consider:

1. Find $\operatorname{arc sec}(2)$.
   If $y = \operatorname{arc sec}(2)$, then $\sec(y) = 2$. Since $\sec(y) = 1/\cos(y)$, then $\cos(y) = 1/2$. Thus, we can find $\operatorname{arc cos}(1/2)$.

   ARCCOS .5
60

   Therefore, $\operatorname{arc sec}(2)$ is 60 degrees or $\pi/3$ radians.
2. Find $\operatorname{arc csc}(2)$.
   If $y = \operatorname{arc csc}(2)$, then $\csc(y) = 2$. Since $\csc(y) = 1/\sin(y)$, then $\sin(y) = 1/2$. Thus, we can find $\operatorname{arc sin}(1/2)$.

   ARCSIN .5
30

   Therefore, $\operatorname{arc csc}(2)$ is 30 degrees or $\pi/6$ radians.[2]

---

[2]The other dyadic uses of the large circle, O, with negative left arguments are: ^-40X yields $(^-1 + X*2)*.5$, ^-50X yields $\operatorname{arcsinh}(X)$, ^-60X yields $\operatorname{arccosh}(X)$, and ^-70X yields $\operatorname{arctanh}(X)$, for those who know about the hyperbolic functions.

1. Evaluate the following:
   (a) $\arcsin(\sqrt{2}/2)$ (e) $\arcsin(0.6293)$
   (b) $\arccos(0)$ (f) $\arccos(0.8290)$
   (c) $\arccos(^-\sqrt{2}/2)$ (g) $\arctan(^-1.2)$
   (d) $\arctan(\sqrt{3}/3)$ (h) $\operatorname{arc\ csc}(\sqrt{2})$

2. Evaluate the following:
   (a) $\cos(\arccos(1))$ (e) $\sec(\arctan(1))$
   (b) $\sin(\arcsin(^-1/2))$ (f) $\arcsin(\sin(5\pi/6))$
   (c) $\cot(\arccos(\sqrt{3}/2))$ (g) $\arcsin(\cos(2\pi/3))$
   (d) $\tan(\arccos(^-3/5))$ (h) $\arccos(\tan(45°))$

3. Use the fact that $\cos(y) = 1/\sec(y)$ to write a program for finding ARCSEC X

12.6 Solving right triangles

To *solve* a triangle means to find the values of any unknown angles and the lengths of any unknown sides. In a triangle with vertices A, B, and C, we shall denote the angles at A, B, and C respectively by the Greek letters α, β, and γ. The sides opposite these vertices will be denoted by the letters a, b, and c respectively. In a right triangle, we shall put the right angle at the vertex C. We can always put a right triangle ABC so that either α or β is in standard position, as in Figure 12.20. Therefore, we can express the trig functions of α and β in terms of the side adjacent to them, the side opposite from them, and the hypoteneuse (the side opposite the right angle γ). This is done as follows: In the triangle in Figure 12.20b,

$$\sin(\alpha) = \frac{a}{c} = \frac{\text{Opposite side}}{\text{Hypotenuse}} \qquad \csc(\alpha) = \frac{c}{a} = \frac{\text{Hypotenuse}}{\text{Opposite side}}$$

$$\cos(\alpha) = \frac{b}{c} = \frac{\text{Adjacent side}}{\text{Hypotenuse}} \qquad \sec(\alpha) = \frac{c}{b} = \frac{\text{Hypotenuse}}{\text{Adjacent side}}$$

$$\tan(\alpha) = \frac{a}{b} = \frac{\text{Opposite side}}{\text{Adjacent side}} \qquad \cot(\alpha) = \frac{b}{a} = \frac{\text{Adjacent side}}{\text{Opposite side}}.$$

In the triangle in Figure 12.20c,

$$\sin(\beta) = \frac{b}{c} = \frac{\text{Opposite side}}{\text{Hypotenuse}} \qquad \csc(\beta) = \frac{c}{b} = \frac{\text{Hypotenuse}}{\text{Opposite side}}$$

$$\cos(\beta) = \frac{a}{c} = \frac{\text{Adjacent side}}{\text{Hypotenuse}} \qquad \sec(\beta) = \frac{c}{a} = \frac{\text{Hypotenuse}}{\text{Adjacent side}}$$

$$\tan(\beta) = \frac{b}{a} = \frac{\text{Opposite side}}{\text{Adjacent side}} \qquad \cot(\beta) = \frac{a}{b} = \frac{\text{Adjacent side}}{\text{Opposite side}}.$$

In either case, the relationships between the trig functions, the sides adjacent to and opposite from the angle, and the hypotenuse remain the

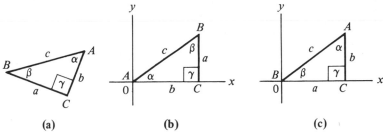

Figure 12.20 Three ways of orienting a right triangle.

same. Using these expressions for the trigonometric functions, we can now solve right triangles.

Example 1

A flagpole stands on ground level. At a point 130 feet from the base of the flagpole, it is found that the angle of elevation of the imaginary line from the ground to the top of the flagpole is 35 degrees (Figure 12.21). Find the height of the flagpole.

$$\tan(35°) = \frac{\text{Opposite side}}{\text{Adjacent side}} = \frac{h}{130}, \quad \text{or } h = 130 \cdot \tan(35°).$$

Using our programs, we get

     130 × *TAN DEGREES* 35
91.02697997

The height of the flagpole is about 91 feet.

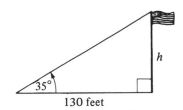

Figure 12.21 Flagpole of Example 1

Example 2

A ladder 24 feet long is placed against a vertical wall. The foot of the ladder is 8 feet from the base of the wall (see Figure 12.22).

(a) Find the angle between the foot of the ladder and the ground.

$$\cos(\alpha) = \frac{\text{Adjacent side}}{\text{Hypotenuse}} = \frac{8}{24} = \frac{1}{3}, \quad \text{so } \alpha = \arccos(1/3).$$

     ARCCOS 1 ÷ 3
70.52877937

The angle is about 70.5 degrees.

355

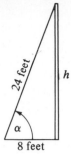

Figure 12.22   Ladder of Example 2.

(b) Find the height of the ladder up the wall.

$$\sin(\alpha) = \frac{\text{Opposite side}}{\text{Hypotenuse}} = \frac{h}{24}, \quad \text{or } h = 24 \cdot \sin(\alpha).$$

24 × SIN DEGREES 70.52877937
22.627417   The ladder is 22.6 feet up the wall.

Example 3

A roof rises 5 inches for each foot measured horizontally.

(a) Find the angle the roof makes with the horizontal.

In the small triangle at the corner of the roof above (see Figure 12.23),

$$\tan(\alpha) = \frac{\text{Opposite side}}{\text{Adjacent side}} = \frac{5}{12}, \quad \text{thus, } \alpha = \arctan\left(\frac{5}{12}\right).$$

ARCTAN 5 ÷ 12
22.61986495   The angle is about 22.6 degrees.

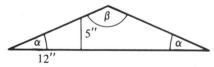

Figure 12.23   Roof of Example 3.

(b) Find the angle at the top of the roof.
Since there are 180 degrees in a triangle, then $\beta = 180 - 2\alpha$.

180 − 2 × ARCTAN 5 ÷ 12
134.76027010

The angle at the top is almost 135 degrees.

356

Example 4

To find the distance across a lake, two trees marked P and Q are located on opposite banks of the lake near the shore. At P, a line is laid off at a right angle to the line PQ and continued a distance of 500 feet to a point S along the bank containing the tree marked P. Thus, we have a right triangle PSQ. The angle at S is measured and found to be 42 degrees. Find the distance across the lake (Figure 12.24).

$$\tan(42°) = \frac{\text{Opposite side}}{\text{Adjacent side}} = \frac{PQ}{500}, \quad \text{or } PQ = 500 \times \tan(42°).$$

500 × *TAN DEGREES* 42
450.2020221

Thus, the distance across the lake is about 450 feet.

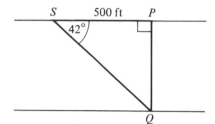

Figure 12.24   Lake of Example 4.

EXERCISES

1. In the triangle below, if $a = 10$, and $c = 12$, find b and the angles.

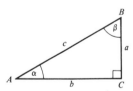

2. In the triangle above, if $a = 45$, and the angle $\beta = 36$ degrees, find the values of b, c, and the angle α.

3. A roof rises 3 inches for each foot measured horizontally. Find the angle the roof makes with the horizontal.

4. A kite is held tightly by a string 1200 feet long. The string makes an angle of 52 degrees with the ground. How high up is the kite?

5. From a point 40 feet high on a building, a line is stretched to the top of a pole 60 feet high. The horizontal distance from the point on the building to the pole is 75 feet. Find the angle this line makes with the horizontal.

357

6. A tree casts a shadow 75 feet long. The angle between the ground and the imaginary line connecting the tip of the shadow and the top of the tree is 41 degrees. Find the height of the tree.

7. The top of a pole on one side of a stream is 25 feet above the eye of an observer. The angle of elevation from the observer's eye to the top of the pole is 6 degrees. How wide is the stream?

8. In an isosceles triangle, the two equal sides are each 35 feet long, and the equal angles at the base are both 54 degrees. Find the length of the base, the length of the altitude, and the area of the triangle.

12.7 Solving oblique triangles

An *oblique triangle* is a triangle with no right angle. Given three parts of an oblique triangle, at least one of which is a side, we can determine the remaining angles and sides. To do this, we will use either the Law of sines or the Law of cosines or both.

The Law of sines

Let ABC be any triangle, such as in Figure 12.25. Then the Law of sines says

$$\frac{a}{\sin(\alpha)} = \frac{b}{\sin(\beta)} = \frac{c}{\sin(\gamma)}.$$

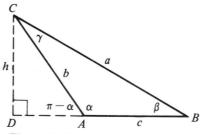

Figure 12.25   An oblique triangle.

The Law of sines can be derived as follows: Drop a perpendicular from the vertex C to the side AB, intersecting it at the point D. Then, in the triangle BDC, $\sin(\beta) = h/a$, or $h = a \cdot \sin(\beta)$.

In triangle ADC,

$$\sin(\pi - \alpha) = \sin(\pi) \cdot \cos(\alpha) - \cos(\pi) \cdot \sin(\alpha)$$
$$= 0 \cdot \cos(\alpha) - {}^{-}1 \cdot \sin(\alpha)$$
$$= \sin(\alpha)$$
$$= \frac{h}{b}.$$

So, $h = b \cdot \sin(\alpha)$.

Setting these two values of h equal, we get

$$h = a \cdot \sin(\beta) = b \cdot \sin(\alpha), \quad \text{or} \quad \frac{a}{\sin(\alpha)} = \frac{b}{\sin(\beta)}.$$

Similarly, by dropping perpendiculars to the other sides of the triangle, it can be shown that

$$\frac{a}{\sin(\alpha)} = \frac{c}{\sin(\gamma)} \quad \text{and} \quad \frac{b}{\sin(\beta)} = \frac{c}{\sin(\gamma)}.$$

Example 1

In the triangle in Figure 12.26, $a = 43$, $\alpha = 54$ degrees, and $\beta = 66$ degrees. Solve for γ, b, and c.

Since there are 180 degrees in a triangle,

$$\gamma = 180 - (54 + 66) = 60 \text{ degrees.}$$

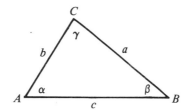

Figure 12.26   Triangle of Example 1.

By the Law of sines,

$$\frac{a}{\sin(\alpha)} = \frac{b}{\sin(\beta)} \quad \text{or} \quad \frac{43}{\sin(54°)} = \frac{b}{\sin(66°)} \quad \text{or} \quad b = 43 \cdot \frac{\sin(66°)}{\sin(54°)}.$$

$43 \times (SIN\ DEGREES\ 66) \div (SIN\ DEGREES\ 54)$
48.55578431

So, $b = 48.56$, approximately.

Also, by the Law of sines,

$$\frac{a}{\sin(\alpha)} = \frac{c}{\sin(\gamma)} \quad \text{or} \quad \frac{43}{\sin(54°)} = \frac{c}{\sin(60°)} \quad \text{or} \quad c = 43 \cdot \frac{\sin(60°)}{\sin(54°)}.$$

$43 \times (SIN\ DEGREES\ 60) \div (SIN\ DEGREES\ 54)$
58

So, $c = 58$.

359

The Law of cosines

Let *ABC* be any triangle, as in Figure 12.25. Then, the Law of cosines says

$$a^2 = b^2 + c^2 - 2 \cdot b \cdot c \cdot \cos(\alpha)$$
$$b^2 = a^2 + c^2 - 2 \cdot a \cdot c \cdot \cos(\beta)$$
$$c^2 = a^2 + b^2 - 2 \cdot a \cdot b \cdot \cos(\gamma).$$

The Law of cosines is derived as follows: Drop a perpendicular from the vertex *C* to the side *AB* intersecting it at the point *D*, as in Figure 12.25. Then, in the triangle *ADC*, $\cos(\pi - \alpha) = DA / b$. Also,

$$\cos(\pi - \alpha) = \cos(\pi) \cdot \cos(\alpha) + \sin(\pi) \cdot \sin(\alpha) = {}^-1 \cdot \cos(\alpha) + 0 \cdot \sin(\alpha)$$

$$= -\cos(\alpha) = \frac{DA}{b}, \quad \text{so that } DA = -b \cdot \cos(\alpha).$$

Now, in triangle *BDC*, $DB = DA + AB = c - b \cdot \cos(\alpha)$. Also, in triangle *ADC*, $\sin(\pi - \alpha) = h / b$. But,

$$\sin(\pi - \alpha) = \sin(\pi) \cdot \cos(\alpha) - \cos(\pi) \cdot \sin(\alpha) = 0 \cdot \cos(\alpha) - {}^-1 \cdot \sin(\alpha) = \sin \alpha,$$

so that $h = b \cdot \sin(\alpha)$.

Now, using the pythagorean theorem in the triangle *BDC*, we get

$$a^2 = h^2 + DB^2 = (b \cdot \sin(\alpha))^2 + (c - b \cdot \cos(\alpha))^2$$
$$= b^2 \cdot \sin^2(\alpha) + c^2 - 2 \cdot b \cdot c \cdot \cos(\alpha) + b^2 \cdot \cos^2(\alpha)$$
$$= b^2 \cdot (\sin^2(\alpha) + \cos^2(\alpha)) + c^2 - 2 \cdot b \cdot c \cdot \cos(\alpha)$$
$$= b^2 + c^2 - 2 \cdot b \cdot c \cdot \cos(\alpha).$$

The other relationships in the Law of cosines are derived similarly. [*Note*: If the triangle is a right triangle, with α the right angle, $\cos(\alpha) = 0$ and the Pythagorean theorem is the result of the Law of cosines.]

Example 2

In Figure 12.27, let $a = 6$, $b = 3$, and $c = 4$. Find the angles α, β, and γ.

By the Law of cosines, $a^2 = b^2 + c^2 - 2 \cdot b \cdot c \cdot \cos(\alpha)$, or $36 = 9 + 16 - 24 \cdot \cos(\alpha)$, or $\cos \alpha = -11/24$. Thus, α is

ARCCOS ‾11 ÷ 24
117.2796127 $\alpha = 117.3$ degrees approximately.

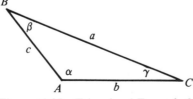

Figure 12.27   Triangle of Example 2.

Also, by the Law of cosines, $c^2 = a^2 + b^2 - 2 \cdot a \cdot b \cdot \cos(\gamma)$, or $16 = 36 + 9 - 36 \cdot \cos(\gamma)$, or $\cos(\gamma) = 29/36$. Thus, γ is

ARCCOS 29 ÷ 36
36.33605751 $\gamma = 36.3$ degrees, approximately.

Finally, since there are 180 degrees in a triangle, the remaining angle is $\beta = 180 - (117.3 + 36.3) = 26.4$ degrees.

Example 3

To find the distance across a lake from a point A on one shore to a point B on the other shore, a line AC is laid out along the shore containing the point A. The distance from A to C is 80 feet. The angle α is measured and found to be 70 degrees. The angle γ is found to be 45 degrees. Find the distance from A to B.

First, $\beta = 180 - (\alpha + \gamma) = 180 - (70 + 45) = 65$ degrees. From Figure 12.28, we need to find c. By the Law of sines,

$$\frac{b}{\sin(\beta)} = \frac{c}{\sin(\gamma)} \quad \text{or} \quad \frac{80}{\sin(65°)} = \frac{c}{\sin(45°)}, \quad \text{so } c = 80 \cdot \frac{\sin(45°)}{\sin(65°)}.$$

80×(*SIN DEGREES* 45) ÷ (*SIN DEGREES* 65)
62.4164807

The distance from A to B is about 62.4 feet.

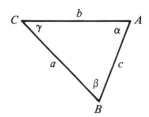

Figure 12.28   Illustration for Example 3.

Example 4

A playground slide is 20 feet long. A ladder 18 feet long reaches from behind the slide to the top. The angle between the ladder and slide is 105 degrees. Find the distance from the foot of the slide to the foot of the ladder.

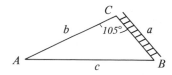

Figure 12.29   Illustration for Example 4.

By the Law of cosines,

$$c^2 = a^2 + b^2 - 2 \cdot a \cdot b \cdot \cos(\gamma)$$
$$= 324 + 400 - 720 \cdot \cos(105°)$$
$$= 724 - 720 \cdot \cos(105°).$$

So, in APL, c is

(724 − 720 × *COS DEGREES* 105) ∗ .5
30.17200213

Thus, the distance from the foot of the slide to the foot of the ladder is about 30 feet.

EXERCISES

1. If $a = 10$, $\alpha = 45$ degrees, $\beta = 30$ degrees, find b, c, and γ.

2. If $a = 70$, $b = 80$, $\alpha = 50$ degrees, find β, γ, and c.

3. If $a = 8$, $b = 10$, $c = 6$, find α, β, and γ.

4. If $a = 3$, $b = 2$, $\gamma = 60$ degrees, find c, α, and β.

5. If $b = 5$, $c = 9$, $\alpha = 62$ degrees, find a, β, and γ.

6. A playground slide is 20 feet long and makes an angle of 55 degrees with the horizontal. A ladder 18 feet long meets the slide at its top. Find the angle the ladder makes with the horizontal.

7. Two straight roads intersect at a point P and make an angle of 35 degrees with each other. On one road, a house is 1250 yards from the point P. On the other road, a house is 1680 yards from the point P. Find the distance between the houses.

8. The area of any triangle is given by $A = \frac{1}{2} b \cdot h$, where b is the length of the base and h is the height. Prove that the area is also given by $A = \frac{1}{2} b \cdot c \cdot \sin(\alpha)$.

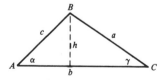

9. Write a program for the area formula in Exercise 8.

10. Find the area of the triangle with sides of lengths $a = 15$, $b = 12$, and $c = 23$. (Use your program you wrote for Exercise 9.)

Appendix

A.0 Using APL on a computer terminal

APL is an interactive computer language in which the user communicates with the computer via an electronic typewriter (called a terminal), a telephone, and an acoustic coupler (a device which holds the telephone handset and is the link between the terminal, the telephone, and the computer). The exact procedure for establishing communication with the computer varies from installation to installation. The reader should check with the computer center for the procedure appropriate for his particular institution. The keyboard for a standard APL terminal is shown in Figure A.1.

To begin with, it is necessary to establish a telephone connection from the terminal to the computer. First, turn on the coupler and the terminal. Also press the key marked COMM (for communicate) on the terminal. (COMM is not shown in the picture of the APL keyboard above.) Then, dial the telephone number for the computer. When you hear a high-pitched sound, this means that the computer has been reached. Place the telephone handset in the acoustic coupler with the cord to the back of the coupler. Usually, at most computer installations, the computer will now respond with an opening message such as the one described below. (At some installations, the user must make some initial response by typing a character or sequence of characters followed by pressing the *RETURN* key.)

75/08/11.　　　10.15.05.　　　　　　The time and date.

　　　USER NUMBER:

The user then types in his user number. Then, he presses the *RETURN* key.

363

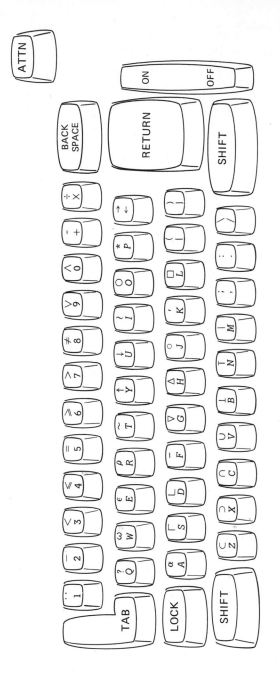

Figure A.1   The APL keyboard.

For example,

>   *USER NUMBER*: A000000

The computer will then ask for a password. If the user has a password, he now enters it. If not, he merely presses the *RETURN* key. The computer then prints out a message such as

>   *TERMINAL 22, COR*
>   *RECOVER/SYSTEM*:

The user must now tell the computer the computer language he wishes to use. Thus, typically, he would enter APL:

>   *RECOVER/SYSTEM*: *APL*

Then, he presses *RETURN* again, and the computer responds with a message such as

>   *CLEARWS*

The message *CLEARWS* indicates that the user has begun with a clear active workspace. He can now begin to type APL expressions. However, if there is a workspace which has been previously *SAVED*,[1] and which the user wishes to have replace the *CLEARWS* as his active workspace, he then types

>   *) LOAD MYWORK*   Assuming the name of the workspace is *MYWORK*.

The computer will then respond with a message such as

>   *SAVED* 75/1/08 02.20.03.   Indicating when this workspace was last saved.

The user can now begin entering APL expressions and use any of the variables or programs previously stored in the workspace *MYWORK*.

The rest of this appendix is intended to aid the beginning student in using APL.

A.1   Introduction to APL

We now discuss some of the fundamental features of the APL language. To begin with, there are two types of data in APL: literal data and numerical data.

---

[1]To save some work in a workspace called *MYWORK*, merely enter

>   *) SAVE MYWORK*

at the conclusion of the work being saved. Programs and variables may be saved in this way.

Literal data

Literal data are enclosed in quotes. For example:

     'WELCOME TO APL'   If this is typed by the programmer, followed by pressing the *RETURN* key, the computer will respond

WELCOME TO APL

Numerical data

The APL terminal can be used to do computations. For example:

     5+3   To add 5 to 3, enter *5+3* and press
8   *RETURN*. The computer responds with the answer 8.

Other computations are done similarly. For example:

     3−5
⁻2

Observe that the symbol for subtraction is distinguished from the negative symbol in APL. The subtraction symbol is located above the + sign on the keyboard, while the negative symbol is a raised minus sign located above the 2.

Assignment

The symbol ← is used to assign a name to a value. Consider the following examples:

     A←5   The name *A* is assigned to 5.
     A   A request for the value of *A*.
5   The computer responds with 5.

     B←7
     B
7

     A−B   The value of *A−B* is requested.
⁻2

     B←A−B   *B* is now assigned the value of *A−B*.
     B   The request for *B*.
⁻2   Notice that the computer responds with the latest value of *B*.

The computer retains only the latest value of any name.

Assignment of literal data

$C \leftarrow$ '*WELCOME*'	Literal data are assigned names by enclosing the characters in quotes.
$D \leftarrow$ '*TO*'	
$E \leftarrow$ '*APL*'	
C,D,E *WELCOMETOAPL*	The comma is used in APL to chain data end to end. Notice that the computer didn't space the words as one would probably like.

In order to get the correct spaces between the words, there are two possibilities:

$C \leftarrow$ '*WELCOME* '	Leave a space between the last letter and the end quotes.
$D \leftarrow$ '*TO* '	
$E \leftarrow$ ' *APL* '	
C,D,E *WELCOME TO APL*	Spaced correctly.
OR $C \leftarrow$ '*WELCOME*' $D \leftarrow$ '*TO*' $E \leftarrow$ '*APL*'	
$C,$' ',$D,$' ',E *WELCOME TO APL*	Tell the computer to leave spaces by typing a space between the quotes.

Mixing literal data and numerical data

Literal and numerical data may be displayed on the same line, provided that they are separated by semicolons. For example:

$X \leftarrow 3$ $Y \leftarrow 5$	*X* and *Y* arc numerical data.
$L \leftarrow$ ' *IS LESS THAN* '	*L* is literal data.
$X;L;Y$ 3 *IS LESS THAN* 5	Numerical and literal data are displayed on the same line.

Order of operations

In conventional mathematics, we are accustomed to a hierarchy of numerical operations. Therefore, if we are presented with the expression $3 \times 2 + 5$, we would probably multiply 3 by 2 and then add 5, obtaining 11. We are

367

used to a rule that multiplication is done before addition. However, if we were to enter this expression on an APL terminal, we would obtain the result 21. This is because in APL there is no hierarchy of operations. APL has too many operations to make this desirable. The rule in APL is to perform the rightmost operation first and then proceed from right to left. Thus, the addition $2+5$ is done first, followed by the product 3×7. This "right to left rule" holds regardless of the operations. The only rule that takes precedence over this right to left rule is that operations within parentheses are done first as they are encountered in going from right to left. Thus, $(3\times2)+5$ would yield 11.

The expression $5\times3+3\times4$ yields 75 in APL, since in APL it is equivalent to $(5\times(3+(3\times4)))$. In order to get the answer of 27, we would have to insert parentheses as follows: $(5\times3)+(3\times4)$. This right to left rule can cause the beginning student problems if he is not careful. He should be sure that the expression he is entering is really the expression he wants evaluated. When in doubt, he should insert parentheses. Consider the following examples:

        3+4×2
11

        2∗3+2   ∗ is the exponentiation operation in
32   APL. 2∗5 is 2 raised to the 5th
                                         power.

        (2∗3)+2
10

        10−3∗2+1
⁻17

        10−(3∗2)+1
0

        ι4+1   *Note:* ιN yields the vector of positive
1 2 3 4 5   integers up to and including N.

        (ι4)+1
2 3 4 5

Monadic and dyadic functions

APL often uses the same symbol in two ways; one monadic and the other dyadic. A *monadic function* has only one argument. In APL, the argument always appears on the right of the function. For example:

        ÷4   The reciprocal of 4.
.25

        |⁻5   The absolute value of ⁻5.
5

368

A *dyadic function* has two arguments; one on each side of the function. For example:

      2÷4   The division function.
.5

      3|7   The residue function. It returns the
1   remainder when 7 is divided by 3.

The "quad" and the "quote quad"

The "quad" □ is used to request input from the person using the computer. This is illustrated below:

      A←□   The user is allowed to enter any data he desires for A. The □: is printed by the computer to prompt the user to enter input.

□:

      5   He has decided to enter 5.
      A   A request for the value of A.
5

      A+□   The user is asked to add the number of his choice to A.
□:

      2   He chooses to add 2.
7   The result is 7.

As seen in the text, the "quad" is very useful in writing programs in which the user is to interact with the computer. It is also possible to allow a student to enter literal data of his choice by using the "quote quad" ⎕. To make this ⎕, type the □, backspace one space and overstrike the '. The following example illustrates the use of the "quote quad":

      A←⎕   A request for literal data to be named A.
HELP

      A A has been assigned the word *HELP*.
HELP

      B←⎕   A request for literal data to be named B. B has been assigned the word *ME*.

ME

      A,' ',B   The computer is requested to print out A space B.
HELP ME

369

Reduction

If α is a function and V is a vector, then the notation α/V is called reduction. It reduces V to a single number by applying the operation α to the successive elements of V (from right to left). For example:

$$V \leftarrow 1\ 2\ 3\ 4$$

$+/V$ This is sum reduction. The elements of V are added.

10

$\times/V$ This is times reduction. The elements of V are multiplied.

24

Compression

If A is a vector consisting entirely of 0's and 1's and if V is another vector, then the notation A/V is called compression. The result of A/V is that the elements of V corresponding to the 1's in A are kept, while the elements in V corresponding to the 0's in A are deleted. In general, A and V must have the same number of elements. For example:

$$0\ 1\ 0\ 1\ 1\ 0/3\ 4\ 7\ 8\ 2\ 9$$
4 8 2

$$1\ 0\ 1\ 1\ 0/\text{'}APPLE\text{'}$$
APL

Outer product

If A and B are vectors and α a dyadic function, an expression of the form $A\circ.\alpha B$ is called an outer product. The result is an array or matrix obtained by performing the function α on every pair of elements of A and B. The small circle, called null, is located above the J on the keyboard. Consider the following examples:

$$A \leftarrow 1\ 2\ 3$$
$$B \leftarrow 4\ 5\ 6$$

$A\circ.+B$ Each element of A is added to each element of B.
5 6 7
6 7 8   1 is added to 4, 5, and 6.
7 8 9   2 is added to 4, 5, and 6.
                             3 is added to 4, 5, and 6.

$A\circ.\times B$ Each element of A is multiplied by each element of B.
4 5 6
8 10 12   1 is multiplied by 4, 5, and 6.
12 15 18   2 is multiplied by 4, 5, and 6.
                             3 is multiplied by 4, 5, and 6.

$A \circ . = A$	Each element of A is compared to
1 0 0	each element of A.
0 1 0	1 is compared to 1, 2, and 3.
0 0 1	2 is compared to 1, 2, and 3.
	3 is compared to 1, 2, and 3.

If two numbers are $=$, a 1 is printed. 1 means can be interpreted to mean "true" and 0 "false."

Inner product

If A and B are arrays of the same length, and if α and ω are dyadic functions, then the expression of the form $A\alpha.\omega B$ is called an inner product. The result is that α is applied to A and B element by element, followed by ω reduction applied to the result. For example:

$A \leftarrow 1\ 2\ 3$
$B \leftarrow 4\ 5\ 6$

	$A + . \times B$	The corresponding elements of A
32		and B are multiplied and then the results added.
	$+ / A \times B$	Another way to accomplish the same
32		result using reduction.
	$A \times . + B$	The corresponding elements of A
315		and B are added and then the results multiplied.
	$\times / A + B$	Another way to accomplish the same
315		result using reduction.

A.2 Program definition

Programs begin with a del, ∇, followed by the name of the program. The computer then prints the line number [1]. The programmer then enters the proper expression on this line. He then presses the **RETURN** key, and the computer responds with line number [2]. This process continues until the program is completed. To end the program, the programmer enters another ∇, on the last line of the program. The program is now ready for execution.

Consider the following program for computing simple interest at 5 percent per year.

	∇ *INTEREST*	The name of the program is *INTEREST*.
[1]	$R \leftarrow .05$	The interest rate is 5 percent.
[2]	$I \leftarrow P \times R \times T$	Interest is principal, P, times rate, R, times time, T.

[3]   *I*   The computer is instructed to print
                                             out *I*.

[4]   ∇   The end of the program.

To execute the program, enter the values of *P* and *T*, and then enter the
name of the program, *INTEREST*. For example:

    P←100
    T←2

    INTEREST
10   The value of *I* is 10.

Instead of entering the values of *P* and *T* separately before running the
program, it is possible to include them in the program header. (We will
consider the techniques for revising a program in Section A.4.)

    ∇ *P INTEREST T*
[1]   *R*←.05
[2]   *I* ←*P*×*R*×*T*
[3]   *I*
[4]   ∇

The program is now executed as follows:

    100 *INTEREST* 2
10

Even better would be to include the result of the program, *I*, in the
header of the program also. This is done as follows:

    ∇ *I*←*P INTEREST T*
[1]   *R*←.05
[2]   *I*←*P*×*R*×*T*
[3]   ∇
    100 *INTEREST* 2
10

[*Note*: If the result *I* is included in the header of the program, it should not
be requested in the body of the program also. Otherwise, the computer will
print out the answer, 10, twice. Thus, the former line 3 which instructed the
computer to print out the value of *I* has been eliminated.]

When the header of the program has an explicit result, as this program
has the result *I*, then the program itself has this result. The importance of
this is that the program can now be thought of as a function just as any of
the keyboard operations are functions. The program *INTEREST* uses the
values of *P* and *T* to produce the value *I*. As a function, this program can
now be used in other programs as a *subprogram*. To illustrate this point, we
shall use the program *INTEREST* as a subprogram in the following program *AMOUNT* which computes the amount of the loan at 5 percent
interest.

```
        ∇ A←P AMOUNT T
[1]     R←.05
[2]     A←P+P INTEREST T
[3]     ∇

        100 AMOUNT 2
110
```

[2] line comment: The amount owed is principal plus interest.

One other point worth considering is that variables included in the header of the program are "local" variables. That is, they are local to the program and do not retain their values outside of the program. Variables in a program that are not included in the header of the program become "global" variables and retain their values outside of the program. They might, therefore, affect later calculations unwantedly. In the final version of the program *INTEREST*, *I*, *P*, and *T* are local variables, while *R* is global, since it is not included in the header. Consider the following example:

```
        R←.07
        P←200
        T←3
        I ←42
        100 INTEREST 2
10

        I

42

        P

200

        T

3

        R
.05
```

Let these represent some values of *R*, *P*, *T*, and *I* stored previously, which the programmer would not like to have changed.

Note that the previous values of *I*, *P*, and *T* have not been altered by executing the program *INTEREST*. However, the value of *R* has been changed from 0.07 to 0.05. This is because *I*, *P*, and *T* are local variables in the program *INTEREST*, whereas *R* is a global variable.

If you now proceeded to do other calculations under the assumption that *R* is still 0.07, you would have problems.

To make *R* a local variable, include it in the header of the program as follows:

```
        ∇ I←P INTEREST T;R
[1]     R←.05
[2]     I ←P×R×T
[3]     ∇
```

Variables are made local to a program by appending them to the program header and separating them from the rest of the header and from each

373

other by semicolons. Consider the effect of including R in the header as a local variable now:

$$R \leftarrow .07$$
$$P \leftarrow 200$$
$$T \leftarrow 3$$
$$I \leftarrow 42$$

100 *INTEREST* 2

10

R   Note that R retains its value of 0.07

.07   that it had prior to running the pro-
                                gram *INTEREST*.

In general, if possible, it is a good idea to include all variables used exclusively in a program as "local" variables.

As another example, consider the following program which computes the amount S accumulated when a principal P is deposited in a savings bank at a yearly interest rate of 5 percent for N years. Such a bank uses compound interest. So the name chosen for the program is *COMPOUND*.

$\nabla\ S \leftarrow P\ COMPOUND\ N; R$
[1] $R \leftarrow .05$
[2] $S \leftarrow P \times (1 + R) * N$
[3] ∇

100 *COMPOUND* 4
121.550625

If \$100 is deposited in a savings bank at 5 percent interest compounded yearly, then in 4 years, it yields \$121.55.

A.3 Branching

Branching is an instruction to change the regular sequence of steps in a program. In APL, there are two types of branching: unconditional branching and conditional branching. In general, branching statements are indicated by a right arrow →. An unconditional branching statement has the form

$$\rightarrow \text{(a designated line)}.$$

This means: "branch to this designated line."

A conditional branching statement has the form

$$\rightarrow \text{(a given condition)}/\text{a designated line}.$$

If the given condition is true, the computer branches to the designated line. Otherwise, it just proceeds to the next line of the program.

Consider the following program *FUNCTION* which prints out the ordered pairs (X, Y) for the function $Y = X^2$ starting at some specified value of X.

	∇ *FUNCTION X*	
[1]	$Y \leftarrow X*2$	Y is assigned the value $X*2$.
[2]	X, Y	The computer is told to print out the pair X, Y.
[3]	$X \leftarrow X+1$	X is increased by 1.
[4]	$\rightarrow 1$	The computer is told to branch to line 1 with this new value of X.
[5]	∇	

In order to run this program, the student chooses an initial value of X, say ⁻5, and enters the following:

```
    FUNCTION ‾5
‾5  25
‾4  16
‾3  9
‾2  4
‾1  1
0 0
1 1
2 4
3 9
4 16
5 25
6 36
```

The trouble is that there is no built-in way to stop this program. It will go on incrementing X and printing out pairs indefinitely unless the programmer does something to stop it. In order to stop an endless program such as this one, press the *ATTN* key. The computer will then stop the program and is ready for new work.

Let us now consider a better way to write this program with a built in stopping condition.

	∇ *FUNCTION X;I*	Note that the local variable I is in the header. The initial value of I is 0.
[1]	$I \leftarrow 0$	
[2]	*COUNTER*: $I \leftarrow I+1$	I is incremented by 1. Note that this
[3]	$Y \leftarrow X*2$	line has a line label *COUNTER*.
[4]	X, Y	(Line labels are discussed below.)
[5]	$X \leftarrow X+1$	
[6]	$\rightarrow (I<11)/COUNTER$	If I is less than 11, the computer is
[7]	∇	sent back to *COUNTER* on line 2.

In the previous program, as soon as I is not less than 11, the computer proceeds to line 7, where the program is terminated.

```
     FUNCTION ̄5
 ̄5   25
 ̄4   16
 ̄3    9
 ̄2    4
 ̄1    1
 0    0
 1    1
 2    4
 3    9
 4   16
 5   25
```

Line labels

In the above program *FUNCTION*, a line label *COUNTER* was used. A line label is a name given to a particular statement in a program. It is always followed by a colon. In programs involving branching, it is a good idea to use line labels for reference—especially if one expects to alter the program. This is because very few programmers can write every program in final form on their first try. Thus, if originally they wanted to branch to the statement on line 5, in a revised form of the program the original line 5 might now be changed to line 7. This would require the programmer to rewrite the branching statement. However, if the statement on line 5 had a line label, there would be no need to rewrite the branching statement. In addition, line labels can make the finished program easier to interpret by users of the program. Line labels are local variables. That is, they do not retain their values outside of the program. However, they do not need to be included in the header of the program.

Another example

We now consider a program for computing the absolute value of a number. The absolute value of a number is the positive value of the number. Thus, if X is positive, then the absolute value of X is X. However, if X is negative, then the absolute value of X is $-X$. Note that the negative of a variable X is denoted by $-X$ rather than $ ̄5$ which is used to represent a negative constant. The absolute value of 0 is 0.

∇ *AV←ABSOLUTE X*	The name of the program is *ABSOLUTE*.
[1] →(*X*<0)/*NEGATIVE*)	If X is less than 0, branch to *NEGATIVE*. Otherwise, proceed to the next line.

[2]	*POSITIVE*: $AV \leftarrow X$	Line label *POSITIVE*, for if X is not less than 0.
[3]	$\rightarrow 0$	Branch to 0 (explained below).
[4]	*NEGATIVE*: $AV \leftarrow -X$	Line label *NEGATIVE*, for if
[5]	∇	X is less than 0.

Examples

        ABSOLUTE ¯5
5

        ABSOLUTE 5
5

        ABSOLUTE 0
0

   If the branching statement →0 is used in a program, this is equivalent to telling the computer to exit from the program. In branching commands, the header is not included as a line in the program. Thus, there is no line 0 for the computer to branch to. So, it exits from the program.
   We conclude this brief discussion of branching with one final example. This program, *SUM*, computes the *SUM* of the first K positive integers.

	$\nabla\ S \leftarrow SUM\ K; N$	
[1]	$S \leftarrow 0$	The initial value of S is 0.
[2]	$N \leftarrow 1$	N is a counter whose initial value is 1.
[3]	*ADDON*: $S \leftarrow S + N$	*ADDON* is a line in which the next value of N is added to the previous value of S.
[4]	$N \leftarrow N + 1$	Increase N by 1.
[5]	$\rightarrow (N \leqslant K)/ADDON$	If N is $\leqslant K$, branch to *ADDON* and add this new value of N to S.
[6]	∇	Otherwise, the program is ended.

        SUM 100
5050

The sum of the first 100 positive integers is 5050.

   Notice that in this program, the incrementing is done after the computation rather than before the computation as was done in the program *FUNCTION*. The reader should study the differences between these two programs.
   It is worth mentioning that in APL there is usually less need for branching than in many other programming languages. This is because of the large number of keyboard functions available in APL and the array handling capabilities in APL. A much easier form of *SUM* using a couple

of these functions follows:

```
      ∇ S←SUM K
[1]   S←+/ιK
[2]   ∇
      SUM 100
5050
```

ιK yields the vector of positive integers up to and including K. +/ιK adds the numbers in ιK.

It is also worth mentioning that there really is no need for a program to add the first K positive integers. It can be accomplished directly as follows:

```
      +/ι 100
5050
```

A.4 Program revision and editing procedures

Once a program has a name, no other program or variable in that workspace can be given the same name. Therefore, a program cannot be revised by simply rewriting it. The computer would interpret this as trying to give another program the same name as the original program. For example, in a previous section, the program INTEREST was changed. Originally, it was as follows:

```
      ∇ INTEREST
[1]   R←.05
[2]   I←P×R×T
[3]   I
[4]   ∇
```

Later, it was changed to the following:

```
      ∇ I←P INTEREST T; R
[1]   R←.05
[2]   I←P×R×T
[3]   ∇
```

This cannot be accomplished by merely starting from the beginning and entering this new version of INTEREST. If one attempted this, upon typing

```
      ∇ I←P INTEREST T; R
```

instead of getting the line number [1], he would get the error message

```
      DEFN ERROR
```

Meaning an error in defining a program.

The original program INTEREST could be erased by typing

```
      )ERASE INTEREST
```

and then the new program *INTEREST* could be entered. However, if only minor revisions are to be made in the program, this would be needlessly time consuming. This is especially true if the program is quite long. In this section, we consider some techniques for program revision. We shall consider revising one line, revising the header, revising several lines, deleting a line, adding lines, and correcting typographical errors.

Revising one line of a program

If we wish to revise line 5 of a program named *PROGRAM*, we would type

∇ *PROGRAM* [5]

[5] (Revise line 5 here)
[6] ∇

Cause the computer to print out the line number 5. Then simply enter the corrected version of line 5. Upon pressing *RETURN*, the computer will print out the line number 6. Since we do not wish to revise line 6, we simply enter ∇. This tells the computer that we are finished with our revisions, for now.

Revising the header of the program

To revise the header of the program *PROGRAM*, type

∇ *PROGRAM* [0]

[0] (Enter new header)
[1] ∇

For the sake of program revision, the header is referred to as line 0.

Revising consecutive lines of a program

To revise lines 5, 6, and 7 of *PROGRAM*, type

∇ *PROGRAM* [5]

[5] (Revise line 5)
[6] (Revise line 6)
[7] (Revise line 7)
[8] ∇

The computer automatically prints out the next line number as an invitation for further revision. Since we do not wish to revise line 8, we type ∇.

Revising nonconsecutive lines of a program

To revise lines 3, 6, and 8 of *PROGRAM*, type

∇ *PROGRAM* [3]

[3] (Revise line 3)
[4] [6] (Revise line 6)
[7] [8] (Revise line 8)
[9] ∇

A line number may be overridden in favor of a new line number by simply entering this new line number, followed by the revision.

379

Deleting a line from a program

There are different methods for deleting a line from a program. The more common method for deleting line 5 from *PROGRAM* is to type

∇ *PROGRAM* [5]	In other words, cause the computer
[5] (Press the *RETURN* key now)	to print out the line number. Then,
[6] ∇	simply press *RETURN*.

At some other computer installations, to delete line 5 from *PROGRAM*, type

∇ *PROGRAM* [Δ5] ∇

Adding lines to a program

If a program *PROGRAM* has 10 lines, and we wish to add two more lines at the end, we type

∇ *PROGRAM* [11]	
[11] (Enter the new line)	
[12] (Enter the new line)	When we are finished adding new
[13] ∇	lines, we enter ∇.

Inserting new lines in a program

To insert 3 new lines between lines 5 and 6 of *PROGRAM*, type

∇ *PROGRAM* [5.1]	
[5.1] (Enter the new line)	Use decimals to denote the first new
[5.2] (Enter the next line)	line to be inserted. The computer will
[5.3] (Enter the next line)	print out the next decimal as an in-
[5.4] ∇	vitation to insert another new line. When all new lines have been inserted, type ∇.

Correcting typographical errors

To correct a typographical error, press the *ATTN* key. This will cause the computer to rotate the typewriter to the next line. Then, the programmer backspaces to the leftmost point where the error was made and retypes the line from that point on. Of course, this is assuming that the *RETURN* key has not been pressed before the typographical error has been detected. If the *RETURN* key has been pressed, then override the next line number by typing the previous line number and retype the previous line correcting the error.

For example, suppose that line 3 of a program should read $Y \leftarrow X * 2$, and we accidentally typed $Y \leftarrow X + 2$, but we have not pressed the *RETURN*

key. Then, we correct the error as follows:

[3] $Y \leftarrow X + 2$
      $*2$

Press *ATTN* key and backspace to the point of error (2 spaces), and then retype the line from the point of the error.

If we had already pressed the *RETURN* key, then we would correct the error as follows:

[3] $Y \leftarrow X + 2$
[4]   [3] $Y \leftarrow X * 2$

Displaying a program

To display a program *PROGRAM*, type

∇ *PROGRAM* [☐] ∇

The computer will then print out the latest version of *PROGRAM*. This procedure may be used to display any program in the active workspace.

Let us now revise the program *INTEREST* to the form indicated at the beginning of this section.

∇ *INTEREST*
[1] $R \leftarrow .05$
[2] $I \leftarrow P \times R \times T$
[3] I
[4] ∇

The old version of *INTEREST* to be revised.

∇ *INTEREST* [0]
[0] $I \leftarrow P$ *INTEREST* $T; R$
[1]   [3]
[3]
[4] ∇

Request to revise the header.
The new header.

To delete line 3, cause the computer to recognize line 3 by overriding line 1. Then enter a blank line 3 by pressing *RETURN*.

∇ *INTEREST* [☐] ∇

Request to display the revised version of *INTEREST*.

∇ $I \leftarrow P$ *INTEREST* $T; R$
[1] $R \leftarrow .05$
[2] $I \leftarrow P \times R \times T$
      ∇

Notice that the final ∇ is not given a line number by the computer. There are just two lines in the program *INTEREST*.

381

In the branching section (Section A.3) we had two versions of a program *FUNCTION*. Let us see how to revise the original form of *FUNCTION* to the form with the built in stopping condition.

```
      ∇ FUNCTION X
[1]    Y←X*2
[2]    X,Y                         The original form of FUNCTION.
[3]    X←X+1
[4]    →1
       ∇
```

```
      ∇ FUNCTION [0]               We have to alter the header to
                                   include I.
[0]    FUNCTION X;I
[1]    [0.1]  I←0                  We want to insert these two lines
                                   before line 1.
[0.2]  COUNTER: I←I+1
[0.3]  [4]  →(I<11)/COUNTER        Alter line 4 to include the stopping
[5]    ∇                           condition.
```

So, the revised form of *FUNCTION* is as follows:

```
      ∇ FUNCTION [□] ∇             Request to display the revised form
                                   of FUNCTION.
```

```
      ∇ FUNCTION X; I
[1]    I←0
[2]    COUNTER: I←I+1
[3]    Y←X*2
[4]    X,Y
[5]    X←X+1
[6]    → (I<11)/COUNTER
       ∇
```

A.5 The trace command

The trace command causes the computer to print out a line-by-line account of its computations in a program. This is often very helpful in determining where errors occur in a program and in trying to understand how a program works. The form of the trace command depends on the APL system being used. Two of the more common forms of the trace command are given here.

If a program called *PROGRAM* has 10 lines, the trace command would be

$$T \, \Delta \, PROGRAM \leftarrow \iota 10$$

To remove the trace, type

    T Δ PROGRAM ← ι0

To illustrate the trace command, let us trace the program *FUNCTION* of the last section. To cut down on the amount of printout, we will revise line 6 of *FUNCTION* so that it will print out 5 pairs instead of 11 pairs.

    ∇ *FUNCTION* [6]
[6]   →(*I*<5)/*COUNTER* ∇

Thus, the program *FUNCTION* now is as follows:

    ∇ *FUNCTION* [☐] ∇

    ∇ *FUNCTION X;I*
[1]   *I*←0
[2]   *COUNTER*: *I*←*I*+1
[3]   *Y*←*X*∗2
[4]   *X,Y*
[5]   *X*←*X*+1
[6]   →(*I*<5)/*COUNTER*

T Δ FUNCTION ← ι6	Trace the 6 lines of *FUNCTION*.
FUNCTION 0	Run the program *FUNCTION* with initial value 0.

FUNCTION[1] 0
FUNCTION[2] 1
FUNCTION[3] 0
0 0

FUNCTION[5] 1	The computer prints out a line by line record of its computations.
FUNCTION[6] 2	

FUNCTION[2] 2
FUNCTION[3] 1
1 1
FUNCTION[5] 2
FUNCTION[6] 2
FUNCTION[2] 3
FUNCTION[3] 4
2 4
FUNCTION[5] 3
FUNCTION[6] 2
FUNCTION[2] 4
FUNCTION[3] 9
3 9
FUNCTION[5] 4
FUNCTION[6] 2
FUNCTION[2] 5

FUNCTION[3] 16
4 16
FUNCTION[5] 5
FUNCTION[6]

  T △ FUNCTION←ι0   Remove the trace.

  FUNCTION 0

0 0   The program is run with the trace
1 1   removed.
2 4
3 9
4 16

Solutions to exercises

Chapter 1

Section 1.1

1. (a) True; (b) False; (c) True; (d) False; (e) True; (f) True; (g) True; (h) False.

2. (a) 1; (b) 0; (c) 1; (d) 0; (e) 1; (f) 1; (g) 1; (h) 0.

3. (a) 3 6 9; (b) *a e i*; (c) January–December; (d) Violet, indigo, blue, green, yellow, orange, red.

4. (a) 1; (b) 0; (c) 1 1 1 1; (d) 1; (e) 1; (f) 0; (g) 1 0 0; (h) 0; (i) 0 (j) 1.

5. There are 16 subsets.

6. 2^n

7. $x=2, y=1$.

8. (a) 1 1 0 0; (b) 1; (c) 0 0 0 0; (d) 0; (e) 1 1 1 1; (f) 1; (g) 0; (h) 1.

Section 1.2

1. (a) *C,D*; (b) *C,D*; (c) 0 3; (d) 5 7 8; (e) 1 2 4 5 7 8; (f) 1 2 4 5 7 8; (g) 0 3 5 6 7 8 9; (h) 6 9; (i) 6 9; (j) ∅; (k) 1 2 3 4 5 7 8 9 0; (l) Same as *k*.

2. (a) *ME*; (b) *AMPLE*; (c) *APL*; (d) *ABDFGHJKLNOPQSUVWXYZ*; (e) *LP*.

3. (a) Ace of spades; (b) All aces or face cards; (c) All clubs, hearts, diamonds; (d) ∅; (e) All spades which are not face cards; (f) Jack, queen, king of spades.

4.   ∇ *S←A SYMMETRIC B*
  [1]   *S←(A DIFFERENCE B) UNION (B DIFFERENCE A)* ∇

Section 1.4

1. (a) True; (b) True; (c) True; (d) True.
3. (a) True; (b) False; (c) True.

Section 1.5

1. (a) 4; (b) 5; (c) 2; (d) 7; (e) 5; (f) 2.
2. (a) 8; (b) 8; (c) 4; (d) 12; (e) 4; (f) 18.
3. (a) 32; (b) 8; (c) 36.
4. 80
5. 20
6. $50
7. (a) 320; (b) 180; (c) 110; (d) 15; (e) 185.
8. (a) 6; (b) 5; (c) 4; (d) 5.
9. (a) 240; (b) 760; (c) 700; (d) 1000.
10. (a) 300; (b) 90; (c) 520; (d) 600.

Chapter 2

Section 2.1

1. $A \leftarrow 1\ 1\ 0\ 0$
$B \leftarrow 1\ 0\ 1\ 0$

(a) $(\sim(A \lor B)) \land B$
0 0 0 0

(b) $(A \not\equiv B) \land B$
0 0 0 0

(c) $B \land \sim(A \lor B)$
0 0 0 0

(d) $\sim B \lor A$
0 0 0 1

(e) $\sim A \land B$
0 1 1 1

(f) $(\sim A) \lor (\sim B)$
0 1 1 1

(g) $\sim(A \land B)$
0 1 1 1

(h) $\sim A \neq B$
1 0 0 1

3. $A \leftarrow 1\ 1\ 1\ 1\ 0\ 0\ 0\ 0$
$B \leftarrow 1\ 1\ 0\ 0\ 1\ 1\ 0\ 0$
$C \leftarrow 1\ 0\ 1\ 0\ 1\ 0\ 1\ 0$

(a) $(A \land B) \lor C$
1 1 1 0 1 0 1 0

(b) $(A \lor C) \land (B \lor C)$
1 1 1 0 1 0 1 0

(c) $\sim((\sim A) \land (\sim B)) \land C$
1 1 1 1 1 1 0 1

(d) $(A \lor B) \land C$
1 0 1 0 1 0 0 0

(e) $((\sim A) \land (\sim B)) \land C$
1 0 1 0 1 0 0 0

(f) $(A \lor B \lor C) \land \sim(A \lor B \lor C)$
0 0 0 0 0 0 0 0

(g) $(A \land B \land C) \lor \sim(A \land B \land C)$
1 1 1 1 1 1 1 1

(h) $A \neq (B \neq C)$
1 0 0 1 0 1 1 0

5. (a) $A \land \sim C$
0 1 0 1 0 0 0 0

(b) $\sim(A \lor C)$
0 0 0 0 0 1 0 1

(c) $A \lor B$
1 1 1 1 1 1 0 0

(d) $C \land B \land \sim A$
0 0 0 0 1 0 0 0

(e) $C \land (A \lor B)$
1 0 1 0 1 0 0 0

(f) $((\sim A) \land (\sim B)) \land C$
0 0 0 0 0 0 1 0

6. (a) T; (b) F; (c) T; (d) F; (e) F; (f) F.

7. (a) $A \wedge C$ (b) $(\sim A) \vee (\sim B)$ (c) $(A \vee B) \wedge \sim C$
     1 0 1 0 0 0 0 0   0 0 1 1 1 1 1 1   0 1 0 1 0 1 0 0

     (d) $C \wedge ((\sim A) \wedge (\sim B))$ (e) $B \wedge (\sim (A \vee C))$
     0 0 0 0 0 0 1 0   0 0 0 0 0 0 1 0

     (f) $A \wedge ((\sim B) \wedge (\sim C))$
     0 0 0 1 0 0 0 0

8. (a) F; (b) F; (c) T; (d) F; (e) F; (f) F.

Section 2.2

1. $A \leftarrow 1\ 1\ 0\ 0$
   $B \leftarrow 1\ 0\ 1\ 0$

     (a) $(A \vee (\sim A)) \leqslant B$ (b) $A \leqslant (B \leqslant A)$ (c) $A = (\sim B)$
     1 0 1 0   1 1 1 1   0 1 1 0

     (d) $((\sim B) \leqslant (\sim A)) \leqslant (A \leqslant B)$ (e) $(A \vee B) = C$
     1 1 1 1   1 0 1 0 1 0 0 1

     (f) $((A \leqslant B) \wedge \sim B) \leqslant \sim A$
     1 1 1 1

3. $A \leftarrow 1\ 1\ 1\ 1\ 0\ 0\ 0\ 0$
   $B \leftarrow 1\ 1\ 0\ 0\ 1\ 1\ 0\ 0$
   $C \leftarrow 1\ 0\ 1\ 0\ 1\ 0\ 1\ 0$

     (a) $C = B$ (b) $C \leqslant (B \wedge \sim A)$
     1 0 0 1 1 0 0 1   0 1 0 1 1 1 0 1

     (c) $A \leqslant ((\sim B) \wedge (\sim C))$ (d) $C = (B \wedge \sim A)$
     0 0 0 1 1 1 1 1   0 1 0 1 1 0 0 1

4. (a) F; (b) T; (c) F; (d) T.

5. (a) $(A \wedge B) \leqslant C$ (b) $(A \wedge B) = C$
     1 0 1 1 1 1 1 1   1 0 0 1 0 1 0 1

     (c) $(A \vee B) \leqslant C$ (d) $((\sim B) \wedge (\sim A)) \leqslant \sim C$
     1 0 1 0 1 0 1 1   1 1 1 1 1 1 0 1

6. (a) F; (b) F; (c) F; (d) T.

7. (a), (e), (g), (h) are implications and (e), (g), (h) are double implications.

Section 2.3

1. All are equivalent.

2. (a) Equivalent; (b) Not equivalent; (c) Not equivalent; (d) Equivalent.

Section 2.4

1. (a) Valid; (b) Not valid; (c) Valid; (d) Not valid; (e) Valid; (f) Valid.

2. $(A \leqslant B) \leqslant ((\sim B) \leqslant (\sim A))$, valid.

3. $((A \vee B) \wedge \sim A) \leqslant B$, valid.

4. $((A \leqslant B) \wedge (B \leqslant C)) \leqslant (A \leqslant C)$, valid.

5. $((A \vee B) \wedge A) \leqslant \sim B$, fallacy.

6. $((A = B) \wedge A) \leqslant B$, valid.

7. $((A \leqslant B) \wedge \sim A) \leqslant \sim B$, fallacy.

8. $((A \leqslant B) \wedge (B \leqslant C)) \leqslant (A \leqslant C)$, valid.

9. $((A \leqslant B) \wedge \sim B) \leqslant \sim A$, valid.

10. $((\sim A) \leqslant B) \leqslant (A \leqslant \sim B)$, fallacy.

11. $((A \leqslant B) \wedge \sim A) \leqslant \sim B$, fallacy.

12. $((\sim A) \leqslant (\sim B)) \leqslant (B \leqslant A)$, valid.

Chapter 3

Section 3.2

1. (a) 18 9 0 9; (b) 10 9 ‾2 13; (c) 8 0 ‾1 ‾10; (d) ‾4 36 49;
(e) 2 3 ‾1 5 ‾2 6 7; (f) 5 6 2 8; (g) *LENGTH ERROR*; (h) ‾4 0 ‾1 2;
(i) ‾3; (j) 9.433981132; (k) ‾1 3; (l) 1.

2. (a) ‾3; (b) ‾9.

3. (a) ρS; (b) $S[4]$; (c) $S[1\ 5]$; (d) $+/S$; (e) $S[3] \leftarrow 38$.

4. (a) $\quad S1 + S2 + S3$ (b) $\quad +/S1 + S2 + S3$
279 199 94 103 195 $\qquad$ 870

(c) $P \leftarrow .10\ .15\ .20\ .10\ .15$

(d) $\quad S1 + . \times P,$ $\qquad S2 + . \times P,$ $\qquad S3 + . \times P.$
42.4 $\qquad$ 39.75 $\qquad$ 33.95

Section 3.3

1. (a) $M \leftarrow 3\ 4\rho 2\ 1\ 4\ 3\ 5\ 0\ 2\ 1\ 3\ ‾1\ 7\ 0$; (b) $\quad \rho\ M$; (c) $\quad (\rho M)[2]$;
$\qquad$ 3 4 $\qquad$ 4

(d) $M[2;3] \leftarrow 8$; (e) $M[;3] \leftarrow 3\ ‾1\ 5$; (f) $M[2;] \leftarrow 1\ 0\ 0\ 0$;

(g) $\quad \lozenge M$; (h) $M,[1]\ 0\ 0\ 0\ 1$; (i) $‾1\ 1 \downarrow M$; (j) $2\ ‾2 \uparrow M$.
2 1 3
1 0 ‾1
3 0 5
3 0 0

2. (a) 3 4; (b) 2; (c) 5 ‾1 0 4; (d) 3 ‾1 5; (e) 0; (f) 3; (g) 4; (h) 0 10 ‾4 6;
(i) 0 ‾16 ‾10 4; (j) 2 1
0 5
‾2 0.

3. (a) $P[2;]$; (b) $P[;4]$; (c) $P[2;3]$; (d) $P[2;4] \leftarrow 3$; (e) $P[;5] \leftarrow P[;5] + 1$;
(f) $P \leftarrow P,[1]\ 4\ 2\ 0\ 5\ 2$; (g) $5 \times P[3;]$.

Section 3.4

1. (a) $\begin{bmatrix} 9 & 2 & 7 \\ ‾2 & 6 & 3 \end{bmatrix}$ (b) $\begin{bmatrix} ‾1 & ‾3 & ‾8 \\ 3 & ‾14 & 3 \end{bmatrix}$ (c) $\begin{bmatrix} 25 & 0 & 1 \\ 0 & 4 & 9 \end{bmatrix}$

(d) $\begin{bmatrix} 7 & 6 & 8 \\ 4 & 9 & 5 \end{bmatrix}$ (e) $\begin{bmatrix} 2 & ‾3 & ‾2 \\ ‾3 & ‾5 & 0 \end{bmatrix}$ (f) $\begin{bmatrix} 5 & 0 & 1 & 2 & 1 & 3 \\ 0 & ‾2 & 3 & ‾1 & 4 & 0 \end{bmatrix}$

(g) $\begin{bmatrix} 2 & 1 & 3 \\ -1 & 4 & 0 \\ 5 & 0 & 1 \\ 0 & -2 & 3 \end{bmatrix}$ (h) [1 5 3] (i) [6 0] (j) 9.

3. $\begin{bmatrix} 7 & 8 \\ 16 & 23 \end{bmatrix}$

5. (a) $\begin{bmatrix} 16 & 1 \\ 20 & -4 \end{bmatrix}$ (b) $\begin{bmatrix} 22 & 26 \\ 11 & 17 \end{bmatrix}$ (c) $\begin{bmatrix} 20 & 12 & 34 \\ 4 & 0 & 20 \\ 6 & 2 & 19 \end{bmatrix}$ (d) $\begin{bmatrix} 19 & 19 \\ 43 & 51 \end{bmatrix}$.

7. (a) $\begin{bmatrix} -56 & -56 \\ -88 & -88 \end{bmatrix}$ (b) $\begin{bmatrix} -60 & -68 \\ -76 & -84 \end{bmatrix}$

8. Both sides of the equation yield $\begin{bmatrix} 2 & 37 & 61 \\ 45 & 16 & 92 \\ 64 & 37 & 131 \end{bmatrix}$.

9.　　　$S+.\times P$
　34 17 11
　33 19 13
　35 16 10

10. (a)　　　0←10 15　　(b)　　　0←5 4 6
　　　　　　0+.×T　　　　　　　(0+.×T)+.×C
　　　17.5 65 25　　　　　497.5

Chapter 4

Section 4.1

1. In each case, see if $(A+.\times X)=B$ yields a 1 (for true). (a),(b),(c) are true and (d) is false.

2. (a) $(0,-2)$, $(1.2,0)$; (b) $(12,0,0)$, $(4,0,2)$; (c) $(1,1,2,1)$, $(0,0,2,0)$ are examples of solutions. These are not unique.

3.　　　∇X←A SOLVE B
　[1]　X←B÷A ∇

4. (a) 2.333333333; (b) 0.7142857142; (c) -4.5; (d) -0.75.

Section 4.2

1. (a) True; (b) True; (c) False; (d) True.

2. (a) $x=1, y=1$; (b) $x=.5, y=.5$; (c) Redundant system; (d) Inconsistent system; (e) $x=11, y=-14$.

3. (a) 3 bookcases, 4 tables (b) 25 nuts, 15 bolts (c) 5 oz A, 10 oz B
　　$12x+16y=100$　　　$4x+6y=190$　　　$100x+80y=1300$
　　$2x+1.5y=12$　　　$x-2y=-5$　　　$8x+6y=100$.

Section 4.4

1. (a) A[1 2;]←A[2 1;]; (b) A[2;]←A[2;]+(-3)×A[1;]; (c) A[3;]←A[3;]+(-7) × A[1;]; (d) A[2;]←(1+14)×A[2;].

4. (a) $x = 1.058823529$, $y = 1.294117647$; (b) Redundant system: $x = z - 2.33333333$, $y = 2.6666667 - 2z$; (c) $x = {}^-0.5$, $y = 0.5$, $z = 2.5$; (d) $x = {}^-4.7375$, $y = 7.9625$, $z = {}^-2.5625$, $w = 0.3$; (e) Inconsistent system, no solutions.

Section 4.5

1. (a) $\begin{bmatrix} 1 & 0 & 0 \\ 0 & 1 & 0 \\ 0 & 0 & 1 \end{bmatrix}$ (b) $\begin{bmatrix} 1 & 0 & {}^-1 \\ 0 & 1 & 2 \\ 0 & 0 & 0 \end{bmatrix}$ (c) $\begin{bmatrix} 1 & 0 & 0 & 0 \\ 0 & 1 & 0 & 0 \\ 0 & 0 & 1 & 0 \\ 0 & 0 & 0 & 1 \end{bmatrix}$

(d) $\begin{bmatrix} 1 & 0 & 0 & 2 \\ 0 & 1 & 0 & 3 \\ 0 & 0 & 1 & 0 \\ 0 & 0 & 0 & 0 \end{bmatrix}$ (e) $\begin{bmatrix} 1 & 2 & 0 & 3.4 \\ 0 & 0 & 0 & 7 \\ 0 & 0 & 1 & {}^-0.6 \end{bmatrix}$.

Section 4.6

1. (a) $\begin{bmatrix} 1 & {}^-2 \\ {}^-0.5 & 1.5 \end{bmatrix}$ (b) $\begin{bmatrix} {}^-0.16666667 & 0.5 & {}^-0.16666667 \\ {}^-1.16666667 & 0.5 & 0.83333333 \\ 0.83333333 & {}^-0.5 & {}^-0.16666667 \end{bmatrix}$

(c) *DOMAIN ERROR*, no inverse

(d) $\begin{bmatrix} {}^-0.09259259259 & 0.1296296296 & 0.03703703704 \\ 0.3240740741 & 0.0462962963 & {}^-0.1296296296 \\ {}^-0.1944444444 & {}^-0.027777778 & 0.2777777778 \end{bmatrix}$ (e) No inverse.

4. Suppose A has two inverses X and Y. Then, $Y = I \cdot X = (X \cdot A) \cdot Y = X \cdot (A \cdot Y) = X \cdot I = X$.

5. $\nabla I \leftarrow IDENTITY\ N$
[1] $I \leftarrow (\iota N) \circ . = (\iota N)\ \nabla$

Section 4.7

1. See the answers to Exercise 1 of Section 4.6.

2. (a) True; (b) False; (c) True.

3. (a) $x = {}^-4$, $y = 4.5$; (b) $x = 0.1666666667$, $y = 1.166666667$, $z = 1.166666667$; (c) $x = 5.625$, $y = {}^-6.5$, $z = {}^-0.25$; (d) *DOMAIN ERROR*, redundant system; (e) $x_1 = 0.916666667$, $x_2 = {}^-1$, $x_3 = {}^-1.58333333$, $x_4 = 0.916666667$, $x_5 = {}^-1$.

4. (a) $4x + 10y = 900$
     $x - 2y = 0$ $x = 100$, $y = 50$.

   (b) $x + 2y + 3z = 1000$
     $20x + 50y + 90z = 10000$ $x = 1500$, $y = 500$, $z = {}^-500$.
     $10x + 30y + 50z = 5000$ Thus, it is physically impossible to meet these conditions.

   (c) $20x + 30y + 10z = 9500$
     $12x + 20y + 4z = 5800$ $x = 100$, $y = 200$, $z = 150$.
     $5x + 9y + 2z = 2600$

Chapter 5

Section 5.1

1. $\nabla D \leftarrow D2\ A$
[1] $D \leftarrow (A[1;1] \times A[2;2]) - (A[1;2] \times A[2;1])$
∇

2. $\nabla D \leftarrow D3\ A$
[1] $D \leftarrow (A[1;1] \times A[2;2] \times A[3;3]) - (A[1;1] \times A[2;3] \times A[3;2])$
[2] $D \leftarrow D + (A[1;2] \times A[2;3] \times A[3;1]) - (A[1;2] \times A[2;1] \times A[3;3])$
[3] $D \leftarrow D + (A[1;3] \times A[2;1] \times A[3;2]) - (A[1;3] \times A[2;2] \times A[3;1])$
[4] ∇

3. (a) 2; (b) 0; (c) 12; (d) $^-$32; (e) $^-$108; (f) $^-$12.

4. (a) 2; (b) $^-$32; (c) 0; (d) $^-$350.

Section 5.2

1. (a) 2; (b) $^-$32; (c) 0; (d) $^-$350.

2. 0.

Section 5.3

1. (a) $\begin{bmatrix} 1 & 2 & 4 \\ 3 & 1 & 3 \\ 8 & 2 & 0 \end{bmatrix}$ and $^-$34
(b) $\begin{bmatrix} 2 & 3 & 4 \\ 0 & 9 & ^-2 \\ 1 & ^-1 & 3 \end{bmatrix}$ and $^-$8

(c) $\begin{bmatrix} 5 & 0 & ^-2 \\ 3 & 1 & 3 \\ 8 & 2 & 0 \end{bmatrix}$ and $^-$26
(d) $\begin{bmatrix} 50 & ^-200 & ^-26 & 8 \\ ^-26 & 104 & ^-34 & ^-20 \\ ^-84 & ^-336 & 12 & ^-156 \\ ^-8 & ^-166 & 20 & 70 \end{bmatrix}$

(e) $^-$396 (f) $^-$396.

2. $\nabla D \leftarrow COF3\ M$
[1] $D \leftarrow M[3;] + . \times (COFACTORS\ M)[3;]\ \nabla$

3. $^-$32.

4. (a) 0; (b) 0.

Section 5.4

1. (a) $^-$32
(b) $\begin{bmatrix} ^-63 & 13 & 35 \\ 17 & ^-3 & ^-13 \\ 36 & ^-12 & ^-20 \end{bmatrix}$

$A + . \times ADJOINT\ A$
(c) $\begin{matrix} ^-32 & 0 & 0 \\ 0 & ^-32 & 0 \\ 0 & 0 & ^-32 \end{matrix}$

(d) $\begin{bmatrix} 1.96875 & ^-0.40625 & ^-1.09375 \\ ^-0.53125 & 0.09375 & 0.40625 \\ ^-1.125 & 0.375 & 0.625 \end{bmatrix}$.

3. (a)   *DET A*   (b)   *DET A*

      ¯6   0

         ADJOINT A   *ADJOINT A*

     9 ¯6   0   0   0   0

   ¯7   4   ¯21 ¯7 21   7

    A +.× ADJOINT A   3   1 ¯3 ¯1

   ¯6 0   3   1 ¯3 ¯1

    0 ¯6   *A +.× ADJOINT A*

       INVERT A   ¯1.136868377 *E* ¯13 0 0 0

 ¯1.5   1   0.000000000 *E* 0   0 0 0

 1.166666667 ¯0.6666666667   0.000000000 *E* 0   0 0 0

                          ¯1.136868377 *E* ¯13 0 0 0

                              INVERT A

                           A IS SINGULAR

4. (a) $\begin{bmatrix} 2 & 1 & 3 \\ 3 & 2 & 2 \\ 1 & 1 & 3 \end{bmatrix} A$, $B[5 \quad 4 \quad 3]$ (b) $\begin{bmatrix} 1 & 0 & ¯1 \\ ¯1.75 & 0.75 & 1.25 \\ 0.25 & ¯0.25 & 0.25 \end{bmatrix}$

(c) $x = 2, y = ¯2, z = 1$.

5. $DET(A \times ADJ(A)) = DET(DET(A) \times I) = DET(A) * N$, since the determinant of a diagonal matrix is the product of the diagonal elements. So, $DET(A) \times DET(ADJ(A)) = DET(A) * N$, or, $DET(ADJ(A))) = (DET(A)) * N - 1$.

Section 5.5

1. $x = 0.16666666667, y = 1.1666666667, z = 1.1666666667$.

3. *THE SYSTEM IS DEPENDENT*.

5.   ∇ *XI ← A CRAMERS B*

[1]   ' *WHAT IS I?* '

[2]   *I ←* ☐

[3]   *D ← DET A*

[4]   *A[;I] ← B*

[5]   *C ← DET A*

[6]   *XI ← C ÷ D*

     ∇

Chapter 6

Section 6.1

1. (a) 64, 0, ¯8; (b) 11, 3, ¯1; (c) 1.5874, 0, ¯1.25992; (d) ¯6, ¯2, ¯12; (e) .4, 2, ¯2.

3. (a) ¯4 ¯64   (b) ¯4 ¯5   (c) ¯4 ¯1.5874   (d) ¯4 ¯30   (e) ¯4 ¯0.666667

       ¯3 ¯27   ¯3 ¯3   ¯3 ¯1.44225   ¯3 ¯20   ¯3 ¯1

       ¯2 ¯8   ¯2 ¯1   ¯2 ¯1.25992   ¯2 ¯12   ¯2 ¯2

       ¯1 ¯1   ¯1 1   ¯1 ¯1   ¯1 ¯6   ¯1 DOMAIN ERROR

        0 0   0 3   0 0   0 ¯2   0 2

        1 1   1 5   1 1   1 0   1 1

        2 8   2 7   2 1.25992   2 0   2 0.666667

        3 27   3 9   3 1.44225   3 ¯2   3 0.5

        4 64   4 11   4 15874   4 ¯6   4 0.4.

5. The ranges are given by the second columns in the solutions to Exercise 3.

6. (a)–(d) All real numbers; (e) All real numbers except for ⁻1.

7.   ∇A←ABS X
  [1]   A←(X∗2)∗.5 ∇

8.   ∇A←B AREA H
  [1]   A←.5×B×H ∇

Section 6.2

1.

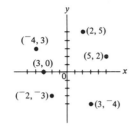

(⁻4, 3)
•(2, 5)
(5, 2)•
(3, 0)
(⁻2, ⁻3)•
•(3, ⁻4)

2. (a)

⁻3	6
⁻2	1
⁻1	⁻2
0	⁻3
1	⁻2
2	1
3	6

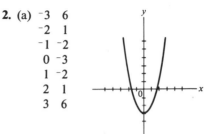

(b)

0	1
1	2
2	2.41
3	2.73
4	3
5	3.24
6	3.45
7	3.64
8	3.83

(c)

⁻2	⁻5
⁻1	⁻2
0	1
1	4
2	7

(d)

⁻5	1.25
⁻4	1.33
⁻3	1.5
⁻2	2
⁻1	*DOMAIN ERROR*
0	0
1	0.5
2	0.667
3	0.75
4	0.8
5	0.833

(e)

⁻4	57
⁻3	34
⁻2	17
⁻1	6
0	1
1	2
2	9
3	22
4	41

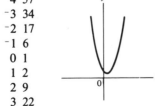

(f)

⁻4	0.0625
⁻3	0.125
⁻2	0.25
⁻1	0.5
0	1
1	2
2	4
3	8
4	16

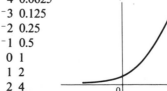

Solutions to exercises

3. (a)
⁻3	12
⁻2	6
⁻1	2
0	0
1	0
2	2
3	6
4	12

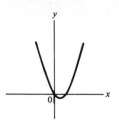

(b)
⁻5	⁻7
⁻4	⁻5
⁻3	⁻3
⁻2	⁻1
⁻1	1
0	3
1	5
2	7
3	9
4	11
5	13

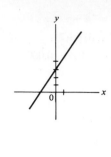

(c)
⁻3	⁻18
⁻2	⁻2
⁻1	2
0	0
1	⁻2
2	2
3	18

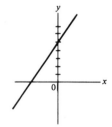

Section 6.3

1. (a) 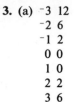 **(b)** 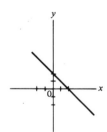 **(c)**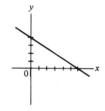

(d) 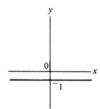 **(e)**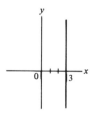

2. (a) 2; (b) ⁻1; (c) ⁻2/3; (d) 0; (e) infinity or no slope.

3. (a) $y = 5x - 1$; (b) $y = ⁻2x + 3$; (c) $y = 5$; (d) $y = 2x - 5$; (e) $y = -x$.

4. (a) 5/2; (b) ⁻7/4; (c) No slope; (d) 0.

5. (a) $y = (5/2)x - (1/2)$; (b) $y = (⁻7/4)x + (1/4)$; (c) $x = 3$; (d) $y = 1$.

6. (a) (0,2); (b) (13/7,4/7); (c) (⁻9/11,3/11); (d) No solution.

7.   ∇ X ← *SOLVE COEFS*
[1]   X ← (*COEFS*[3] − *COEFS*[2]) ÷ *COEFS*[1] ∇

394

8. $C = 3x + 5$.

9. $D = 10x + 25$.

10. $V = {}^-10x + 10{,}000$.

11. $C = 15x + 100$.

12. $C = 0.08x + 10$.

Section 6.4

1. $\nabla V \leftarrow VERTEX\ COEFS; X; Y$
   [1] $X \leftarrow {}- COEFS[2] \div 2 \times COEFS[1]$
   [2] $Y \leftarrow COEFS[3] - (COEFS[2] * 2) \div (4 \times COEFS[1])$
   [3] $V \leftarrow X, Y\ \nabla$

2. (a) $(0, {}^-2)$; (b) $(\frac{1}{2}, {}^-2\frac{1}{4})$; (c) $(3, 0)$; (d) $(4, 10)$; (e) $(\frac{1}{4}, {}^-3\frac{1}{8})$; (f) $({}^-1, 2)$.

3. (a) $\pm\sqrt{2} = \pm 1.4.4$; (b) ${}^-1, 2$; (c) $3, 3$; (d) $4 \pm \sqrt{10} = 0.84$ and 7.16;
   (e) ${}^-1, \frac{3}{2}$; (f) No real roots.

5. (a) 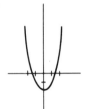   (b)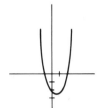

    (c)    (d)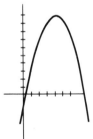

    (e)    (f)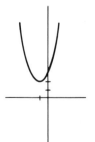

7. (a) 250; (b) \$22,500; (c) 100, 400.

8. (a) $R = {}^-20x^2 + 400x + 16000$; (b) 10; (c) \$300, 20.

9. (a) 4 seconds; (b) 262 feet.

10. (a) 1000; (b) \$10,000; (c) 0, 2000.

Section 6.5

1.

	(a)		(b)		(c)		(d)		(e)
-10	-1025	-10	-1670	-10	9506	-10	10988	-10	-120035
-9	-752	-9	-1189	-9	6162	-9	7279	-9	-72203
-8	-533	-8	-810	-8	3782	-8	4598	-8	-40989
-7	-362	-7	-521	-7	2162	-7	2735	-7	-21635
-6	-233	-6	-310	-6	1122	-6	1504	-6	-10391
-5	-140	-5	-165	-5	506	-5	743	-5	-4395
-4	-77	-4	-74	-4	182	-4	314	-4	-1553
-3	-38	-3	-25	-3	42	-3	103	-3	-419
-2	-17	-2	-6	-2	2	-2	20	-2	-75
-1	-8	-1	-5	-1	2	-1	-1	-1	-11
0	-5	0	-10	0	6	0	-2	0	-5
1	-2	1	-9	1	2	1	-1	1	-3
2	7	2	10	2	2	2	8	2	1
3	28	3	59	3	42	3	55	3	85
4	67	4	150	4	182	4	194	4	519
5	130	5	295	5	506	5	503	5	1885
6	223	6	506	6	1122	6	1084	6	5197
7	352	7	795	7	2162	7	2063	7	12021
8	523	8	1174	8	3782	8	3590	8	24595
9	742	9	1655	9	6162	9	5839	9	45949
10	1015	10	2250	10	9506	10	9008	10	80025.

2. (a)

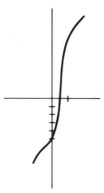

(b)

(c)

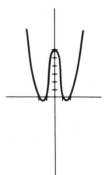

(d)

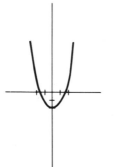

(e)

3. (a) 1.328; (b) 1.625; (c) 1.414; (d) 1.309; (e) 1.942.

5. ⁻0.481, 1.311, 3.170.

7. $A = {}^-2$, $B = 3$, $C = 0$.

8. $\nabla V \leftarrow VOLUME\ X$
[1] $V \leftarrow (4 \div 3) \times (\circ 1) \times (X * 3)\ \nabla$

Section 6.6

1. (a) No roots, $x = 2$ is an asymptote;
(b) 5 is a root, $x = 1$ is an asymptote;
(c) ⁻1 and 1 are roots, no asymptotes;
(d) No roots or asymptotes;
(e) No roots, $x = 0$ and $x = 1$ are asymptotes.

2.

(a)		(b)		(c)		(d)	
⁻5	⁻1.42857	⁻5	⁻1.666667	⁻5	0.923077	⁻5	0.03846
⁻4	⁻1.66667	⁻4	⁻1.8	⁻4	0.882353	⁻4	0.05882
⁻3	⁻2	⁻3	⁻2	⁻3	0.8	⁻3	0.833333
⁻2	⁻2.5	⁻2	⁻2.333333	⁻2	0.6	⁻2	0.2
⁻1	⁻3.33333	⁻1	⁻3	⁻1	0	⁻1	0.5
0	⁻5	0	⁻5	0	⁻1	0	1
1	⁻10	1	$9.99E999$	1	0	1	0.5
2	$9.99E999$	2	3	2	0.6	2	0.2
3	10	3	1	3	0.8	3	0.1
4	5	4	0.3333333	4	0.822353	4	0.05882
5	3.333333	5	0	5	0.923077	5	0.03846

(e)
⁻5	0.866667
⁻4	0.85
⁻2	
⁻1	1
0	$9.99E999$
1	$9.99E999$
2	2.5
3	1.666667
4	1.41667
5	1.3.

3.

(a)						(e)	
1.1	⁻11.111111	2.5	20	0.8	⁻21	⁻0.9	1.05848
1.2	⁻12.5	2.6	16.666667	0.9	⁻41	⁻0.8	1.13889
1.3	⁻14.2857	2.7	14.2857	1	$9.99E999$	⁻0.7	1.2521
1.4	⁻16.66667	2.8	12.5	1.1	39	⁻0.6	1.41667
1.5	⁻20	2.9	11.11111	1.2	19	⁻0.5	1.66667
1.6	⁻25			1.3	12.3333	⁻0.4	2.07143
1.7	⁻33.33333			1.4	9	⁻0.3	2.79487
1.8	⁻50	(b) 0.1	⁻5.44444	1.5	7	⁻0.2	4.33333
1.9	⁻100	0.2	⁻6	1.6	5.66667	⁻0.1	9.18182
2	$9.99E999$	0.3	⁻6.71429	1.7	4.71429	0	$9.99E999$
2.1	100	0.4	⁻7.666667	1.8	4	0.1	⁻11.2222
2.2	50	0.5	⁻9	1.9	3.44444	0.2	⁻6.5
2.3	33.33333	0.6	⁻11			0.3	⁻5.19048
2.4	25	0.7	⁻14.33333				

cont.

3. (e) *cont.*

0.4	⁻4.83333	0.8	⁻10.25	1.2	10.16667	1.6	3.70833
0.5	⁻5	0.9	⁻20.1111	1.3	6.89744	1.7	3.26891
0.6	⁻5.66667	1.0	9.99E999	1.4	5.28571	1.8	2.94444
0.7	⁻7.09524	1.1	20.0909	1.5	4.33333	1.9	2.69591.

4. (a)

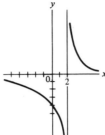

(b)

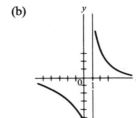

(c)

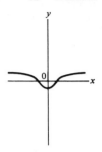

(d)

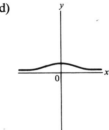

(e)

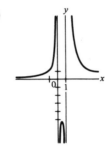

5. (a) \$277; **(b)** \$2500; **(c)** \$22,500.

6. (a) 410; **(b)** 355.

Chapter 7

Section 7.1

1. (a)

⁻5	0.0009765625
⁻4	0.00390625
⁻3	0.015625
⁻2	0.625
⁻1	0.25
0	1
1	4
2	16
3	64
4	256
5	1024

(b)

-5	243
-4	81
-3	27
-2	9
-1	3
0	1
1	0.333333333
2	0.111111111
3	0.037037037
4	0.012345679
5	0.004115226

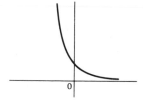

(c)

-5	148.4131591
-4	54.59815003
-3	20.08553692
-2	7.389056099
-1	2.718281828
0	1
1	0.3678794412
2	0.1353352832
3	0.04978706837
4	0.01831563889
5	0.00673794669

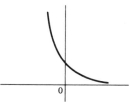

(d)

-5	33554432
-4	65536
-3	512
-2	16
-1	2
0	1
1	2
2	16
3	512
4	65536
5	33554432

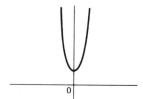

(e)

-5	0.00326776364
-4	0.01026598225
-3	0.03225153443
-2	0.1013211836
-1	0.3183098862
0	1
1	3.141592654
2	9.869604401
3	31.00627668
4	97.40909103
5	306.0196848

(f)

⁻5	32
⁻4	16
⁻3	8
⁻2	4
⁻1	2
0	1
1	2
2	4
3	8
4	16
5	32

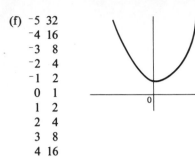

2. (a) 5; (b) 8; (c) 0.04; (d) 1; (e) 8; (f) 4; (g) 3; (h) 625.

4. $a = 5$, $b = 4$.

Section 7.2

1. ∇ *A*←*P INTEREST X*
  [1] *A*←*P*×(1+.05÷2)*2×*X* ∇

2. $839.03

3. $842.87.

4. $845.48.

5. $846.80.

6. $610.27.

7. 984150 weeds.

8. 3.75 square inches.

9. ∇ *A*←*P GROWTH KX*
  [1] *A*←*P*× *K*[1]× *K*[2] ∇

10. (a) $760.98; (b) $492.79.

11. 164.87 bacteria.

12. 16.15 grams.

13. 273.57 pieces.

14. 15.8 pieces.

16. (1+*I*÷*N*)*N×*X* yields 1.349737374; **I*×*X* yields 1.349858808.

Section 7.3

1. (a)

1	0	0.1	⁻2.095903274
2	0.6309297536	0.2	⁻1.464973521
3	1	0.3	⁻1.095903274
4	1.261859507	0.4	⁻0.834043767
5	1.464973521	0.5	⁻0.630929754
6	1.630929754	0.6	⁻0.464973521
7	1.771243749	0.7	⁻0.324659525
8	1.892789261	0.8	⁻0.203114014
9	2	0.9	⁻0.095903274
10	2.095903274	1	0

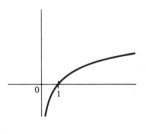

(b) 1 0   0.1 ⁻1
     2 0.3010299957   0.2 ⁻0.6989700043
     3 0.4771212547   0.3 ⁻0.5228787453
     4 0.6020599913   0.4 ⁻0.3979400087
     5 0.6989700043   0.5 ⁻0.3010299957
     6 0.7781512504   0.6 ⁻0.2218487496
     7 0.84509804   0.7 ⁻0.15490196
     8 0.903089987   0.8 ⁻0.096910013
     9 0.9542425094   0.9 ⁻0.0457574906
    10 1   1   0

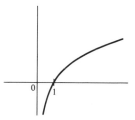

(c) 1 0   0.1 ⁻4.605170186
     2 1.386294361   0.2 ⁻3.218875825
     3 2.197224577   0.3 ⁻2.407945609
     4 2.772588722   0.4 ⁻1.832581464
     5 3.218875825   0.5 1.386294361
     6 3.583518938   0.6 ⁻1.021651248
     7 3.891820298   0.7 ⁻0.713349887
     8 4.158883083   0.8 ⁻0.446287103
     9 4.394449155   0.9 ⁻0.210721031
    10 4.605170186   1   0

2. (a) 2; (b) 4; (c) $1/3$; (d) 1; (e) ⁻2; (f) $\sqrt{2}$.

4. (a) 0.3710678623; (b) 2.371067862; (c) 4.371067862; (they all have the same decimal part).

5. (a) 0.8613531161; (b) 1.113282753; (c) 1.974635869; ($c = a + b$).

6. (a) 1.386294361; (b) 2.708050201; (c) 4.094344562; (d) 2.772588722; ($c = a + b, d = 2a$).

Section 7.4

2. $x = (\ln 9)/(\ln 5) = 1.365212389.$

3. $\nabla\ X \leftarrow A\ SOLVE\ B$
   [1] $X \leftarrow (\circledast B) \div (\circledast A)\,\nabla$

4. (a) $\ln 40 = \ln 5 + \ln 8 = 3.688879454;$
(b) $\ln 1.6 = \ln 8 - \ln 5 = 0.4700036392;$
(c) $\ln 25 = 2 \cdot \ln 5 = 3.218875825;$
(d) $\ln 200 = \ln 8 + 2 \cdot \ln 5 = 5.298317367$
(e) $\ln \sqrt{5} = (\ln 5)/2 = 0.8047189562.$

5. (a) 55751.454; (b) 5.819112628; (c) 169.0522737; (d) 18.46618529.

6. 8.66 years.

7. 13.86 years.

8. 21.97 years.

9. \$3258.10.

10. $\log_b(b^x) = x \cdot \log_b b = x \cdot 1 = x.$

Chapter 8

Section 8.1

1. (a) 5; (b) $^-0.2$; (c) 6; (d) 4; (e) Infinity; (f) Infinity; (g) 2; (h) 1.

Section 8.2

1. (a) $^-4$, $^-2.5$, $^-2.1$, $^-2.01$, $^-2.001$; (b) $^-1$, $^-1.5$, $^-1.9$, $^-1.99$, $^-1.999$; (c) $^-2$; (d) $^-2$.

2. (a) 2; (b) 0; (c) 5; (d) 24.

3. (a) $y=2x-5$; (b) $y=2$; (c) $y=5x+1$; (d) $y=24x-54$.

4. (a) 2; (b) 7; (c) 2.

Section 8.4

1. Decreasing if $x<2$, increasing if $x>2$.

2. Increasing if $x<1$ or $x>3$, decreasing if $1<x<3$.

3. (a) Increasing; (b) Decreasing; (c) 127.7.

4. $10\pi=31.4159$ square inches per inch.

5. $y=2x+8$.

6. (a) 32 feet/second; (b) 12 feet/second; (c) 52 feet/second; (d) 8 feet/second^2.

7. (a) $^-160$ feet/second; (b) $^-320$ feet/second (however, it will have already hit the ground); (c) $^-32$ feet/second^2.

8. \$3,999,950, which is ridiculous.

9. (a) \$1/item; (b) \$2/item; (c) 100.

10. $x=6$ (5.997 to be more precise).

11. After 2 hours.

12. 0.000000002 pounds/mile.

Section 8.3

1. (a) $F'(x)=3$; (b) $F'(x)=9x^2-2x+4$; (c) $F'(x)=4x^4+6x^2-8x+5$; (d) $F'(x)=(3/\sqrt{x})+(8/x^5)$; (e) $F'(x)=3x^{1/2}+6x^{-3}$.

3. (a) 4; (b) 1; (c) 0; (d) 11; (e) $^-1/4$.

5. (a) $y=2x-2$; (b) $y=-x+3$.

6. $(^-1,16),(2,^-11)$.

Section 8.5

1. (a) $dy/dx=(8x^3-4)/16x^2$; (b) $dx/dx=42x^6+50x^4+42x$;
(c) $dy/dx=(x+2)+(2x+2)\cdot\ln(3x)$; (d) $dy/dx=(10^{x^3+5x})\cdot(3x^2+5)\cdot(\ln 10)$;
(e) $dy/dx=(e^{x^3+5x})\cdot(3x^2+5)$; (f) $dy/dx=((1/x)-\ln x)/e^x)$;
(g) $dy/dx=(1/(\ln 10))\cdot(2x+5)/(x^2+5x+1)$; (h) $dy/dx=(2x+5)/(x^2+5x+1)$; (i) $dy/dx=(8x+20)(x^2+5x+1)^3$; (j) $dy/dx=5(\ln x)^4/x$.

2. (a) ‾42; (b) $21/32 = 0.65625$; (c) 0; (d) 2; (e) *DOMAIN ERROR.*

4. $y = x - 1$.

5. $4e = 10.873$.

Section 8.6

1. (a) Maximum (4, 10); (b) Minimum (1.25, ‾6.125); (c) Maximum (1, 10), minimum (3,6); (d) Maximum (‾1.46, ‾3.99), minimum (.457, ‾11.10); (e) Maximum (1, ‾3), minimum (1.36, ‾3.11).

2. (a) Minimum (1,1), maximum (4,10); (b) Minimum (1.25, ‾6.125), maximum (5,22); (c) Minimum (3,6), maximum (5,26); (d) Minimum (1, ‾9), maximum (5,295); (e) Minimum (1.36, ‾3.11), maximum (5,1885).

Section 8.7

1. $x = 2500$, $P = \$62,400$.

2. 6 weeks.

3. 45 by 45 by 30 by 30 by 30.

4. 156.25 feet by 208.333 feet.

5. 4 by 4 by 4.

6. 4.08 by 4.08 by 4.08.

7. Cut out squares 1.569 on a side. The volume is about 67.6 cubic inches.

8. \$300, 20 empty apartments, \$18,000.

9. $r = 1.68$ inches, $h = 3.37$ inches.

10. $t = 3\frac{1}{3}$ days.

Section 8.8

1. (1,4) maximum
No minimum
No inflection point
Increasing if $x < 1$
Decreasing if $x > 1$
Always concave downward.

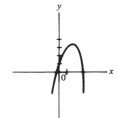

2. (0,4) maximum
(2,0) minimum
(1,2) inflection point
Increasing if $x < 0$ or $x > 2$
Decreasing if $0 < x < 2$
Concave upward if $x > 1$
Concave downward if $x < 1$.

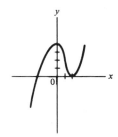

3. $(^-3, 145)$ maximum;
$(4, ^-198)$ minimum;
$(0.5, ^-26.5)$ inflection point
Increasing if $x < ^-3$ or $x > 4$
Decreasing if $^-3 < x < 4$
Concave upward if $x < 0.5$
Concave downward if $x > 0.5$.

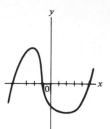

4. $(0.694, 6.71)$ maximum
$(0.36, 4.37)$ minimum
$(^-0.167, 4.57)$ inflection point
Increasing if $x < ^-0.694$ or $x > 0.36$
Decreasing if $^-0.694 < x < 0.36$
Concave upward if $x > ^-0.167$
Concave downward if $x < ^-0.167$.

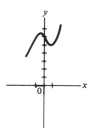

5. $(1/e, ^-1/e) = (0.368, ^-0.368)$ minimum
No maximum or inflection point
Increasing if $x > 1/e$
Decreasing if $0 < x < 1/e$
Concave upward for all $x > 0$.

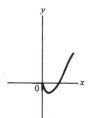

6. $(0, 1)$ maximum
No minimum
$(\pm 1, .607)$ inflection points
Increasing if $x < 0$
Decreasing if $x > 0$
Concave upward if $x < ^-1$ or $x > 1$
Concave downward if $^-1 < x < 1$.

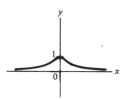

Chapter 9

Section 9.1

1. (a) $2x^3 - 4x^2 + 2x + c$; (b) $\frac{1}{4}x^4 + \frac{1}{3}x^3 + \frac{1}{2}x^2 + x + c$; (c) $\frac{2}{5}x^{5/2} + x^{-2} + c$;
(d) $12\sqrt{x} + c$; (e) $4\ln x + c$; (f) $^-\frac{1}{5}e^{-5x} + c$.

2. $y = x^4 + 3x^2 + 1$.

3. $C = 2x^2 - 200x + 100$.

4. $S = 2t^3 - 12t^2 + 12t$.

5. $h = ^-20t - 16t^2 + 500$.

6. $y = e^{3x} - 1$.

7. $A = \pi x^2$, circle.

Section 9.2

1. (a) $2x^4 - 3x^2 + 2x + c$; (b) $\frac{1}{6}x^6 + \frac{1}{4}x^4 + x + c$; (c) $2e^{2x} - 4 \cdot \ln x + c$;
   (d) $4\sqrt{x} + (3/x) + c$; (e) $\frac{1}{6}(4x+1)^{3/2} + c$; (f) $\frac{1}{2}(2x^2+1)^7 + c$; (g) $x + c$;
   (h) $2x \cdot \ln x - 2x + c$; (i) $\frac{1}{2}\ln(2x+1) + c$; (j) $\frac{1}{2}(\ln x)^2 + c$.
2. (a) $v = t^2 - 4t + 5$; (b) $s = \frac{1}{3}t^3 - 2t^2 + 5t$.
3. $y = (-1/(4x+2)) + \frac{1}{2}$.
4. $C = 2e^{5x} + 98$.

Section 9.3

1. (a) 11.25; (b) 12.12; (c) 10.
2. (a) 8; (b) 11; (c) 10.666667.
3. (a)

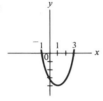

(b) $^-$10.666667; (c) $3c$ is the negative of $2c$; (d) The curve in 3 lies entirely below the x axis. Area below the x axis is negative.
4. (a)

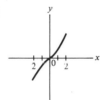

(b) 0; (c) The curve is below the x axis and above the x axis in equal amounts. Therefore, the negative and positive areas cancel out.
5. $\nabla S \leftarrow SUM\ I$
   [1] $S \leftarrow + /(\iota I) * 2\ \nabla$

       SUM 100
   338350

Section 9.4

1. (a) 11; (b) $^-$2; (c) 4.6; (d) 2.545
2. (a)   (b) 0; (c) 8.

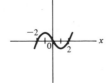

3. (a) 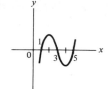 (b) 0; (c) 8.

4. (a) 0.5; (b) 2.666667; (c) 36.

Section 9.5

1. (a) 27; (b) $143\frac{1}{3}$; (c) $e-1$; (d) 1.75; (e) $8\frac{2}{3}$; (f) $\frac{1}{2}\cdot\ln 3$.
2. 18.
3. $21\frac{1}{3}$.
4. (a) 0; (b) $\frac{1}{2}$.
5. See Exercise 1, Part (c).

Section 9.6

1. 4800 feet.
2. $40\frac{2}{3}$.
3. $10,000(e-1)=\$17,182.82$.
4. 10,202.
5. 9 months.
6. First, write an **FN** program:

      ∇ *Y←FN X*
[1]   *Y←*(○1)×(   )∗2 ∇   (Insert $F(x)$ within the parentheses.)

Then, you can use the program **INTEGRAL** as

    A INTEGRAL B.

7. $30\pi = 94.248$.
8. $4\pi/3 = 4.189$.
9. $512\pi/15 = 107.23$.
10. $(8/27)(10\sqrt{10} - 1) = 9.073$.

Chapter 10

Section 10.1

1. (a) 1/2; (b) 1/4; (c) 1/6; (d) 1/3.
2. (a) 3/16, 3/16; (b) 9/16.
3. (a) 1/13; (b) 1/4; (c) 1/52; (d) 4/13; (e) 1/2.
4. $P(A)=2/5, P(B)=2/5, P(C)=1/5$.
5. $k = 120/274$.

6. (a) $\{(H,H,H),\ (H,H,T),\ (H,T,H),\ (H,T,T),\ (T,H,H),\ (T,T,H),\ (T,H,T),$ $(T,T,T)\}$; (b) 1/8; (c) 7/8; (d) 3/8.

Section 10.2

1. (a) 2/13; (b) 7/13; (c) 1/13; (d) 6/13.

2. (a) 11/36; (b) 1/6; (c) 2/9; (d) 5/6.

3. (a) 1/8; (b) 1/8; (c) 7/8; (d) 1/4.

4. (a) 5/14; (b) 25/28; (c) 15/56; (d) 5/7.

5. (a) 25/64; (b) 55/64; (c) 15/64; (d) 5/8.

6. (a) 21/50; (b) 32/50; (c) 12/30; (d) 9/21.

7. (a) 3/5; (b) 0.6; (c) 0.4; (d) Dependent.

8. (a) 0.42; (b) 0.88; (c) 0.28; (d) 0.6.

9. (a) 0.2401; (b) 0.7599; (c) 0.0081; (d) 0.4116.

10. (a) 0.06; (b) 0.44; (c) 0.56; (d) 0.3.

Section 10.3

1. 2000.

2. 40,320.

3. 624.

4. 1024.

5. 243.

6. (a) 1,816,214,400; (b) 5005.

7. 2100.

8. 210.

9. 480.

10. (a) 252; (b) 105; (c) 126; (d) 126.

11. (a) 1287; (b) 48; (c) 22,308; (d) 123,552.

Section 10.4

1. (a) $\dfrac{\binom{6}{2}\binom{6}{2}}{\binom{12}{4}}=0.4545$; (b) $\dfrac{\binom{6}{4}\binom{6}{0}}{\binom{12}{4}}=0.0303$; (c) $1-\dfrac{\binom{6}{4}\binom{6}{0}}{\binom{12}{4}}=0.9697$;

(d) $\dfrac{\binom{6}{0}\binom{6}{4}}{\binom{12}{4}}+\dfrac{\binom{6}{1}\binom{6}{3}}{\binom{12}{4}}=0.2727$.

2. (a) $\dfrac{\binom{6}{0}\binom{4}{3}}{\binom{10}{3}}=0.0333$; (b) $\dfrac{\binom{6}{3}\binom{4}{0}}{\binom{10}{3}}=0.1667$;

(c) $\dfrac{\binom{6}{3}\binom{4}{0}+\binom{6}{2}\binom{4}{1}}{\binom{10}{3}}=0.6667$; (d) $\dfrac{\binom{6}{2}\binom{4}{1}+\binom{6}{1}\binom{4}{2}}{\binom{10}{3}}=0.8.$

3. (a) $\dfrac{\binom{4}{4}\binom{48}{3}}{\binom{52}{7}}=0.00013$; (b) $13\cdot$Part (a)$=0.00168$; (c) $\dfrac{\binom{13}{7}}{\binom{52}{7}}=0.000013$;

(d) $4\cdot$Part (c)$=0.00005.$

4. (a) $\dfrac{\binom{9}{6}}{\binom{20}{6}}=0.00217$; (b) $\dfrac{\binom{9}{3}\binom{6}{2}\binom{5}{1}}{\binom{20}{6}}=0.1625$; (c) $\dfrac{\binom{9}{2}\binom{6}{2}\binom{5}{2}}{\binom{20}{6}}=0.1393$;

(d) 0.

Section 10.5

1. (a) $\binom{4}{2}(0.3)^2(0.7)^2=0.2646$; (b) $\binom{4}{0}(0.3)^0(0.7)^4=0.2401$;

(c) $1-$Part (b)$=0.7599$; (d) $\binom{4}{4}(0.3)^4(0.7)^0=0.0081.$

2. (a) $\binom{10}{10}(0.8)^{10}(0.2)^0=0.1074$; (b) $\binom{10}{8}(0.8)^8(0.2)^2+\binom{10}{9}(0.8)^9(0.2)^1+$

$\binom{10}{10}(0.8)^{10}(0.2)^0=0.6778$; (c) $\binom{10}{10}(0.8)^{10}(0.2)^0+\binom{10}{9}(0.8)^9(0.2)^1=0.3758.$

3. (a) $\binom{6}{6}(0.5)^6(0.5)^0=0.015625$; (b) $\binom{6}{3}(0.5)^3(0.5)^3=0.3125$;

(c) $1-$Part (a)$=0.984375$; (d) 0.5.

4. (a) $\binom{5}{2}(0.1)^2(0.9)^3=0.0729$; (b) $\binom{5}{0}(0.1)^0(0.9)^5=0.59049$; (c) $4\cdot(0.1)^2=0.04.$

5. (a) $\binom{12}{0}(0.3)^0(0.7)^{12}+\binom{12}{1}(0.3)^1(0.7)^{11}+\binom{12}{2}(0.3)^2(0.7)^{10}+\binom{12}{3}(0.3)^3(0.7)^9$

$=0.4925$;

(b) $\binom{12}{7}(0.3)^7(0.7)^5+\binom{12}{8}(0.3)^8(0.7)^4+\binom{12}{9}(0.3)^9(0.7)^3+\binom{12}{10}(0.3)^{10}(0.7)^2+$

$\binom{12}{11}(0.3)^{11}(0.7)^1+\binom{12}{12}(0.3)^{12}(0.7)^0=0.0386.$

Section 10.6

1. (a) $\dfrac{(e^{-4})(4^0)}{!0!}=0.0183$; (b) $\dfrac{(e^{-4})(4^2)}{!2}=0.1465$;

(c) $1-\left(\dfrac{(e^{-4})(4^0)}{!0}+\dfrac{(e^{-4})(4^1)}{!1}\right)=0.9084$;

(d)

$\dfrac{(e^{-4})(4^0)}{!0}+\dfrac{(e^{-4})(4^1)}{!1}+\dfrac{(e^{-4})(4^2)}{!2}+\dfrac{(e^{-4})(4^3)}{!3}+\dfrac{(e^{-4})(4^4)}{!4}+\dfrac{(e^{-4})(4^5)}{!5}=$

0.7851.

2. (a) $\dfrac{(e^{-10})(10^0)}{!0} = 0.000045$;

(b) $1 - \left(\dfrac{(e^{-10})(10^0)}{!0} + \dfrac{(e^{-10})(10^1)}{!1} + \dfrac{(e^{-10})(10^2)}{!2} \right) = 0.9972$;

(c) $\dfrac{(e^{-10})(10^0)}{!0} + \dfrac{(e^{-10})(10^1)}{!1} + \dfrac{(e^{-10})(10^2)}{!2} + \dfrac{(e^{-10})(10^3)}{!3} + \dfrac{(e^{-10})(10^4)}{!4} = $
0.0292;

(d) $\dfrac{(e^{-10})(10^1)}{!1} = 0.000454$.

3. Done as in Problem 2, except that $u = 2.5$ instead of 10.

(a) 0.0821; (b) 0.4562; (c) 0.8912; (d) 0.2052.

4. (a) $\dfrac{(e^{-3})(3^0)}{!0} = 0.0498$; (b) $\dfrac{(e^{-12})(12^0)}{!0} + \dfrac{(e^{-12})(12^1)}{!1} + \ldots + \dfrac{(e^{-12})(12^{10})}{!10}$
$= 0.3472$;

(c) $1 - \left(\dfrac{(e^{-9})(9^0)}{!0} + \dfrac{(e^{-9})(9^1)}{!1} + \ldots + \dfrac{(e^{-9})(9^7)}{!7} \right) = 0.6761$.

5. (a) $\dfrac{(e^{-10})(10^0)}{!0} + \dfrac{(e^{-10})(10^1)}{!1} + \ldots + \dfrac{(e^{-10})(10^4)}{!4} = 0.0293$;

(b) $1 - \left(\dfrac{(e^{-10})(10^0)}{!0} + \dfrac{(e^{-10})(10^1)}{!1} + \ldots + \dfrac{(e^{-10})(10^{20})}{!20} \right) = 0.0016$.

6. (a) $1 - \left(\dfrac{(e^{-25})(25^0)}{!0} + \dfrac{(e^{-25})(25^1)}{!1} + \ldots + \dfrac{(e^{-25})(25^{19})}{!19} \right) = 0.8664$;

(b) $\dfrac{(e^{-25})(25^0)}{!0} + \dfrac{(e^{-25})(25^1)}{!1} + \ldots + \dfrac{(e^{-25})(25^{10})}{!10} = 0.0006$.

Chapter 11

Section 11.1

1. (a) 5 ? 12; (b) ? (5ρ12); (c) (5 ? 10)+5; (d) (5 ? 11)−6.

3. No.

4. (b) 3 F's, 6 D's, 3 C's, 10 B's, 6 A's.

5. ∇ *DESCENDING ← REARRANGE SAMPLE*
[1] *DESCENDING ← SAMPLE[⇩ SAMPLE]* ∇

6. 1 17, 5 18's, 9 19's, 8 20's, 2 21's, 3 23's, 1 24, 1 30.

7. 3 in 350–399, 2 in 400–449, 5 in 450–499, 9 in 500–549, 3 in 550–599, 4 in 600–649, 2 in 650–699, 2 in 700–749, 2 in 750–800.

8. * * *
 * *
 * * * * *
 * * * * * * * *
 * * *
 * * * *
 * *
 * *
 * *

Section 11.2

1.
       ∇ *MD←MEDIAN SAMPLE*
[1]   →((2|ρ *SAMPLE*)=0)/*EVEN*
[2]   *ODD*: *MD←*(*SORT SAMPLE*)[(1 + ρ *SAMPLE*)÷2]
[3]   →0
[4]   *EVEN*: *MD←*((*SORT SAMPLE*)[(ρ *SAMPLE*)÷2+(*SORT SAMPLE*)
    [(1 + ρ *SAMPLE*)÷2])÷2
    ∇

2.
       ∇ *M←MODE SAMPLE*
[1]   *FREQUENCIES←+*/(*SAMPLE* ∘.= *SAMPLE*)
[2]   *MOST←*⌈/*FREQUENCIES*
[3]   *M←SAMPLE* [(*FREQUENCIES*= *MOST*)/ιρ *FREQUENCIES*] ∇

3.
       ∇ *R←RANGE SAMPLE*
[1]   *R←*(⌈*SORT SAMPLE*)−(⌊ *SORT SAMPLE*) ∇

4. Mean, 77.39; median, 82.5; mode, 83; range, 45; variance, 175.3; standard deviation, 13.24.

5. Mean, 20.1; median, 19.5; mode, 19; range, 13; variance, 6.09; standard deviation, 2.47.

6. Mean, 546.7; median, 525; modes, 500, 510, 520, 540, 610, 650, 720; range, 440; variance, 11332; standard deviation, 106.45.

7. Mean 2.706, standard deviation 0.606789142

8. (a) Mean, 20.44; median, 14; mode, 14, 15; the median
(b) Mean, 5.1; median, 5; mode, 5; the mean
(c) Mean, 5.875; median, 6.5; modes, 4, 7; the mean.

9. Select a random sample of students and find their mean weight.

10. Keep the machine oiled to reduce the variance.

11.
       ∇ *V←VAR SAMPLE*
[1]   *V←*((+ / *SAMPLE*∗2)÷ρ *SAMPLE*)−(*MEAN SAMPLE*)∗2 ∇

12. 2 F's, 2 D's, 18 C's, 8 B's, 0 A's.

13. 2.3547, 3.1416, 2.7183, ⁻7.3891.

14.
       ∇ *Q←P ROUND N*
[1]   *Q←*(10∗ − *P*)×⌊0.5+ *N*×10∗*P* ∇

Section 11.3

1. 0.0231.

2. 0.6826.

3. 0.0914.

4. 0.0651.

5. (a) 0.9191; (b) 459.

6. 2.31 percent A's, 13.58 percent B's, 68.26 percent C's, 13.58 percent D's, 2.31 percent F's.

7. 0.0067.

8. About 2 students.

Section 11.4

1. (a) 0.1583; (b) 0.5; (c) 0.5332.

2. (a) 0.0001; (b) 0.0013.

3. 0.9997.

4. (a) 0.9769; (b) 0.9538.

5. 0.8633.

Chapter 12

Section 12.1

1. (a)  (b) 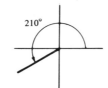 (c)

(d) (e)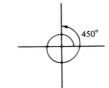

2. (a) $\pi/4 = 0.7854$; (b) $\pi/10 = 0.3142$; (c) $^-\pi/20 = ^-0.1571$; (d) $4\pi/3 = 4.1888$; (e) $^-\pi/9 = ^-0.3491$.

3. (a) 108; (b) $^-240$; (c) 114.59; (d) 15; (e) $^-85.94$.

5. $\nabla \; S \leftarrow R \; LENGTH \; T$
[1] $S \leftarrow R \times T \; \nabla$

Section 12.2

2. $\sin(\theta) = 5/13$; $\cos(\theta) = ^-12/13$; $\tan(\theta) = ^-5/12$; $\cot(\theta) = ^-12/5$; $\sec(\theta) = ^-13/12$; $\cos(\theta) = 13/5$.

3. $\sin(\theta) = ^-3/5$; $\cos(\theta) = 4/5$; $\tan(\theta) = ^-3/4$; $\cot(\theta) = ^-4/3$; $\sec(\theta) = 5/4$; $\csc(\theta) = ^-5/3$.

4. $\sin(\theta) = 3/5$; $\tan(\theta) = ^-3/4$; $\cot(\theta) = ^-4/3$; $\sec(\theta) = ^-5/4$; $\csc(\theta) = 5/3$.

5. $\sin(\theta) = ^-4/5$; $\cos(\theta) = ^-3/5$; $\cot(\theta) = 3/4$; $\sec(\theta) = ^-5/3$; $\csc(\theta) = ^-5/4$.

6. $\sin(60°) = \sqrt{3}/2 = 0.8660$; $\cos(60°) = 1/2 = 0.5$; $\tan(60°) = \sqrt{3} = 1.7321$; $\cot(60°) = 1/\sqrt{3} = 0.5774$; $\sec(60°) = 2$; $\csc(60°) = 2/\sqrt{3} = 1.1547$.

7. $\sin(0°) = 0$; $\cos(0°) = 1$; $\tan(0°) = 0$; $\cot(0°) = $ undefined; $\sec(0°) = 1$; $\csc(0°) = $ undefined.

8. $\sin(45°)=1/\sqrt{2}=0.7071$; $\cos(45°)=1/\sqrt{2}=0.7071$; $\tan(45°)=1$;
$\cot(45°)=1$; $\sec(45°)=\sqrt{2}=1.4142$; $\csc(45°)=\sqrt{2}=1.4142$.

9. $\sin 150°=1/2=0.5$; $\cos(150°)=^-\sqrt{3}/2=^-0.8660$; $\tan(150°)=^-1/\sqrt{3}=$
$^-0.5774$; $\cot(150°)=^-\sqrt{3}=^-1.7321$; $\sec(150°)=^-2/\sqrt{3}=^-1.1547$;
$\csc(150°)=2$.

10. $\sin(225°)=^-1/\sqrt{2}=^-0.7071$; $\cos(225°)=^-1/\sqrt{2}=^-0.7071$; $\tan(225°)=1$;
$\cot(225°)=1$; $\sec(225°)=^-\sqrt{2}=^-1.4142$; $\csc(225°)=^-\sqrt{2}=^-1.4142$.

11.   ∇ *S←X SINE Y*
    [1]   *S←Y÷((X∗2)+(Y∗2))∗5* ∇

12.   ∇ *C←X COSINE Y*
    [1]   *C←X÷((X∗2)+(Y∗2))∗.5* ∇

13.   ∇ *T←X TANGENT Y*
    [1]   *T←(X SINE Y)÷(X COSINE Y)* ∇

14.   ∇ *C←X COTANGENT Y*
    [1]   *C←(X COSINE Y)÷(X SINE Y)* ∇

    ∇ *S←X SECANT Y*
    [1]   *S←1÷(X COSINE Y)* ∇

    ∇ *C←X COSECANT Y*
    [1]   *C←1÷(X SINE Y)* ∇

Section 12.3

1. (a) 0.9975; (b) 0; (c) $^-0.4161$; (d) 2; (e) 3.732; (f) Undefined.
2. (a) 3.732; (b) 0.9659; (c) 0.5; (d) 3.0407; (e) $^-2$; (f) $^-0.4142$.
3. (a) 0.5832; (b) 0.9664; (c) $^-2.6466$; (d) 0.9051.

Section 12.4

Degrees	Radians	$\cos X$	$\cot X$	$\csc X$
0	0.	1.	9.9E99	9.9E99
15	0.2618	0.9659	3.7321	3.8637
30	0.5236	0.8660	1.7321	2.
45	0.7854	0.7071	1.	1.4142
60	1.0472	0.5	0.5774	1.1547
75	1.3090	0.2588	0.2679	1.0353
90	1.5708	0.	0.	1.
105	1.8326	$^-0.2588$	$^-0.2679$	1.0353
120	2.0944	$^-0.5$	$^-0.5774$	1.1547
135	2.3562	$^-0.7071$	$^-1.$	1.4142
150	2.6180	$^-0.8660$	$^-1.7321$	2.
165	2.8798	$^-0.9659$	$^-3.7321$	3.8637
180	3.1416	$^-1.$	$^-9.9E99$	9.9E99
195	3.4034	$^-0.9659$	3.7321	$^-3.8637$
210	3.6652	$^-0.8660$	1.7321	$^-2.$
225	3.9270	$^-0.7071$	1.	$^-1.4142$

240	4.1888	⁻0.5	0.5774	⁻1.1547
255	4.4506	⁻0.2588	0.2679	⁻1.0353
270	4.7124	0.	0.	⁻1.
285	4.9742	0.2588	⁻0.2679	⁻1.0353
300	5.2360	0.5	⁻0.5774	⁻1.1547
315	5.4978	0.7071	⁻1.	⁻1.4142
330	5.7596	0.8660	⁻1.7321	⁻2.
345	6.0214	0.9659	⁻3.7321	⁻3.8637
360	6.2832	1.	⁻9.9E99	⁻9.9E99

Ranges

$$^{-}1 \leqslant \cos x \leqslant 1$$
$$^{-}\infty \leqslant \cot x \leqslant \infty$$
$$\csc x \geqslant 1 \text{ or } \csc x \leqslant {}^{-}1$$

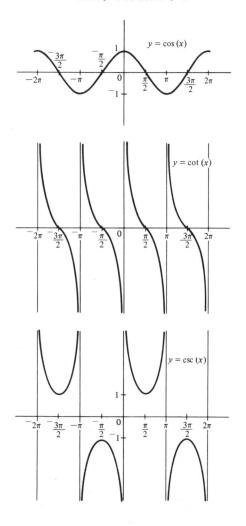

413

Section 12.5

1. (a) 45°; (b) 90°; (c) 135°; (d) 30°; (e) 39°; (f) 34°; (g) ⁻50.2°; (h) 45°.

2. (a) 1; (b) ⁻0.5; (c) $\sqrt{3}$; (d) ⁻4/3; (e) $\sqrt{2}$; (f) 30°; (g) ⁻30°; (h) 0.

3.   ∇ *Y←ARCSEC X*
   [1]   *Y←⁻2○(1÷X)* ∇

Section 12.6

1. $b = \sqrt{44} = 6.63$, $\alpha = 56.44°$, $\beta = 33.56°$.

2. $b = 32.69$, $c = 55.62$, $\alpha = 54°$.

3. 14.04°.

4. 945.6 feet.

5. 14.93°.

6. 65.2 feet.

7. $t = 237.86$ feet.

8. $b = 41.14$ feet, $h = 28.32$ feet, $A = 582.5$ square feet.

Section 12.7

1. $b = 7.07$, $c = 13.66$, $\gamma = 105°$.

2. $\beta = 61.1°$, $\gamma = 68.9°$, $c = 85.25$.

3. $\alpha = 53.13°$, $\beta = 90°$, $\gamma = 36.87°$.

4. $c = \sqrt{7} = 2.65$, $\alpha = 79.1°$, $\beta = 40.9°$.

5. $a = 7.98$, $\beta = 33.6°$, $\gamma = 84.4°$.

6. 65.53°.

7. 971.83 yards.

9.   ∇ *A←SIDES AREA ANGLE*
   [1]   *A←.5× SIDES* [1]× *SIDES* [2]× *SINDEGREES ANGLE* ∇

10. 80.62 square units.

Program Index

Subject index

417